Collins

White Rose Maths
AQA GCSE 9-1
Higher
Student Book 1

Caroline Hamilton and Ian Davies

Published by Collins
An imprint of HarperCollins*Publishers*
The News Building
1 London Bridge Street
London
SE1 9GF

HarperCollins*Publishers*
Macken House
39/40 Mayor Street Upper
Dublin 1
D01 C9W8
Ireland

Browse the complete Collins catalogue at
collins.co.uk

© HarperCollins*Publishers* Limited 2024

10 9 8 7 6 5 4 3 2 1

ISBN: 978-0-00-866959-1

British Library Cataloguing-in-Publication Data
A catalogue record for this publication is available from the British Library.

Series editors: Ian Davies and Caroline Hamilton
Authors: Matthew Ainscough, Rob Clasper, Rhiannon
  Davies and Sahar Shillabeer
Publisher: Katie Sergeant
Product manager: Richard Toms
Development editor: Karl Warsi
Editorial: Richard Toms, Amanda Dickson and
  Deborah Dobson
Proofreading and answer checking: Steven Matchett,
  Anne Stothers, Amanda Dickson, Eric Pradel,
  Trevor Senior and Anna Cox
Cover designer: Sarah Duxbury
Typesetter: Jouve India Private Limited
Production controller: Alhady Ali
Printed and bound in India

This book contains FSC™ certified paper and other controlled sources to ensure responsible forest management.

For more information visit: www.harpercollins.co.uk/green

Text acknowledgements
The publishers gratefully acknowledge the permission granted to reproduce the copyright material in this book. Every effort has been made to trace copyright holders and to obtain their permission for the use of copyright material.

Contents

Number Algebra Ratio, proportion and rates of change

Contents

Number   Algebra   Ratio, proportion and rates of change

How to use this book

Welcome to the **Collins White Rose Maths AQA GCSE 9–1 Higher tier** course.

There are two Student Books in the series:

- **Student Book 1** covers Number, Algebra, and Ratio, proportion and rates of change.
- **Student Book 2** covers Geometry and measures, Probability, and Statistics.

Sometimes you will need some knowledge of a different area of mathematics within the topic you are studying. For example, you may need to set up and solve an algebraic equation when solving a geometry problem. You will often be able to use your earlier knowledge and skills from Key Stage 3 to help you do this.

Here is a short guide to how to get the most out of this book. We hope you enjoy continuing your learning journey.

Caroline Hamilton and Ian Davies, series editors

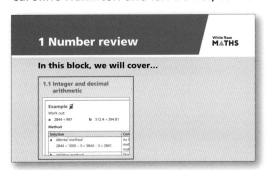

Block overviews Each block of related chapters starts with a visual introduction to the key concepts and learning you will encounter.

Are you ready? Before you start each part of a chapter, remind yourself of the maths you should already know with these questions. If you need more practice, refer to the *Collins White Rose Maths Key Stage 3* course or the series of *Collins White Rose Maths AQA GCSE 9–1 Foundation Student Books*.

Explanatory text Key words and concepts are explained before moving on to worked examples.

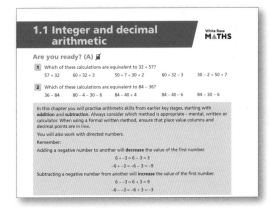

Using your calculator Where appropriate, you are given advice on how to use the features of your calculator to find or check answers. Not all calculators work in the same way, so make sure you know how your model works.

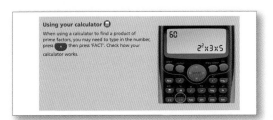

Worked examples Learn how to approach different types of questions with worked examples that clearly walk you through the process of answering. Visual representations are provided to help when necessary.

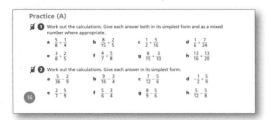

Practice Put what you have just learned into practice. Sometimes symbols are used in questions or whole sections to show when you should, or should not, use a calculator. If there is no symbol, the question or section can be approached in either way.

Many of the Practice sections conclude with a **What do you think?** exercise to encourage further exploration.

Consolidate Reinforce what you have learned in the chapter with additional practice questions.

Stretch Take your learning further and challenge yourself to apply it in new ways or different areas of maths.

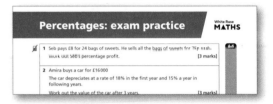

Exam practice At the end of each block, you will find exam-style questions to practise your learning. These are organised into grade bands of 4–6 and 7–9. There is extra practice at the end of the three main parts of the book.

Glossary Look up the meanings of any key words or phrases you are not sure about.

Answers Check your work using the answers provided at the back of the book.

1 Number review

In this block, we will cover...

1.1 Integer and decimal arithmetic

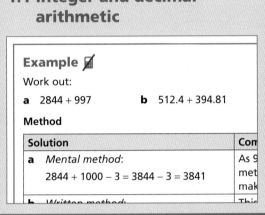

Example ✏

Work out:

a 2844 + 997   **b** 512.4 + 394.81

Method

Solution	Com
a *Mental method*: 2844 + 1000 − 3 = 3844 − 3 = 3841	As 9 met mak
b *Written method:*	This

1.2 Fraction arithmetic

Practice (B)

✏ **1** Work out the calculations. Give each answ

    a $\frac{1}{8} \times \frac{1}{6}$ **b** $\frac{1}{5} \times \frac{3}{7}$

    e $\frac{6}{11} \times \frac{3}{22}$ **f** $\frac{4}{9} \times \frac{5}{8}$

✏ **2** Work out the calculations. Give each answ

    a $3 \times \frac{1}{5}$ **b** $\frac{3}{8} \times 2$

✏ **3** Write the reciprocal of each number.

    a $\frac{1}{?}$ **b** $\frac{1}{?}$

1.3 Factors, multiples and primes

Consolidate – do you need more

1 Write the numbers as a product of prime f

    a 50   **b** 90

2 Use powers to show that:

    a 576 is a square number   **b** 1296 is a

3 Work out the HCF of:

    a 32 and 80   **b** 60 and 150

Are you ready? (A) 🗒

1 Which of these calculations are equivalent to 32 + 57?

$57 + 32$ $60 + 32 + 3$ $50 + 7 + 30 + 2$ $60 + 32 - 3$ $30 - 2 + 50 + 7$

2 Which of these calculations are equivalent to 84 − 36?

$36 - 84$ $80 - 4 - 30 - 6$ $84 - 40 + 4$ $84 - 40 - 6$ $84 - 30 - 6$

In this chapter you will practise arithmetic skills from earlier key stages, starting with **addition** and **subtraction**. Always consider which method is appropriate – mental, written or calculator. When using a formal written method, ensure that place value columns and decimal points are in line.

You will also work with directed numbers.

Remember:

Adding a negative number to another will **decrease** the value of the first number.

$$6 + -3 = 6 - 3 = 3$$

$$-6 + -3 = -6 - 3 = -9$$

Subtracting a negative number from another will **increase** the value of the first number.

$$6 - -3 = 6 + 3 = 9$$

$$-6 - -3 = -6 + 3 = -3$$

Example 🗒

Work out:

a $2844 + 997$ **b** $512.4 + 394.81$ **c** $37 - 12.11$

Method

Solution	Commentary
a *Mental method:* $2844 + 1000 - 3 = 3844 - 3 = 3841$	As 997 is close to 1000, you can use a mental method. 997 can be rewritten as 1000 − 3, making the calculation easier to complete.
b *Written method:* $512.4 + 394.81$ $\begin{array}{r} 5\,1\,2\,.\,4\,0 \\ +\,3\,9\,4\,.\,8\,1 \\ \hline 9\,0\,7\,.\,2\,1 \\ _1_1 \end{array}$	This calculation has too many digits to hold in your head, so a written method is more appropriate. Ensure that all place value columns are aligned correctly. Using a placeholder by writing 512.4 as 512.40 helps with this.

c *Written method:*

37 − 12.11

Again, there is too much to remember to do this calculation mentally.

Writing 37 as 37.00 is essential to support the written method here.

Practice (A)

Work out the calculations using mental or written methods as appropriate. Compare your choices with a partner.

1 **a** 346 + 231
   b 584 + 1233
   c 65.7 + 27.6
   d 0.7 + 0.23 + 1.2
   e three hundred + 274
   f £3.12 + £2.99 + £5.90

2 **a** 6284 − 358
   b 62.8 − 6.84
   c 1 thousand − 284
   d 0.84 − 0.513
   e £10 − £3.99 − £2.50
   f £300 − (£215 + £52.34)

3 **a** −4 + −3
   b 4 + −3
   c −4 + 3
   d −4 − −3
   e 7 + −9
   f −7 + 9
   g −7 + −9
   h −7 − −9
   i 63 − 96
   j −63 − 96
   k −63 − −96
   l −96 − −63

4 Given that $a = 6.4$, $b = -7$, $c = 32$ and $d = 9.6$, work out the value of the expressions.

   a $a + d$
   b $a + c$
   c $b - c$
   d $c - d$
   e $b + a$
   f $d - b$

5 Write the next two terms of each of the linear sequences.

   a 62, 84, 106, ___, ___
   b 10.2, 9.3, 8.4, ___, ___
   c 40, 24, 8, ___, ___
   d −1.9, −1.3, −0.7, ___, ___

What do you think? (A)

1 Using the fact that 543 + 78 = 621, write the answers to the calculations.

   a 543 + 80
   b 540 + 78
   c 5.43 + 0.78
   d 643 + 178
   e 540 + 81
   f 545 + 80

2 Work out the missing digits in the additions.

   a
```
    □ 5 □
  + 3 □ 4
  ─────────
  1 1 2 1
```

   b
```
    □ 5 · 3
    5 □ · 0 □
  + 3 0 · □ 4
  ───────────
  1 4 9 · 0 7
```

Are you ready? (B)

1 Given that $32 \times 19 = 608$, write the answers to the calculations.

a 3.2×19 **b** 0.32×0.19 **c** 320×1900 **d** 3.2×0.19

e $608 \div 32$ **f** $60.8 \div 1.9$ **g** $6.08 \div 3.2$ **h** $6080 \div 1.9$

2 Given that $3060 \div 85 = 36$, write the answers to the calculations.

a $306 \div 85$ **b** $306 \div 8.5$ **c** $30.6 \div 3.6$ **d** 8.5×3.6

In this section you will review **multiplication** and **division**. Again, always consider which method is appropriate – mental, written or calculator. You need to be confident in multiplying and dividing by powers of 10, for example:

$61.34 \times 1000 = 61\,340$ (move the digits three places to the left and add a placeholder)

$61.34 \div 1000 = 0.06134$ (move the digits three places to the right and add placeholders)

You should also know how to multiply and divide with directed numbers.

Remember:

- The product of a negative and a positive value is always negative.
- The result of a division with one positive and one negative value is always negative.
- The product of two negative numbers is always positive.
- The result of dividing with two negative values is always positive.

For example: $4 \times -3 = -12$ $-4 \times -3 = 12$ $12 \div -4 = -3$ $-12 \div -3 = 4$

Example

Work out each calculation using a written method.

a 52×27 **b** $3462 \div 8$ **c** 5.2×2.7 **d** 324×98 **e** $6 \div 0.2$

Solution	Commentary
a Method A / Method B / Method C (see working below)	There are many different methods for multiplying two 2-digit numbers. Here is a reminder of some of these. Use the one you are most comfortable with.

a **Method A**

		5	2
	×	2	7
	3	₁6	4
1	0	4	0
1	4	0	4
	₁		

Method B

×	50	2
20	1000	40
7	350	14

1	0	0	0
	3	5	0
		4	0
		1	4
1	4	0	4
	₁		

Method C

```
      5       2
   1 /1   0 /0
 1 / /0   / /4  2
 4 /3   1 /1
   / /5   / /4  7
      0       4
= 1404
```

b   **Method A**

		4	3	2	.	7	5
8	3	4	^{2}6	^{2}2	.	^{6}0	^{4}0

Method B

		4	3	2	.	7	5
8	3	4	6	2			
	3	2					
		2	6	2			
		2	4				
			2	2			
			1	6			
				6			

etc.

When dividing, you can use either short division or long division. Again, use the method that you are most comfortable with. You can copy and complete the long division shown in Method B yourself.

An alternative method for dividing by 8 could be to halve the number, halve the result and then halve the second result.

---

c

$$52 \times 27 = 1404$$

$\div 10 \qquad \div 10 \quad \div 100$

$$5.2 \times 2.7 = 14.04$$

Notice that the digits in this calculation are the same as those in part **a**, so you can use your knowledge of place value to find the answer.

Here, 52 is divided by 10 to give 5.2, and 27 is divided by 10 to give 2.7

Overall, the calculation has been divided by 100, so the answer needs to be divided by 100

---

d $324 \times 98 = (324 \times 100) - (324 \times 2)$

$\qquad = 32\,400 - 648$

$\qquad = 31\,752$

98 can be rewritten as $100 - 2$, and both of these can be easily multiplied by 324 to help find the answer.

You could use the formal method for the subtraction or work it out in stages:

$32\,400 - 600 = 31\,800$, $31\,800 - 40 = 31\,760$, $31\,760 - 8 = 31\,752$

---

e $6 \div 0.2 = \dfrac{6}{0.2} = \dfrac{60}{2} = 30$

When dividing by decimals, rewrite the calculation as a fraction. Then use equivalent fractions to make the denominator an integer, and divide.

Practice (B) 🖩

1   Use an appropriate method to work out:

    a 32×6 **b** 46×9 **c** 98×9 **d** 321×7

    e 64×19 **f** 96×83 **g** 321×28 **h** 452×98

2   Choose the most appropriate method and work out:

    a 3.6×9 **b** 6×9.4 **c** 19.6×7.2 **d** 7.2×9.31

    e 0.4×35 **f** 0.8×1.8 **g** 0.35×0.6 **h** 0.07×0.584

3 Work out the calculations. Give non-integer answers in decimal form.

a $688 \div 8$

b $978 \div 6$

c $8406 \div 8$

d $9842 \div 9$

e $805 \div 15$

f $6435 \div 25$

g $965.4 \div 8$

h $30055 \div 40$

4 Work out the calculations.

a $5 \div 0.5$

b $7 \div 0.14$

c $18 \div 0.6$

d $25 \div 0.4$

e $0.8 \div 0.2$

f $0.45 \div 0.09$

g $0.008 \div 0.04$

h $0.0032 \div 0.08$

5 The areas of the shapes are equal.

Work out the value of x.

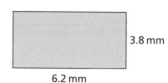
3.8 mm
6.2 mm

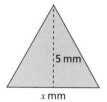
5 mm
x mm

6 Write the next two terms of each geometric sequence.

a $6, 18, 54, \_\_\_, \_\_\_$

b $80, 40, 20, \_\_\_, \_\_\_$

c $5, 12.5, 31.25, \_\_\_, \_\_\_$

d $800, 160, 32, \_\_\_, \_\_\_$

What do you think? (B) 💡

1 Use an efficient method to work out:

a 23×9

b 35×99

c 572×9.9

2 What is the relationship between the digits in a multiplication and the final digit of the answer?

Are you ready? (C) 📝

Work out the following calculations.

1 a 7^2

b 2^4

c 3^3

d $(-4)^2$

2 a $8 + 5 - 2$

b $9 - 4 + 8$

c $12 - 7 + 19$

d $32 + 18 - 14$

3 a $3 \times 6 \div 2$

b $40 \times 3 \div 10$

c $8 \div 4 \times 6$

d $18 \div 4 \times 6$

When a calculation involves more than one operation, you use the **order of operations** to decide which one should be done before another.

( ) ← Brackets first

^{2} & $\sqrt{}$ ← Then any powers (indices) or roots

× & ÷ ← Then multiplication and division at the same time; work from left to right

+ & − ← Finally, addition and subtraction at the same time; work from left to right

Example

Work out:

a $8 + 2 \times 7$ **b** $30 \div (2 + 3)$ **c** $3^2 \times 4 - 8 \div 2$ **d** $21 - (2 \times 3)^2$

Solution	Commentary
a $8 + \underline{2 \times 7}$ $= 8 + 14$ $= 22$	Multiplication takes priority over addition, so work out 2×7 first. Then calculate the addition.
b $30 \div \underline{(2 + 3)}$ $= 30 \div 5$ $= 6$	Calculate the brackets first. Then calculate the division.
c $\underline{3^2} \times 4 - 8 \div 2$ $= \underline{9 \times 4} - \underline{8 \div 2}$ $= 36 - 4$ $= 32$	Indices take priority here. Multiplication and division take priority over subtraction. Now calculate the subtraction.
d $21 - \underline{(2 \times 3)}^2$ $= 21 - \underline{6^2}$ $= 21 - 36$ $= -15$	Calculate the brackets first. The indices take priority over the subtraction. Finally, complete the subtraction.

Practice (C)

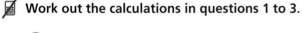 **Work out the calculations in questions 1 to 3.**

1 **a** $5 \times 3 + 4$ **b** $5 + 3 \times 4$ **c** $5^2 \times 3 + 4$ **d** $5 + 3^2 \times 4$

2 **a** $7 + 4^2$ **b** $8^2 \times 3 - 4$ **c** $(7 - 3)^2$ **d** $(3 + 4) \times 6$

   e $20 \div (3^2 + 1)$ **f** $10 \times (4^2 - 6)$ **g** $\sqrt{25} + 4^2$ **h** $(2 \times 4)^2 - 8 \div 2$

3 **a** $9 - 3 \times -4$ **b** $(-3 - 6) \div 3$ **c** $(-5 + 2)^2$ **d** $-3 \times 4 - 6$

4 Write brackets to make the calculations correct.

   a $2 + 5 \times 3 - 1 = 20$ **b** $6^2 - 4 \times 3 + 6 = 0$

   c $5 - 3 \times 8 \div 2 = 8$ **d** $18 \times 4^2 \div 4 \times 8 = 9$

5 Given $a = 3$, work out the value of each expression.

   a $3a$ **b** $3a^2$ **c** $(3a)^2$ **d** $(3 + a)^2$

6 Given $x = -3$, $y = 5$ and $z = 0.4$, work out the value of each expression.

   a xy **b** $x + yz$ **c** $x^2 - 2y$ **d** xyz

What do you think? (C) 💭

1 **a** Use the digits 3, 4, 6 and 7 to make the statement true.

$$\square \times \square^2 + \square \times \square = 165$$

b What different answers can you find using 3, 4, 6 and 7 and the same operations as the calculation in part **a**?

Consolidate – do you need more?

Work out the calculations using appropriate methods.

1 **a** $637 + 842$ **b** $15\,477 + 23\,485$ **c** $843.55 + 432.8$ **d** $0.216 + 0.084$

2 **a** $1653 - 964$ **b** $15\,574 - 3656$ **c** $3489.2 - 321.86$ **d** $0.0089 - 0.000\,31$

3 **a** 684×26 **b** 68×348 **c** 3.4×6.32 **d** 0.052×0.84

4 **a** $5464 \div 8$ **b** $34\,174 \div 5$ **c** $84 \div 1.2$ **d** $55.25 \div 0.85$

5 **a** $-8 - 3$ **b** -8×-3 **c** $8 - -3$ **d** $-63 \div 9$

     e $-63 - -52$ **f** 16×-9 **g** $-162 - 345$ **h** $-135 \div -5$

6 **a** $13 + 6 \times 3$ **b** $6^2 - 2 \times 4$ **c** $5 \times 4 - 7^2$ **d** $2^3 + 3^2$

     e $(5 \times 3^2 - 4) \times 9$ **f** $\sqrt{16} \times (7 + 8)$ **g** $10 - (3 + (-2)^2)$ **h** $(\sqrt{9} + 3 \times 7)^2$

7 Write the next two terms in each sequence.

     a $9, 27, 81,$ _____ , _____   **b** $-86, -79, -72,$ _____ , _____

     c $0.96, 0.79, 0.62,$ _____ , _____   **d** $3000, 600, 120,$ _____ , _____

Stretch – can you deepen your learning?

1 **a** Using the digits 5, 6, 7, 8 and 9, what is the greatest product that can be made?

b Investigate the different possible greatest sums using numbers with different amounts of digits.

2 How many ways can you use the digits 1, 2, 3 and 4 exactly once, the four operations and brackets to get the answer 24?

Investigate with other sets of digits.

Are you ready? (A)

1 Simplify each fraction.

   a $\dfrac{18}{24}$ **b** $\dfrac{27}{30}$ **c** $\dfrac{48}{72}$ **d** $\dfrac{90}{108}$

2 Work out:

   a $\dfrac{1}{7} + \dfrac{3}{7}$ **b** $\dfrac{9}{11} - \dfrac{3}{11}$ **c** $3\dfrac{2}{7} + 1\dfrac{3}{7}$ **d** $4\dfrac{3}{8} - 2\dfrac{1}{8}$

3 Write as a mixed number in its simplest form.

   a $\dfrac{72}{5}$ **b** $6\dfrac{13}{10}$ **c** $\dfrac{8}{12}$ less than 9   **d** $12\dfrac{15}{9}$

Using your calculator 🔲

Calculators will give you a fraction answer in its simplest form.

When typing a mixed number into a calculator, you often need to press <kbd>SHIFT</kbd> before the fraction button.

In some models, to convert an answer given as an improper fraction to a mixed number, press <kbd>SHIFT</kbd> then <kbd>S⇌D</kbd>.

To add or subtract fractions with the same denominator, you only have to add or subtract their numerators. If the denominators are different, convert the fractions so that they have a common denominator. If you use the lowest common denominator (the LCM of the denominators), you will be working with smaller numbers.

Always give your answer in the form specified in the question; for example, simplest form, as a mixed number, or both.

Example 1

Work out:

   a $\dfrac{3}{8} + \dfrac{1}{6}$ **b** $2\dfrac{2}{5} + 1\dfrac{3}{10}$

Method

Solution	Commentary
a $\overset{\times 3}{}\dfrac{3}{8}\overset{\times 3}{} + \overset{\times 4}{}\dfrac{1}{6}\overset{\times 4}{}$ $= \dfrac{9}{24} + \dfrac{4}{24} = \dfrac{13}{24}$	To add fractions together, the denominators need to be the same. The lowest common multiple of 8 and 6 is 24 (you will revisit LCM in Chapter 1.3). Convert the given fractions to equivalent fractions with the denominator 24, then add. As 13 and 24 have no common factor, the answer is in simplest form.

b   Method A

$2\frac{2}{5} + 1\frac{3}{10}$

$\frac{2}{5} + \frac{3}{10} = \frac{4}{10} + \frac{3}{10} = \frac{7}{10}$

$2 + 1 + \frac{7}{10} = 3\frac{7}{10}$

One method for adding mixed numbers is to add the fractional parts and the integers separately.

Notice that the LCM to use with 5 and 10 is 10 itself.

Method B

$2\frac{2}{5} + 1\frac{3}{10}$

$= \frac{12}{5} + \frac{13}{10} = \frac{24}{10} + \frac{13}{10} = \frac{37}{10} = 3\frac{7}{10}$

A different method is to convert the mixed numbers to improper fractions, then add these by making the denominators the same.

Once added, convert the improper fraction back to a mixed number.

This method is not suitable if working with numbers with large integer parts.

Example 2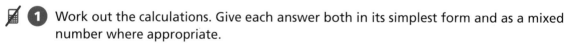

Work out $5\frac{2}{5} - 2\frac{8}{15}$

Give your answer as a mixed number in its simplest form.

Method

Solution	Commentary
$5\frac{2}{5} - 2\frac{8}{15}$ $5 - 2 = 3$ $\frac{2}{5} - \frac{8}{15} = \frac{6}{15} - \frac{8}{15} = -\frac{2}{15}$	Subtract the integer parts and fractional parts separately. Notice that the answer for the fractional parts is negative.
So $5\frac{2}{5} - 2\frac{8}{15} = 3 - \frac{2}{15} = 2\frac{15}{15} - \frac{2}{15} = 2\frac{13}{15}$	Now recombine the integer and fractional parts. To deal with the negative fraction, think of 3 as $2 + 1 = 2 + \frac{15}{15}$, or $2\frac{15}{15}$

Practice (A)

1 Work out the calculations. Give each answer both in its simplest form and as a mixed number where appropriate.

a $\frac{5}{8} + \frac{1}{4}$ b $\frac{8}{15} + \frac{2}{5}$ c $\frac{1}{2} + \frac{5}{16}$ d $\frac{1}{6} + \frac{7}{24}$

e $\frac{3}{8} + \frac{2}{5}$ f $\frac{4}{7} + \frac{5}{8}$ g $\frac{8}{15} + \frac{3}{10}$ h $\frac{13}{16} + \frac{13}{20}$

2 Work out the calculations. Give each answer in its simplest form.

a $\frac{5}{36} - \frac{2}{9}$ b $\frac{9}{16} - \frac{3}{4}$ c $\frac{7}{12} - \frac{5}{6}$ d $-\frac{1}{2} + \frac{5}{9}$

e $\frac{2}{7} - \frac{5}{9}$ f $\frac{5}{6} - \frac{3}{4}$ g $\frac{8}{9} - \frac{5}{6}$ h $\frac{5}{12} - \frac{5}{8}$

3 Work out the calculations. Give each answer both in its simplest form and as a mixed number where appropriate.

a $\dfrac{6}{8} + \dfrac{5}{12} - \dfrac{5}{6}$ 　　**b** $\dfrac{5}{18} - \dfrac{2}{3} + \dfrac{11}{15}$ 　　**c** $\dfrac{1}{8} + \dfrac{2}{5} + \dfrac{4}{15}$ 　　**d** $1 - \dfrac{5}{7} - \dfrac{8}{9}$

4 Work out the calculations. Give each answer as a mixed number in its simplest form.

a $2\dfrac{8}{9} - 1\dfrac{3}{8}$ 　　**b** $3\dfrac{5}{6} - 1\dfrac{3}{8}$ 　　**c** $2\dfrac{5}{12} - 1\dfrac{7}{8}$ 　　**d** $6\dfrac{2}{7} + \dfrac{3}{8}$

e $1\dfrac{8}{9} - 2\dfrac{3}{4}$ 　　**f** $-3\dfrac{6}{11} - 2\dfrac{1}{5}$ 　　**g** $6\dfrac{2}{9} + 2\dfrac{3}{5} - \dfrac{9}{15}$ 　　**h** $19\dfrac{2}{15} + 13\dfrac{8}{45}$

5 Work out the perimeters of the shapes.

a

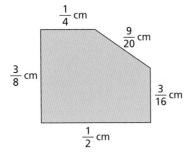

b

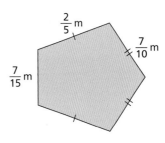

c

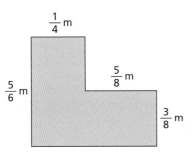

d

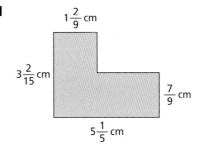

6 Data is collected on how far students in a class can throw a ball.

The shortest distance is $2\dfrac{5}{8}$ m.

The greatest distance is $8\dfrac{2}{9}$ m.

a Work out the range of the distances that the ball is thrown.

b For another class, the range of the data is $\dfrac{7}{12}$ m.

The greatest distance is $1\dfrac{2}{9}$ m.

Work out the shortest distance that the ball is thrown.

7 Solve the equations.

a $r + \dfrac{2}{5} = \dfrac{4}{9}$ 　　　　**b** $p - 1\dfrac{3}{8} = 5\dfrac{2}{7}$

8 Use your calculator to check your answers to questions 1–7

What do you think? (A)

1 Give an example to support and a counterexample to contradict each statement.

  a When adding two fractions, the answer will be a mixed number.

  b When subtracting two fractions, the answer will **not** be a mixed number.

  Make up some more statements about adding and subtracting fractions, and challenge a partner to decide if they are **always**, **sometimes** or **never** true.

2 Would you approach these calculations by converting both numbers to decimals or by converting both numbers to fractions?

$$\frac{3}{5} + 0.1 \qquad 2.5 - \frac{1}{5} \qquad \frac{3}{4} + 0.6 \qquad 0.4 + 1\frac{1}{3} \qquad 3.6 - 1\frac{1}{5}$$

Compare your choices and your answers with a partner.

Are you ready? (B)

1 Convert the mixed numbers to improper fractions.

  a $3\frac{2}{7}$ **b** $2\frac{5}{9}$ **c** $9\frac{3}{10}$ **d** $5\frac{7}{8}$

2 Convert the improper fractions to mixed numbers.

  a $\frac{19}{8}$ **b** $\frac{17}{6}$ **c** $\frac{63}{8}$ **d** $\frac{54}{7}$

3 Match each calculation in the top row to its equivalent in the bottom row.

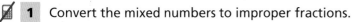

| $16 \div 2$ | 0.75×16 | 16×2 | 0.4×16 | 0.25×16 |

| $16 \div 0.5$ | $16 \div 10 \times 4$ | 0.5×16 | $16 \times 3 \div 4$ | $16 \div 4$ |

To multiply fractions, you multiply the numerators and multiply the denominators:

$$\frac{4}{5} \times \frac{6}{7} = \frac{24}{35}$$

Look for factors to make the calculations easier: $\dfrac{\cancel{7}^{1}}{\cancel{12}_{4}} \times \dfrac{\cancel{9}^{3}}{\cancel{14}_{2}} = \dfrac{3}{8}$

The **reciprocal** of a fraction is found by swapping the numerator and denominator.

For example, the reciprocal of $\frac{3}{5}$ is $\frac{5}{3}$

Notice that the product of a fraction and its reciprocal is always 1

For example: $\dfrac{3}{5} \times \dfrac{5}{3} = \dfrac{15}{15} = 1$

To **divide by a fraction**, you multiply by its reciprocal.

$$\frac{4}{7} \div \frac{5}{9} = \frac{4}{7} \times \frac{9}{5} = \frac{36}{35} = 1\frac{1}{35}$$

Example 1

Work out:

a $\dfrac{2}{7} \times \dfrac{3}{8}$ **b** $\dfrac{5}{9} \div \dfrac{2}{3}$

Method

Solution	Commentary
a **Method A** $\dfrac{2}{7} \times \dfrac{3}{8} = \dfrac{2 \times 3}{7 \times 8}$ $= \dfrac{6}{56} = \dfrac{3}{28}$	To multiply proper fractions, multiply the numerators and multiply the denominators. Simplify the answer if possible.
Method B $\dfrac{\overset{1}{\cancel{2}}}{7} \times \dfrac{3}{\underset{4}{\cancel{8}}}$ $= \dfrac{3}{28}$	Look for common factors; in this example, 2 is a factor of 2 and 8 Once all numbers have been divided by any common factors, multiply the numerators and multiply the denominators.
b **Method A** $\dfrac{5}{9} \div \dfrac{2}{3}$ $\dfrac{5}{\underset{3}{\cancel{9}}} \times \dfrac{\overset{1}{\cancel{3}}}{2} = \dfrac{5}{6}$	Dividing by a number is the same as multiplying by its reciprocal. The reciprocal of $\dfrac{2}{3}$ is $\dfrac{3}{2}$ You can use the common factor 3 to make the multiplication easier.
Method B $\dfrac{5}{9} \div \dfrac{2}{3} = \dfrac{5}{9} \div \dfrac{6}{9} = \dfrac{5}{6}$	Another method for dividing fractions is to convert both fractions to equivalent fractions with the same denominator. Then, divide the numerators.

Example 2

Work out:

a $\dfrac{3}{5} \times 4$ **b** $\dfrac{2}{3} \div 5$ **c** $2\dfrac{3}{7} \times 1\dfrac{2}{9}$ **d** $3\dfrac{1}{8} \div 2\dfrac{3}{4}$

Method

Solution	Commentary
a $\dfrac{3}{5} \times 4 = \dfrac{12}{5}$ $= 2\dfrac{2}{5}$	To multiply a fraction by an integer, just multiply the numerator by the integer. In this case, you can think of 4 as $\dfrac{4}{1}$ $\dfrac{3}{5} \times \dfrac{4}{1} = \dfrac{12}{5} = 2\dfrac{2}{5}$
b $\dfrac{2}{3} \div 5 = \dfrac{2}{3} \times \dfrac{1}{5}$ $= \dfrac{2}{15}$	To divide a fraction by an integer, multiply by the reciprocal of the integer.

c $2\frac{3}{7} \times 1\frac{2}{9}$ $\frac{17}{7} \times \frac{11}{9} = \frac{187}{63}$ $= 2\frac{61}{63}$	When multiplying mixed numbers, the first step is to convert the mixed numbers to improper fractions. Then multiply them in the same way as you would with proper fractions. Convert the answer back to a mixed number and simplify if possible.
d $3\frac{1}{8} \div 2\frac{3}{4}$ $\frac{25}{8} \div \frac{11}{4}$ $\frac{25}{{}_2\cancel{8}} \times \frac{\cancel{4}^{1}}{11} = \frac{25}{22} = 1\frac{3}{22}$	Use the same approach when dividing mixed numbers, remembering to multiply by the reciprocal after converting to improper fractions.

Practice (B)

1 Work out the calculations. Give each answer in its simplest form.

   a $\frac{1}{8} \times \frac{1}{6}$ **b** $\frac{1}{5} \times \frac{3}{7}$ **c** $\frac{5}{8} \times \frac{3}{10}$ **d** $\frac{3}{10} \times \frac{5}{6}$

   e $\frac{6}{11} \times \frac{3}{22}$ **f** $\frac{4}{9} \times \frac{5}{8}$ **g** $\frac{1}{2} \times \frac{3}{8} \times \frac{4}{7}$ **h** $\frac{5}{9} \times \frac{6}{15} \times \frac{1}{2}$

2 Work out the calculations. Give each answer in its simplest form.

   a $3 \times \frac{1}{5}$ **b** $\frac{3}{8} \times 2$ **c** $6 \times \frac{2}{5}$ **d** $4 \times 1\frac{1}{5}$

3 Write the reciprocal of each number.

   a $\frac{1}{5}$ **b** $\frac{1}{7}$ **c** $\frac{3}{4}$ **d** $\frac{5}{9}$

   e 6 **f** -8 **g** 0.6 **h** 0.875

Work out the calculations in questions 4–7. Give each answer in its simplest form and as a mixed number where appropriate.

4 **a** $\frac{1}{8} \div \frac{1}{6}$ **b** $\frac{1}{5} \div \frac{3}{7}$ **c** $\frac{5}{8} \div \frac{3}{4}$ **d** $\frac{3}{10} \div \frac{2}{5}$

   e $6 \div \frac{1}{5}$ **f** $\frac{9}{20} \div 4$ **g** $\frac{5}{8} \div \frac{3}{4} \div \frac{1}{2}$ **h** $\frac{5}{9} \times \frac{6}{15} \div \frac{5}{6}$

5 **a** $1\frac{2}{5} \times \frac{1}{8}$ **b** $\frac{5}{6} \times 2\frac{2}{5}$ **c** $1\frac{2}{7} \times 3\frac{2}{5}$ **d** $1\frac{2}{3} \times 3\frac{3}{5}$

6 **a** $1\frac{2}{5} \div \frac{3}{8}$ **b** $3\frac{3}{4} \div 2\frac{3}{8}$ **c** $3\frac{1}{9} \div 2\frac{1}{8}$ **d** $4\frac{2}{7} \div 3\frac{1}{3}$

7 **a** $\frac{2}{7} \times \frac{3}{8} \div \frac{3}{7}$ **b** $2\frac{3}{5} \div \frac{2}{7} \times \frac{5}{8}$ **c** $\left(\frac{2}{7}\right)^2 \div \frac{3}{4}$ **d** $1\frac{5}{7} \div \left(2\frac{1}{4}\right)^2$

8 Calculate the area of each shape.

a

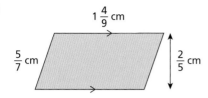

b

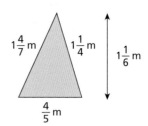

c

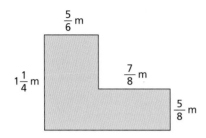

d

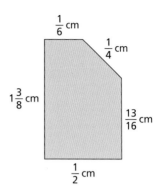

9 Work out:

a $\dfrac{5}{6} + \dfrac{2}{7} \times \dfrac{5}{6}$　　　**b** $2\dfrac{2}{5} \times \left(\dfrac{2}{7} - \dfrac{5}{9}\right)$　　　**c** $\dfrac{8}{11} \div \dfrac{5}{8} + \dfrac{2}{5}$　　　**d** $\left(2\dfrac{3}{7} + 1\dfrac{5}{8}\right) \div 1\dfrac{2}{5}$

10 Use your calculator to check your answers to questions 1–9

What do you think? (B) 💭

1 Use the digits 3 to 8 exactly once each so the calculation has the greatest possible value.

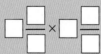

2 Would you approach these calculations by converting both numbers to decimals or by converting both numbers to fractions?

$\dfrac{3}{5} \times 0.1$　　$2.5 \div \dfrac{1}{5}$　　$\dfrac{3}{4} \div 0.6$　　$0.4 \div 1\dfrac{1}{3}$　　$3.6 \times 1\dfrac{1}{5}$

Compare your choices and your answers with a partner.

Consolidate – do you need more?

Work out the calculations. Give each answer both in its simplest form and as a mixed number where appropriate.

1 a $\frac{3}{7} + \frac{6}{11}$ b $\frac{7}{9} - \frac{2}{5}$ c $\frac{1}{6} + \frac{3}{8}$ d $\frac{3}{8} - \frac{5}{12}$

e $\frac{19}{30} - \frac{5}{12}$ f $\frac{29}{100} + \frac{9}{20}$ g $\frac{11}{13} + \frac{1}{4}$ h $\frac{5}{16} - \frac{9}{14}$

2 a $1\frac{2}{9} + 2\frac{3}{5}$ b $3\frac{5}{6} - 2\frac{7}{9}$ c $4\frac{11}{15} - 2\frac{2}{3}$ d $2\frac{7}{8} - 1\frac{3}{5}$

e $3\frac{1}{15} + 2\frac{5}{8}$ f $5\frac{3}{8} - 2\frac{7}{9}$ g $3\frac{2}{15} + 5\frac{5}{6}$ h $8\frac{2}{3} - 5\frac{7}{10}$

3 a $\frac{3}{7} \times \frac{2}{5}$ b $\frac{3}{8} \times \frac{4}{9}$ c $\frac{10}{11} \times 7$ d $\left(\frac{3}{8}\right)^2$

e $2\frac{3}{7} \times \frac{2}{5}$ f $2\frac{2}{5} \times 3\frac{2}{9}$ g $2\frac{3}{8} \times 1\frac{5}{7}$ h $\left(2\frac{3}{5}\right)^2$

4 a $\frac{3}{8} \div \frac{11}{16}$ b $\frac{17}{20} \div \frac{7}{10}$ c $\frac{7}{12} \div \frac{8}{15}$ d $\frac{9}{11} \div 6$

e $1\frac{5}{7} \div \frac{2}{5}$ f $3\frac{2}{9} \div 1\frac{2}{5}$ g $2\frac{5}{8} \div 3\frac{6}{7}$ h $4\frac{3}{10} \div 4\frac{5}{8}$

Stretch – can you deepen your learning?

1 The volume of a cube is $\frac{27}{64}$ cm³.

Calculate the total surface area of the cube.

2 Write three fractions with different denominators that have:

a a total of 1 **b** a product of 1

3 **a** Use fraction multiplication to show that $0.6 \times 1.25 = \frac{3}{4}$

b Further investigate using fraction multiplication or division to solve decimal calculations.

Are you ready? (A)

1 Which of these numbers are prime?

   1   2   5   8   12   19   23   27   29

2 Here is a list of numbers:

   3   6   10   15   18   35   53   108   180

   a Which of the numbers are multiples of 3?

   b Which of the numbers are multiples of 5?

   c Which of the numbers are multiples of both 3 and 5?

Using your calculator

When using a calculator to find a product of prime factors, you may need to type in the number, press **=** then press 'FACT'. Check how your calculator works.

Factors are numbers that exactly divide another number. For example, the factors of 10 are 1, 2, 5 and 10

A **prime number** has exactly two factors, 1 and itself.

Note that 1 is not a prime number as it only has one factor, and 2 is the only even prime number.

Any number can be written as a **product** of its **prime factors** in exactly one way. This is often called its **prime factor decomposition**.

Multiples of a number are found by multiplying the number by a positive integer.
For example, the multiples of 8 start 8, 16, 24, 32, 40 …

Example 1

Write 180 as the product of its prime factors.

Method

Solution	Commentary
180 factor tree: 180 = 2 × 90 90 = 2 × 45 45 = 5 × 9 9 = 3 × 3 (with 2, 2, 5, 3, 3 circled as primes)	To start a prime factor tree, first identify a pair of factors. $180 = 2 \times 90$ 2 is prime, so you can circle this and that branch ends there. 90 is not prime, so you need to continue. $90 = 2 \times 45$ Again, 2 is prime so you can circle this and end that branch. 45 is not prime, so you need to continue. $45 = 5 \times 9$ 5 is prime, but 9 is not. $9 = 3 \times 3$ Now that all branches are ended, you can write 180 as the product of its prime factors.
$180 = 2 \times 2 \times 3 \times 3 \times 5$	Write the prime factors in ascending order, so $180 = 2 \times 2 \times 3 \times 3 \times 5$ You can check your answer by working out $2 \times 2 \times 3 \times 3 \times 5$. It should give you an answer of 180 This is the prime factor decomposition of 180
$180 = 2^2 \times 3^2 \times 5$	This is the result in **index form**, rather than writing out all the prime factors individually.

In the above example, you can start with any factor pair for 180 because the answer will always be the same.

Example 2

Given that $300 = 2^2 \times 3 \times 5^2$, work out the smallest number that 300 can be multiplied by to give a square number.

Method

Solution	Commentary
$(2^2 \times 3 \times 5^2) \times 3 = 2^2 \times 3^2 \times 5^2$ $2^2 \times 3^2 \times 5^2 = (2 \times 3 \times 5)^2$, so the smallest number is 3	When a square is written as a product of prime factors, each prime will have an even power. This is because squaring involves multiplying a number by itself. Only the 3 has an odd power, so 300 must be multiplied by 3 to give a square number.

Practice (A)

1 Write each number as a product of its prime factors. Give your answers in index form.

    a 40   **b** 48   **c** 84   **d** 150

    e 196   **f** 245   **g** 1050   **h** 729

2 Use the fact that $60 = 2^2 \times 3 \times 5$ to write each number as a product of its prime factors.

    a 120   **b** 30   **c** 240   **d** 6000

3 Use the fact that $P = 3^2 \times 5^3 \times 7^3$ to write each expression as a product of prime factors.

    a $3P$ **b** $8P$ **c** $18P$ **d** $100P$

4 **a** Write 450 as a product of its prime factors.

    b What is the smallest number that 450 can be multiplied by to give a square number?

5 What is the smallest number that 540 can be multiplied by to give a square number?

6 What is the smallest number that 864 can be multiplied by to give a cube number?

7 $1260 = 2^2 \times 3^2 \times 5 \times 7$

    Work out if each number below is a factor of 1260. Justify your answers.

    a 35   **b** 20   **c** 24   **d** 90   **e** 50   **f** 135

What do you think? (A) 💡

1 $x = a^2 \times b^4 \times c^6$, $y = a^4 \times b^2 \times c^3$ where a, b and c are prime.

    a Which of these are square numbers?

         x y xy

    b Which of these are cube numbers?

         x y xy

    c Investigate whether other sums or products of x and y are square or cube numbers.

Are you ready? (B)

1 List all of the factors of each number.

    a 20   **b** 24   **c** 36   **d** 84

2 List the first four multiples of each number.

    a 8   **b** 12   **c** 20   **d** 24

3 Write each number as a product of its prime factors.

    a 20   **b** 72   **c** 100   **d** 120

The **highest common factor (HCF)** of a set of numbers is the greatest number that is a factor of all the numbers in the set.

The **lowest common multiple (LCM)** of a set of numbers is the smallest number that is a multiple of all the numbers in the set.

Example 1

Work out the highest common factor (HCF) of 45 and 60

Method

Solution	Commentary
The factors of 45 are 1, 3, 5, 9, 15, 45	List the factors of each number.
The factors of 60 are 1, 2, 3, 4, 5, 6, 10, 12, 15, 20, 30, 60	
The common factors of 45 and 60 are 1, 3, 5, 15	The **common** factors are the numbers that appear in both of the lists.
The highest common factor of 45 and 60 is 15	The HCF is the greatest number that appears in both lists.

Example 2

Work out the lowest common multiple (LCM) of 8 and 10

Method

Solution	Commentary
The multiples of 8 start 8, 16, 24, 32, 40, 48 … The multiples of 10 start 10, 20, 30, 40, 50 …	You could take a listing approach similar to that taken in Example 1, this time listing multiples.
The LCM of 8 and 10 is 40	Keep listing multiples of each number until you have a number that appears in both lists.

The listing approach can be inefficient. Example 3 that follows shows how you can use prime factorisation and a Venn diagram to find the HCF and LCM of a pair of numbers.

Example 3

You are given that $24 = 2^3 \times 3$ and $100 = 2^2 \times 5^2$

Work out:

a the HCF of 24 and 100   **b** the LCM of 24 and 100

Method

Solution	Commentary
Prime factors of 24 — Prime factors of 100 24: 2, 2, 3 — intersection: 2, 2 — 100: 5, 5	Both 24 and 100 have two 2s as prime factors so these go in the intersection.
a $2 \times 2 = 4$ So the HCF of 24 and 100 is 4	Multiply the common factors to find the HCF.
b $2 \times 3 \times 2 \times 2 \times 5 \times 5 = 600$ So the LCM of 24 and 100 is 600	To find the LCM, find the product of all the numbers in the Venn diagram.

Practice (B)

1 Work out the HCF and the LCM of each set of numbers.

    a 40 and 60   **b** 48 and 60   **c** 90 and 36   **d** 12, 18 and 30

2 **a** Write 60 and 84 as a product of their prime factors.

    b Copy and complete the Venn diagram.

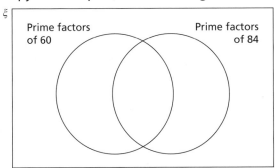

    c Use your Venn diagram to work out:

        i the HCF of 60 and 84   **ii** the LCM of 60 and 84

3 Work out the HCF and LCM of each pair of numbers.

    a 84 and 120   **b** 48 and 200   **c** 180 and 560   **d** 65 and 975

4 Work out the HCF and LCM of 120, 300 and 450

5 $P = 2^2 \times 3^2 \times 7$ and $Q = 2^3 \times 5^2 \times 7$

Work out:

   a   the HCF of P and Q **b**   the LCM of P and Q.

6 $R = 2^3 \times 5^2 \times 11$ $S = 5^3 \times 7^2$ $T = 2^2 \times 3^2 \times 5$

Identify:

   a   the HCF of R and S **b**   the HCF of R, S and T

   c   the LCM of S and T **d**   the LCM of R, S and T.

7 The HCF of two numbers is 12

The LCM of the same two numbers is 360

What might the two numbers be? How many answers can you find?

8 The LCM of X and Y is 30

The HCF of X and Y is greater than 5

Work out two possible pairs of values for X and Y.

9 $A = 2^2 \times 3^2 \times 5^3$

$B = 2^3 \times 5$

$C = 2^2 \times 3 \times 5^2$

Work out the LCM of A, B and C.

What do you think? (B) 💭

1 The LCM of A and B is 720

   a   If $A = 72$, explain why:

       i B cannot be 25   **ii** B cannot be 36   **iii** B could be 180

   b   Work out the other possible values of B.

Consolidate – do you need more?

1 Write the numbers as a product of prime factors in index form.

   a 50   **b** 90   **c** 130   **d** 300

2 Use powers to show that:

   a 576 is a square number   **b** 1296 is a square number   **c** 5832 is a cube number.

3 Work out the HCF of:

   a 32 and 80   **b** 60 and 150   **c** 42 and 105   **d** 240 and 144

4 Work out the LCM of:

    a 24 and 36   **b** 16 and 24   **c** 18 and 54   **d** 36 and 48

5 $A = 2^2 \times 3 \times 5$ $B = 2^3 \times 5^2 \times 7$

    a Work out the HCF of A and B.

    b Work out the LCM of A and B.

    c What is the least number that A can be multiplied by to give a square number?

6 $C = 2^3 \times 3 \times 7^2 \times 11$ $D = 2 \times 3^3 \times 7 \times 11$

    a Work out the HCF of C and D.

    b Work out the LCM of C and D.

    c What is the least number that C can be multiplied by to give a cube number?

Stretch – can you deepen your learning?

1 A bus from Halifax to Leeds leaves at 6:20 am and every 16 minutes after that.

A bus from Halifax to Bradford leaves at 6:20 am and every 18 minutes after that.

When is the next time after 6:20 am that both buses leave at the same time?

2 Marta has two lengths of wood.

One length of wood is 18 m long and the other length is 21 m long.

Marta wants to cut lengths of wood into shorter lengths that are of equal length.

Marta does not want any wood left over.

What is the greatest possible length of each of the shorter lengths of wood?

3 a is a common factor of 72 and 96

b is a common multiple of 8 and 15

    a Investigate the greatest value of $\dfrac{b}{a}$

    b Calculate the smallest value of ab.

4–6

1 Write 600 as a product of powers of its prime factors. **[3 marks]**

2 **(a)** Work out $2\frac{3}{5} + 3\frac{1}{4}$

   Give your answer as a mixed number. **[2 marks]**

   (b) Work out $4\frac{2}{3} \div 6$ **[2 marks]**

3 Work out 0.006×0.34 **[2 marks]**

4 Two numbers a and b are such that

   a is a multiple of 5

   b is a multiple of 3

   the highest common factor (HCF) of a and b is 7

   Write down a possible value for a and a possible value for b. **[2 marks]**

5 Work out $80\,000\,000 \div 200$ **[2 marks]**

6 Work out the lowest common multiple (LCM) of 108 and 120 **[3 marks]**

7 **(a)** Work out $-3 \times -\frac{5}{9}$

   Give your answer as a mixed number. **[2 marks]**

   (b) Work out the missing value.

   $0.407 = \frac{2}{5} + \boxed{}$ **[2 marks]**

8 Work out $74.88 \div 1.6$ **[3 marks]**

9 What is 0.18 as a fraction of 0.3?

   Give your answer in its simplest form. **[2 marks]**

10 Work out $\left(\frac{9}{10} - \frac{7}{15}\right) \div \frac{2}{3}$ **[3 marks]**

11 The first two terms of a sequence are 8 and −3

   The next terms of the sequence are found by adding the previous two terms.

   Work out the next **three** terms of the sequence. **[2 marks]**

12 Show that 3375 is a cube number. **[2 marks]**

2 Indices

White Rose
M▲THS

In this block, we will cover...

2.1 Powers and roots

Example 📝

Evaluate:

a 5^0 **b** 7^{-1}

Method

Solution	Commentary
a $5^0 = 1$	
b $7^{-1} = \dfrac{1}{7}$	A power of –1 is the of 7 is $\dfrac{1}{7}$
c 6^{-3}	This is the reciprocal

2.2 Standard form

Practice (A) 📝

1 Write each number in standard form.

   a 8000   **b** 200 000

   e 980 000   **f** 584 000

   i one million   **j** seven hundred th

   k 200^2 **l** 50^3

2 Write each number in standard form.

   a 0.03   **b** 0.000 08

   e 0.000 634   **f** 0.000 003 684

2.3 Fractional indices

Consolidate – do you need more

1 Work out:

   a $100^{\frac{1}{2}}$ **b** $8^{\frac{1}{3}}$

   e $125^{\frac{1}{3}}$ **f** $64^{\frac{1}{3}}$

2 Work out:

   a $8^{\frac{2}{3}}$ **b** $1000^{\frac{2}{3}}$

   e $16^{0.75}$ **f** $\left(\dfrac{16}{81}\right)^{\frac{3}{4}}$

2.1 Powers and roots

Are you ready? (A)

1 Write these in index form. For example, $3 \times 3 = 3^2$

    a $2 \times 2 \times 2 \times 2$ **b** $0.9 \times 0.9 \times 0.9$

    c $6 \times 6 \times 6 \times 6 \times 6 \times 6$ **d** $4 \times 4 \times 5 \times 5$

2 From the list: 9, 6, 27, 64, 125, 20, 80, 100, 1, 35, 216

    a identify the square numbers   **b** identify the cube numbers.

3   **a** List the first 10 square numbers.   **b** List the first 10 cube numbers.

Index form is a short way of writing repeated multiplication. For example, $3^5 = 3 \times 3 \times 3 \times 3 \times 3$ and $6^4 = 6 \times 6 \times 6 \times 6$

The opposite of repeated multiplication is finding roots.

You know from earlier learning that, for example, $6 \times 6 = 36$ links to $\sqrt{36} = 6$ and $2 \times 2 \times 2 = 8$ links to $\sqrt[3]{8} = 2$. Higher powers and roots work in the same way.

$5^4 = 625$ and $\sqrt[4]{625} = 5$. These statements are read as '5 to the power 4 is equal to 625' and 'The fourth root of 625 is equal to 5'.

Using your calculator

All scientific calculators have a key to help you work out powers that you may not know. This button may be labelled ⬭$x^\blacksquare$ or ⬭x^y or ⬭y^x. Check you can find the 'power' button on your own calculator.

To find 2^{15}, you press ⬭2 ⬭$x^\blacksquare$ (or your different index key) ⬭1 ⬭5. You should get the answer 32 768

To find the 8th root of 6561, you press ⬭SHIFT ⬭$x^\blacksquare$ and see the display $\sqrt[\square]{\square}$. Enter the required digits to see $\sqrt[8]{6561}$. You should get the answer 3

You can **estimate** powers of numbers by considering the powers of the closest integers to the number.

You can use a **number line** to help your reasoning:

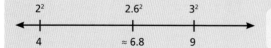

This number line shows that $2.6^2 \approx 6.8$ using the fact that $2^2 = 4$ and $3^2 = 9$

Roots can be estimated by considering roots that have integer values which are above and below the number that you are trying to find the root of.

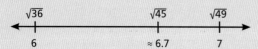

This number line shows that $\sqrt{45} \approx 6.7$ using the fact that $\sqrt{36} = 6$ and $\sqrt{49} = 7$

Example 1

Work out the value of:

a 9^3 **b** $\sqrt{144}$ **c** $\sqrt[3]{64}$

Method

Solution	Commentary

a $9^3 = 9 \times 9 \times 9$
      $= 81 \times 9$
      $= 729$
	The power is 3, which means you are working out the product of three 9s.
	 Work this out in parts, rather than multiplying all at once.

b $\sqrt{144} = 12$	When calculating the square root, look for the number that when multiplied by itself gives the answer 144 You should know at least the first 12 square numbers.
c $\sqrt[3]{64} = 4$	To find the cube root of 64, you need to find the number that multiplied by itself three times gives the answer 64. This is 4 since $4 \times 4 \times 4 = 64$. It is useful to know the first 10 cube numbers.

Example 2

Estimate the value of:

a 3.8^2 **b** $\sqrt{130}$

Method

Solution	Commentary
a $3.8^2 \approx 4^2$ $= 16$ $3.8^2 \approx 16$	When estimating, round the number to 1 significant figure. As 3.8 is less than 4, you know the actual value of 3.8^2 will be less than 16, so 16 is an **overestimate**.
b $\sqrt{121} < \sqrt{130} < \sqrt{144}$ $11 < \sqrt{130} < 12$ $\sqrt{130} \approx 11.4$	Think about the two square numbers that 130 lies between (121 and 144). As 130 is closer to 121 than 144, $\sqrt{130}$ is likely to be less than 11.5, so 11.4 is a good estimate. You can use a number line to help you if you like.

Number line:
$\sqrt{121}$ (11) $\sqrt{130}$ (≈ 11.4) $\sqrt{144}$ (12)

Practice (A)

1 Work out:

     a 3^4 **b** 4^3 **c** 8^3 **d** 5^4

     e 2^6 **f** 4^4 **g** 2.5^2 **h** 1.5^3

2 Use a calculator to work out:

     a 3^7 **b** 5^5 **c** 15^3 **d** 120^3

     e 0.8^3 **f** 3.2^4 **g** 92^4 **h** 23.4^3

3 Use a calculator to work out:

a $\sqrt{289}$ b $\sqrt{361}$ c $\sqrt{2209}$ d $\sqrt{6889}$

e $\sqrt{46.24}$ f $\sqrt{8611.84}$ g $\sqrt{0.0025}$ h $\sqrt{0.6724}$

4 Use a calculator to work out:

a $\sqrt[3]{216}$ b $\sqrt[4]{6561}$ c $\sqrt[3]{551.368}$ d $\sqrt[4]{1785.0625}$

5 Estimate the values of:

a 9.8^2 b 103.4^3 c 3.496^4 d 1.98^7

6 Estimate the roots to 1 decimal place.

a $\sqrt{27}$ b $\sqrt{150}$ c $\sqrt{200}$ d $\sqrt{394}$

e $\sqrt[3]{30}$ f $\sqrt[3]{150}$ g $\sqrt[3]{9}$ h $\sqrt{0.85}$

What do you think? (A)

1 Use different combinations of integers to complete the boxes below.

a $\square^{\square} = 64$ $\square^{\square} = 64$ $\square^{\square} = 64$

b $\square^{\square} = 81$ $\square^{\square} = 81$

2 Estimate the value of $\sqrt{14.8^2 - 2.4 \times 3.95}$

3 Challenge a partner to estimate powers and roots of different numbers. Compare your strategies and how close your estimates are to the actual answers.

Are you ready? (B)

1 Work out the value of:

a 3^4 b 3^3 c 3^2 d 3^1

2 Write the reciprocal of each number.

a 8 b 15 c $\frac{1}{4}$

d $\frac{5}{7}$ e $2\frac{1}{9}$ f 0.25

In this section you will explore raising numbers to the power zero and negative numbers. You should know that:

- any number raised to the power of 0 is 1
- negative powers are the reciprocal of the same positive power.

These facts are illustrated in this example:

$$3^3 = 27$$
$$\begin{array}{c}\\ \div 3\end{array}$$
$$3^2 = 9$$
$$\begin{array}{c}\\ \div 3\end{array}$$
$$3^1 = 3$$
$$\begin{array}{c}\\ \div 3\end{array}$$
$$3^0 = 1$$
$$\begin{array}{c}\\ \div 3\end{array}$$
$$3^{-1} = \frac{1}{3}$$
$$\begin{array}{c}\\ \div 3\end{array}$$
$$3^{-2} = \frac{1}{3^2} \text{ or } \frac{1}{9}$$
$$\begin{array}{c}\\ \div 3\end{array}$$
$$3^{-3} = \frac{1}{3^3} \text{ or } \frac{1}{27}$$

As the power reduces by 1 each time, you divide by 3 again.

As a result, for example, 3^{-2} is the reciprocal of 3^2

Example 📱

Evaluate:

a 5^0 **b** 7^{-1} **c** 6^{-3} **d** $\left(\dfrac{3}{7}\right)^{-2}$

Method

Solution	Commentary
a $5^0 = 1$	
b $7^{-1} = \dfrac{1}{7}$	A power of -1 is the reciprocal of the number. The reciprocal of 7 is $\dfrac{1}{7}$
c 6^{-3} $= \dfrac{1}{6^3} = \dfrac{1}{216}$	This is the reciprocal of 6^3
d $\left(\dfrac{3}{7}\right)^{-2}$ $= \left(\dfrac{7}{3}\right)^2 = \dfrac{49}{9}$ $= 5\dfrac{4}{9}$	This is the reciprocal of $\left(\dfrac{3}{7}\right)^2$ When squaring a fraction, square both the numerator and the denominator. Where necessary, write your answer as a mixed number.

Practice (B) ✍

1 Evaluate, giving your answer as a fraction.

   a 5^{-1} **b** 9^0 **c** 8^{-1} **d** 636^{-1}

   e 2^{-3} **f** 4^{-2} **g** 7^{-2} **h** 3^{-4}

2 Write in index form.

   a $\dfrac{1}{7}$ **b** $\dfrac{1}{6}$ **c** $\dfrac{1}{2^3}$ **d** $\dfrac{1}{8^4}$

3 Write in the form 2^n

   a $\dfrac{1}{2}$ **b** $\dfrac{1}{16}$ **c** $\dfrac{1}{64}$ **d** $\dfrac{1}{8}$

4 Evaluate, giving your answer as a fraction.

   a $\left(\dfrac{1}{5}\right)^{-2}$ **b** $\left(\dfrac{3}{5}\right)^{-3}$ **c** $\left(\dfrac{2}{3}\right)^{-4}$ **d** $\left(1\dfrac{3}{8}\right)^{-2}$

5 Work out:

   a $5^{-2} \times 10$ **b** $\dfrac{3}{8} \times 4^{-1}$ **c** $10^{-1} \div 5^{-2}$ **d** $6^{-2} \times \left(\dfrac{2}{3}\right)^{-2}$

What do you think? (B) 💡

1 Sven says that $5^2 \times 5^{-2} = 5$

   Do you agree? Give a reason for your answer.

2 Write the correct symbol, <, > or =, to make the statement correct.

   $(3^3)^3$ ☐ $3^{(3^3)}$

   Investigate other powers with brackets.

3 Work out the value of $2^{20} \div (2^2)^{12}$

Consolidate – do you need more? ✍

1 Work out:

   a 3^4 **b** 7^3 **c** 5^3

   d 9^3 **e** 16^2 **f** 13^3

2 Work out:

   a $\sqrt{121}$ **b** $\sqrt{196}$ **c** $\sqrt[3]{64}$

   d $\sqrt{64}$ **e** $\sqrt[3]{125}$ **f** $\sqrt[4]{10\,000}$

3 Estimate the roots to 1 decimal place.

   a $\sqrt{50}$ **b** $\sqrt{80}$ **c** $\sqrt[3]{10}$

   d $\sqrt{180}$ **e** $\sqrt[3]{100}$ **f** $\sqrt[3]{85}$

4 Work out:

    a 3^{-1} **b** 5^{-1} **c** $\left(\dfrac{4}{5}\right)^{-1}$

    d 5^{-3} **e** 6^{-2} **f** $\left(1\dfrac{2}{3}\right)^{-2}$

5 Write each set of numbers in ascending order.

    a 6^{-2} 5^{-2} 2^2 5^0 2^{-8}

    b $2 + 4^{-1}$ 2×4^{-1} $2 - 4^{-1}$ $2 \div 4^{-1}$ 4×2^{-1}

Stretch – can you deepen your learning?

1 Use each of the digits once to make the statements true.

    -3 -2 0 1 2 3 3 4 5 6

    $\square^{\square} = 0$ $\left(\dfrac{\square}{\square}\right)^{\square} = \dfrac{1}{216}$ $\square^{\square} = \dfrac{1}{4}$ $\left(\dfrac{\square}{\square}\right)^{\square} = 2\dfrac{10}{27}$

2 Investigate the values for a and b for which the statements are true.

    $ab > b^a$ $a^b > b^a$

Are you ready? (A)

1 Work out:

   a 10^4 **b** 10^2 **c** 10^5 **d** 10^{-1}

2 Write as a single number:

   a 4×1000 **b** $6.3 \times 10\,000$ **c** 3.03×10^3 **d** 3.63×10^5

3 Write as a single power:

   a $3^6 \times 3$ **b** $p^{-4} \times p$ **c** $10^7 \div 10$ **d** $w^4 \div w$

Standard form is where a number is written in the form $A \times 10^n$, where A can be any number from 1 up to, but not including, 10 and n is an integer (whole number).

$8000 = 8 \times 1000 = 8 \times 10^3$ $0.008 = 8 \times \frac{1}{1000} = 8 \times 10^{-3}$

Th	H	T	O	t	h
10^3	10^2	10^1	10^0	10^{-1}	10^{-2}
			8 •		
8	0	0	0 •		

H	T	O	t	h	th
10^2	10^1	10^0	10^{-1}	10^{-2}	10^{-3}
		8 •			
		0 •	0	0	8

Standard form is a way of expressing numbers that is used particularly with very large or very small numbers, to make them easier to understand and to calculate with. It is used commonly in science, technology and engineering.

Using your calculator

You can use your calculator to input and calculate with numbers in standard form. For example, to write 3.4×10^5, type (depending on your calculator model) something like

It should appear something like this: 3.4×10^5

Example 1

Convert:

a 32 500 to standard form

b 6.58×10^6 to an ordinary number.

Method

Solution	Commentary
a 32 500 $= 3.25 \times 10\,000$ $= 3.25 \times 10^4$	A number in standard form is written as $A \times 10^n$, where A is between 1 and 10, so in this case A must be 3.25. You can use your knowledge of multiplication by powers of 10 to work out $n = 4$ or you could use a place value chart to help you. <table><tr><th>TTh</th><th>Th</th><th>H</th><th>T</th><th>O</th><th>t</th><th>h</th></tr><tr><td>10^4</td><td>10^3</td><td>10^2</td><td>10^1</td><td>10^0</td><td>10^{-1}</td><td>10^{-2}</td></tr><tr><td>3</td><td>2</td><td>5</td><td>0</td><td>0 •</td><td></td><td></td></tr><tr><td></td><td></td><td></td><td></td><td>3 •</td><td>2</td><td>5</td></tr></table>
b 6.58×10^6 $= 6.58 \times 1\,000\,000$ $= 6\,580\,000$	This time you multiply A by 10 repeatedly to get the number in ordinary form.

Example 2

Convert:

a 0.0354 to standard form

b 6.58×10^{-3} to an ordinary number.

Method

Solution	Commentary
a 0.0354 $= 3.54 \times 0.01$ $= 3.54 \times 10^{-2}$	To be in the form $A \times 10^n$, A must be 3.54. You would need to divide 3.54 by 10 twice, which is the same as multiplying by 0.01, to get 0.0354 You can think of this in a place value chart: <table><tr><th>H</th><th>T</th><th>O</th><th>t</th><th>h</th><th>th</th><th>tth</th></tr><tr><td>10^2</td><td>10^1</td><td>10^0</td><td>10^{-1}</td><td>10^{-2}</td><td>10^{-3}</td><td>10^{-4}</td></tr><tr><td></td><td></td><td>0 •</td><td>0</td><td>3</td><td>5</td><td>4</td></tr><tr><td></td><td></td><td>3 •</td><td>5</td><td>4</td><td></td><td></td></tr></table>
b 6.58×10^{-3} $= 6.58 \times 0.001$ $= 0.006\,58$	You could also think of this as $6.58 \div 1000$

Example 3

Write 0.215×10^4 in standard form.

Method

Solution	Commentary
Method A	
$\underline{0.215} \times 10^4$ $= 2.15 \times 10^{-1} \times 10^4$ $= 2.15 \times 10^3$	This number is not in standard form as A is less than 1 0.215 can be rewritten as 2.15×10^{-1} To simplify $10^{-1} \times 10^4$, add the indices together as the bases are the same. This is covered in Block 6.
Method B	
$0.215 \times 10^4 = 2150$ $= 2.15 \times 10^3$	Alternatively, you can convert the given number to an ordinary number, then convert it into correct standard form.

Practice (A)

1 Write each number in standard form.

    a 8000   **b** 200 000   **c** 32 000   **d** 5600

    e 980 000   **f** 584 000   **g** 952   **h** 1 651 000

    i one million   **j** seven hundred thousand

    k 200^2 **l** 50^3

2 Write each number in standard form.

    a 0.03   **b** 0.000 08   **c** 0.0038   **d** 0.006 08

    e 0.000 634   **f** 0.000 003 684   **g** 0.003 065   **h** 0.005 38

    i $\dfrac{1}{2}$ **j** three-hundredths   **k** $\dfrac{9}{50}$ **l** $\dfrac{7}{80}$

3 Write each of these as an ordinary number.

    a 3×10^4 **b** 6×10^5 **c** 9.2×10^4 **d** 6.355×10^5

    e 6.805×10^7 **f** 3.84×10^6 **g** 1.265×10^5 **h** 3.054×10^8

4 Write each of these as an ordinary number.

    a 3×10^{-2} **b** 6×10^{-4} **c** 3.8×10^{-3} **d** 6.82×10^{-4}

    e 5.08×10^{-5} **f** 3.98×10^{-7} **g** 2.865×10^{-6} **h** 2.0805×10^{-4}

5 These numbers have been written incorrectly.

    Write them in standard form.

    a 29×10^3 **b** 865×10^{-4} **c** 63.8×10^{-2} **d** 0.0085×10^3

    e 0.384×10^3 **f** 36.8×10^{-3} **g** 8468×10^2 **h** $0.000 58 \times 10^{-3}$

6 Write each set of numbers in order, starting with the least.

   a 6×10^3 3.5×10^3 3×10^6 3.5×10^4

   b 1.32×10^4 1.23×10^4 3.2×10^3 3.21×10^4

   c 8.2×10^{-3} 8.02×10^{-2} 2.8×10^{-2} 2.08×10^{-3}

   d 3.3×10^{-6} 3.3×10^{-7} 2.3×10^{-6} 3.2×10^{-2}

What do you think? (A) 💡

1 Which number is the greatest? Justify your answer.

   0.35 $\dfrac{2}{5}$ 3.8×10^{-2} 0.35×10^2 3.06×10^{-1}

2 $p \times 10^x > q \times 10^y$

Decide if each statement is **always true**, **sometimes true** or **never true**.

     $p > q$ $x > y$ $p = q$

Are you ready? (B) ▨

1 Convert to standard form.

   a $32\,000$ **b** 0.086 **c** $846\,200$ **d** $0.000\,354$

2 Simplify:

   a $10^3 \times 10^4$ **b** $10^7 \div 10^3$ **c** $10^{-2} \times 10^4$ **d** $10^{-4} \div 10^{-2}$

3 Work out:

   a 2.8×3 **b** $8.4 \div 2.1$ **c** 6.18×2.7 **d** $3.6 \div 7.2$

4 Write in correct standard form.

   a 32×10^3 **b** 0.6×10^3 **c** 19.2×10^4 **d** 0.32×10^{-3}

You need to be able to perform calculations with numbers in standard form both with and without a calculator. In this section you will explore multiplication and division.

Using your calculator 🖩

When using a calculator, input each number in standard form in brackets as this will help to prevent errors.

Example 1

Work out:

a $(2 \times 10^3) \times (3.5 \times 10^4)$ **b** $(3.2 \times 10^{-2}) \times (2 \times 10^{-3})$

c $(6 \times 10^5) \div (4 \times 10^3)$ **d** $(4.2 \times 10^6) \div (6 \times 10^2)$

Method

Solution	Commentary
a Method A $(2 \times 10^3) \times (3.5 \times 10^4)$ $= (2 \times 3.5) \times (10^3 \times 10^4)$ $= 7 \times 10^7$	As multiplication is commutative, you can rewrite the calculations in a different order. Then work out each part of this new rewritten calculation in standard form. $$2 \times 3.5 = 7$$ $$10^3 \times 10^4 = 10^7$$
Method B $(2 \times 10^3) \times (3.5 \times 10^4)$ $= 2000 \times 35\,000$ $= 70\,000\,000$ $= 7 \times 10^7$	An alternative method is to convert each number to ordinary form, then complete the multiplication. This answer then needs to be converted back to standard form, unless the question states otherwise.
b Method A $(3.2 \times 10^{-2}) \times (2 \times 10^{-3})$ $= (3.2 \times 2) \times (10^{-2} \times 10^{-3})$ $= 6.4 \times 10^{-5}$	$$10^{-2} \times 10^{-3} = 10^{-2 + -3} = 10^{-5}$$
Method B $(3.2 \times 10^{-2}) \times (2 \times 10^{-3})$ $= 0.032 \times 0.002$ $= (32 \times 2) \div 1\,000\,000$ $= 0.000\,064$ $= 6.4 \times 10^{-5}$	You can see that this method is not as efficient, particularly when working with decimals as you have to think very carefully about the place value of each number in the calculation.
c Method A $(6 \times 10^5) \div (4 \times 10^3)$ $= (6 \div 4) \times (10^5 \div 10^3)$ $= 1.5 \times 10^2$	Again you can consider the powers of 10 and the rest of the numbers separately.
Method B $(6 \times 10^5) \div (4 \times 10^3)$ $= 600\,000 \div 4000$ $= 150$ $= 1.5 \times 10^2$	

d Method A	
$(4.2 \times 10^6) \div (6 \times 10^2)$	
$= (4.2 \div 6) \times (10^6 \div 10^2)$	
$= 0.7 \times 10^4$	This time the calculation results in a number that is not in standard form, so this needs to be adjusted.
$= 7 \times 10^{-1} \times 10^4$	
$= 7 \times 10^3$	
Method B	
$(4.2 \times 10^6) \div (6 \times 10^2)$	
$= 4\,200\,000 \div 600$	
$= 7000$	
$= 7 \times 10^3$	

Example 2 🖩

Work out:

a $(5.6 \times 10^4) \times (2.9 \times 10^5)$ **b** $(3.2 \times 10^{-5}) \div (4 \times 10^{-2})$

Method

Solution	Commentary
a $(5.6 \times 10^4) \times (2.9 \times 10^5)$ $= 1.624 \times 10^{10}$	When keying this into the calculator, either type in each symbol, or use the "$\times 10^x$" button. If your calculator gives the answer in ordinary form, it will need writing in standard form. This is because the question is in standard form.
b $(3.2 \times 10^{-5}) \div (4 \times 10^{-2})$ $= 8 \times 10^{-4}$	

Practice (B)

For questions 1 to 6, work out the calculations. Give your answers in standard form.

1   **a** $(3 \times 10^4) \times (2 \times 10^2)$ **b** $(2 \times 10^5) \times (2.5 \times 10^3)$

     c $(5 \times 10^2) \times (1.5 \times 10^3)$ **d** $(3.2 \times 10^4) \times (2 \times 10^{-2})$

     e $(2.6 \times 10^{-2}) \times (3 \times 10^4)$ **f** $(6.2 \times 10^{-5}) \times (1.5 \times 10^3)$

2   **a** $(6 \times 10^5) \div (2 \times 10^2)$ **b** $(9 \times 10^4) \div (4.5 \times 10^2)$

     c $(7 \times 10^9) \div (2 \times 10^3)$ **d** $(6.5 \times 10^{-2}) \div (2 \times 10^2)$

     e $(9.3 \times 10^5) \div (3 \times 10^{-2})$ **f** $(4.5 \times 10^{-4}) \div (3 \times 10^{-3})$

3   **a** $(3 \times 10^4) \times (6 \times 10^3)$ **b** $(5 \times 10^5) \times (7 \times 10^{-2})$

       c $(3.6 \times 10^5) \times (3 \times 10^2)$ **d** $(3.7 \times 10^4) \times (5.6 \times 10^7)$

       e $(7.2 \times 10^4)^2$ **f** $(4.9 \times 10^4)^3$

       g $(7 \times 10^{-3}) \times (6 \times 10^5)$ **h** $(8 \times 10^{-5}) \times (2.6 \times 10^7)$

       i $(3.25 \times 10^{-4}) \times (6.2 \times 10^{-3})$ **j** $(7.4 \times 10^{-4})^2$

 4   **a** $(3 \times 10^5) \div (4 \times 10^2)$ **b** $(5 \times 10^7) \div (8 \times 10^3)$

       c $(2.5 \times 10^6) \div (5 \times 10^3)$ **d** $(3 \times 10^{-4}) \div (5 \times 10^{-2})$

       e $(3 \times 10^{-4}) \div (7.5 \times 10^{-3})$ **f** $(5.6 \times 10^{-4}) \div (8 \times 10^{-7})$

 5   **a** $\dfrac{(9 \times 10^5) \times (2 \times 10^2)}{3 \times 10^3}$ **b** $(6 \times 10^5) \div (2 \times 10^2) \times (3 \times 10^4)$

       c $\dfrac{6 \times 10^5}{(2 \times 10^{-2}) \times (5 \times 10^{-2})}$ **d** $(2.5 \times 10^{-4}) \div (2 \times 10^2) \times (7 \times 10^{-2})$

6   **a** $(3.7 \times 10^4) \times (2.6 \times 10^7)$ **b** $(2.3 \times 10^{-4}) \times (2.8 \times 10^{-2})$

       c $(2.8 \times 10^4)^2$ **d** $(1.5 \times 10^{-3})^3$ You can use

       e $(6.8 \times 10^4) \div (1.7 \times 10^{-3})$ **f** $(8.96 \times 10^{-2}) \div (2.8 \times 10^{-5})$ a calculator

       g $(3.45 \times 10^5) \div (6 \times 10^2)$ **h** $(4.25 \times 10^{-5}) \div (2.5 \times 10^3)$ to check your answers.

7   The area of Australia is $7.7 \times 10^6 \, \text{km}^2$.

     The area of Austria is $8.4 \times 10^4 \, \text{km}^2$.

     How many times larger is Australia than Austria? Give your answer to 1 significant figure.

8   The mass of one hydrogen atom is 1.67×10^{-24} grams.

     What is the mass of 700 hydrogen atoms? Give your answer in standard form.

9   The distance between Manchester and Wellington (New Zealand) is 1.16×10^4 miles.

     A plane travels at an average speed of 500 mph.

     How long would a flight from Manchester to Wellington take?

What do you think? (B)

1   Here are some calculations:

> $(3 \times 10^{-3}) \times (x \times 10^2) = 7.5 \times 10^y$

> $\dfrac{6.8 \times 10^x}{1.7 \times 10^y} = z \times 10^6$

> $\dfrac{x \times 10^{-2}}{y \times 10^z} = 6 \times 10^4$

> $(x \times 10^4) \times (y \times 10^z) = 8.4 \times 10^{24}$

Which calculations are only true for single values of x, y and/or z?

Which have more than one solution? Give at least two possibilities in each case.

Are you ready? (C)

1 Convert to standard form.

    a 72 000   **b** 0.034   **c** 98 400   **d** 0.000 058

2 Work out:

    a $5.8 + 2.7$ **b** $6.6 - 3.9$ **c** $3.8 + 0.056$ **d** $4.32 - 0.068$

3 Work out the missing numbers.

    a $3.2 \times 10^3 = 0.32 \times 10^{\square}$ **b** $6.89 \times 10^4 = \square \times 10^3$

    c $8.3 \times 10^{-4} = 0.0083 \times 10^{\square}$ **d** $3.5 \times 10^{-2} = \square \times 10^{-4}$

You can also work out additions and subtractions with numbers in standard form both with and without a calculator.

Example

Work out:

a $(3.2 \times 10^3) + (4.1 \times 10^3)$ **b** $(4.8 \times 10^5) - (2.6 \times 10^4)$ **c** $(9.2 \times 10^{-3}) + (9.1 \times 10^{-2})$

Method

Solution	Commentary
a **Method A** $(3.2 \times 10^3) + (4.1 \times 10^3)$ $= (3.2 + 4.1) \times 10^3$ $= 7.3 \times 10^3$	As the indices are the same, simply add together the leading numbers (3.2 and 4.1). As the resulting leading number is still between 1 and 10, the answer is in the correct format.
Method B $(3.2 \times 10^3) + (4.1 \times 10^3)$ $= 3200 + 4100$ $= 7300$ $= 7.3 \times 10^3$	An alternative method is to convert each number to ordinary form then add them together. Then, convert this back to standard form.
b **Method A** $(4.8 \times 10^5) - (2.6 \times 10^4)$ $= (4.8 \times 10^5) - (0.26 \times 10^5)$ $= (4.8 - 0.26) \times 10^5$ $= 4.54 \times 10^5$	As the powers aren't the same, adjust them to be the same. Both numbers need to be written to the largest power. This will be 10^5
Method B $(4.8 \times 10^5) - (2.6 \times 10^4)$ $= 480 000 - 26 000$ $= 454 000$ $= 4.54 \times 10^5$	Alternatively, convert each number to an ordinary number, add them and then convert the result.

c **Method A** $(9.2 \times 10^{-3}) + (9.1 \times 10^{-2})$ $= (0.92 \times 10^{-2}) + (9.1 \times 10^{-2})$ $= (0.92 + 9.1) \times 10^{-2}$ $= 10.02 \times 10^{-2}$ $= 1.002 \times 10^{1} \times 10^{-2}$ $= 1.002 \times 10^{-1}$	As this answer is not in correct standard form, adjust it.
Method B $(9.2 \times 10^{-3}) + (9.1 \times 10^{-2})$ $= 0.0092 + 0.091$ $= 0.1002$ $= 1.002 \times 10^{-1}$	Rewriting as ordinary numbers is usually more efficient for addition and subtraction.

Practice (C)

Work out the calculations in each question below. Give your answers in standard form.

1   **a** $(3.2 \times 10^{3}) + (5.4 \times 10^{3})$ **b** $(7.8 \times 10^{-7}) - (4.2 \times 10^{-7})$

    c $(6.84 \times 10^{-5}) + (2.1 \times 10^{-5})$ **d** $(8.23 \times 10^{4}) - (5.4 \times 10^{4})$

    e $(5.8 \times 10^{4}) - (3.4 \times 10^{4})$ **f** $(5.8 \times 10^{-2}) + (4.1 \times 10^{-2})$

2   **a** $(3.2 \times 10^{3}) + (2.4 \times 10^{4})$ **b** $(2.8 \times 10^{-3}) + (4.8 \times 10^{-4})$

    c $(6.84 \times 10^{-4}) - (2.1 \times 10^{-5})$ **d** $(8.4 \times 10^{4}) - (2.6 \times 10^{3})$

    e $(8.8 \times 10^{4}) - (3.4 \times 10^{2})$ **f** $(3.8 \times 10^{-2}) + (4.1 \times 10^{-4})$

3   **a** $(8.7 \times 10^{-3}) + (4.8 \times 10^{-2})$ **b** $(1.8 \times 10^{4}) - (9.2 \times 10^{3})$

    c $(1.25 \times 10^{-4}) - (3.8 \times 10^{-5})$ **d** $(8 \times 10^{6}) + (5.4 \times 10^{4})$

    e $(1.1 \times 10^{5}) - (9.4 \times 10^{4})$ **f** $(9.3 \times 10^{-4}) + (7.1 \times 10^{-3})$

4   **a** $(8.7 \times 10^{-3}) + (2.3 \times 10^{-3}) - (6.3 \times 10^{-3})$ **b** $(3 \times 10^{5}) - (4.8 \times 10^{5}) + (8.6 \times 10^{5})$

    c $(4.6 \times 10^{3}) - (2 \times 10^{3}) + (6.8 \times 10^{3})$ **d** $(3 \times 10^{5}) - (4.8 \times 10^{3}) - (8.7 \times 10^{4})$

What do you think? (C)

1   Write the digits 5, 6, 7 and 8 in each calculation to give the greatest possible answers.

    a $\left(\square \times 10^{\square}\right) + \left(\square \times 10^{\square}\right)$ **b** $\left(\square \times 10^{\square}\right) - \left(\square \times 10^{\square}\right)$

Consolidate – do you need more?

1 **a** Write each number in standard form.

    i 3000   **ii** 800 000   **iii** 52 000

    iv 68 500   **v** 203 400

  b Write each of these as an ordinary number.

    i 3×10^3 **ii** 8.4×10^4 **iii** 6.84×10^7

    iv 6.08×10^3 **v** 5.045×10^4

2 **a** Write each number in standard form.

    i 0.06   **ii** 0.000 84   **iii** 0.002 508

    iv 0.003 052   **v** 0.000 546

  b Write each of these as an ordinary number.

    i 5×10^{-4} **ii** 2.3×10^{-3} **iii** 3.52×10^{-4}

    iv 3.05×10^{-6} **v** 4.352×10^{-4}

3 Work out the calculations. Give your answers in standard form.

  a $(2 \times 10^3) \times (4 \times 10^4)$ **b** $(7 \times 10^5) \div (2 \times 10^3)$

  c $(2.6 \times 10^{-2}) \times (3 \times 10^6)$ **d** $(4.5 \times 10^{-3}) \times (2 \times 10^{-4})$

  e $(9.6 \times 10^{-3}) \div (3 \times 10^3)$ **f** $(8.4 \times 10^{-3}) \div (4 \times 10^{-5})$

  g $(6 \times 10^2) \times (4 \times 10^4)$ **h** $(2.7 \times 10^{-3}) \div (9 \times 10^3)$

  i $(2.4 \times 10^{-4}) \div (8 \times 10^{-2})$ **j** $(8.2 \times 10^3)^2$

4 Work out the calculations. Give your answers in standard form.

  a $(2 \times 10^4) + (4 \times 10^4)$ **b** $(7.4 \times 10^{-5}) - (2.8 \times 10^{-5})$

  c $(8.6 \times 10^2) + (4.2 \times 10^2)$ **d** $(7.5 \times 10^{-4}) + (4.8 \times 10^{-4})$

  e $(9.6 \times 10^{-3}) - (3 \times 10^{-4})$ **f** $(8.4 \times 10^{-3}) + (4 \times 10^{-5})$

  g $(6.2 \times 10^3) + (9.6 \times 10^4)$ **h** $(1.4 \times 10^{-3}) - (8.7 \times 10^{-4})$

  i $(9.4 \times 10^{-3}) + (9.4 \times 10^{-2})$ **j** $(8.4 \times 10^{-3}) - (4 \times 10^{-5})$

5 $A = 3 \times 10^4$, $B = 2.5 \times 10^3$, $C = 5.1 \times 10^4$

  Work out the value of each expression. Give your answers in standard form.

  a AB   **b** A + B   **c** B^2 **d** AB – C

6 Work out the mean of the five numbers, giving your answer in standard form.

  5×10^3 5.2×10^3 2.8×10^4 9.6×10^2 3.2×10^5

Stretch – can you deepen your learning?

1 Calculate the volume of the cuboid. Give your answer in standard form.

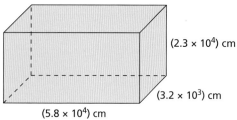

(2.3×10^4) cm

(3.2×10^3) cm

(5.8×10^4) cm

2 Calculate the area of the trapezium. Give your answer in standard form.

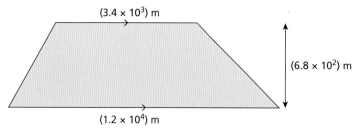

(3.4×10^3) m

(6.8×10^2) m

(1.2×10^4) m

3 Work out the value of each letter.

    a p and q are integers. $(p \times 10^6) + (q \times 10^4) = 3\,070\,000$

    b n is a positive number. $(n \times 10^{-2}) - (n \times 10^{-4}) - 0.060\,39$

    c $(y \times 10^3) \times (y \times 10^4) = 1.225 \times 10^8$

4 **a** $p = 9 \times 10^{4x}$

        Write an expression for p^2 in terms of x. Give your answer in standard form.

    b $(a \times 10^{-7}) + (b \times 10^{-5}) = c \times 10^{-5}$

        Write an expression for a in terms of b and c.

2.3 Fractional indices

Are you ready? (A)

1 Work out:

   a $\sqrt{9} \times \sqrt{9}$ **b** $\sqrt{25} \times \sqrt{25}$ **c** $\sqrt[3]{8} \times \sqrt[3]{8} \times \sqrt[3]{8}$ **d** $\sqrt[3]{64} \times \sqrt[3]{64} \times \sqrt[3]{64}$

2 Simplify:

   a $(x^3)^2$ **b** $(y^6)^2$ **c** $x^{0.5} \times x^{0.5}$ **d** $y^{\frac{1}{3}} \times y^{\frac{1}{3}} \times y^{\frac{1}{3}}$

3 Work out:

   a 5^{-1} **b** 3^{-2} **c** $\left(\dfrac{1}{7}\right)^{-1}$ **d** $\left(\dfrac{3}{5}\right)^{-2}$

When you apply the multiplication rule for indices, you can see that:

$a^{\frac{1}{2}} \times a^{\frac{1}{2}} = a^1$, therefore $a^{\frac{1}{2}} = \sqrt{a}$

Similarly, $a^{\frac{1}{3}} \times a^{\frac{1}{3}} \times a^{\frac{1}{3}} = a^1$, therefore $a^{\frac{1}{3}} = \sqrt[3]{a}$

You can then generalise to give $a^{\frac{1}{n}} = \sqrt[n]{a}$

Example 1

Work out:

a $25^{\frac{1}{2}}$ **b** $27^{\frac{1}{3}}$ **c** $64^{-\frac{1}{3}}$

Method

Solution	Commentary
a $25^{\frac{1}{2}} = \sqrt{25} = 5$	To raise a number to the power half, find its square root.
b $27^{\frac{1}{3}} = \sqrt[3]{27} = 3$	$x^{\frac{1}{3}}$ means the cube root of x.
c $64^{-\frac{1}{3}} = \left(\dfrac{1}{64}\right)^{\frac{1}{3}}$ $\dfrac{1}{\sqrt[3]{64}} = \dfrac{1}{4}$	Remember that a negative power means the reciprocal of the positive power. Then look at the fractional power.

Example 2

Simplify $\left(16x^4\right)^{\frac{1}{2}}$

Method

Solution	Commentary
$\left(16x^4\right)^{\frac{1}{2}}$ $= \left(16\right)^{\frac{1}{2}} \times \left(x^4\right)^{\frac{1}{2}}$ $= 4x^2$	Consider the coefficient and the variable separately: 16 to the power half is the square root of 16 You can use the power of a power rule of indices, which states $(a^b)^c = a^{bc}$, to deal with the variable. In this case, $\left(x^4\right)^{\frac{1}{2}} = x^{4 \times \frac{1}{2}} = x^2$

Practice (A)

1 Work out:

a $36^{\frac{1}{2}}$ b $81^{\frac{1}{2}}$ c $144^{\frac{1}{2}}$ d $225^{\frac{1}{2}}$ e $125^{\frac{1}{3}}$

f $216^{\frac{1}{3}}$ g $81^{\frac{1}{4}}$ h $128^{\frac{1}{7}}$ i $1^{\frac{1}{15}}$ j $625^{\frac{1}{4}}$

2 Work out:

a $25^{-\frac{1}{2}}$ b $8^{-\frac{1}{3}}$ c $(-8)^{\frac{1}{3}}$ d $36^{-\frac{1}{2}}$ e $(-125)^{-\frac{1}{3}}$

3 Write the expressions in index form.

a $\sqrt{a}$ b $\sqrt{b}$ c $\sqrt[3]{c}$ d $\sqrt[7]{d}$ e $\sqrt[n]{e}$

4 Simplify:

a $\left(9x^4\right)^{\frac{1}{2}}$ b $\left(27a^9\right)^{\frac{1}{3}}$ c $\left(25b^{12}\right)^{\frac{1}{2}}$ d $\left(36y^6z^4\right)^{\frac{1}{2}}$

e $\left(64h^6k^9\right)^{-\frac{1}{3}}$ f $\left(81x^8y^4\right)^{-\frac{1}{4}}$ g $\left(-125d^6e^9f^{-3}\right)^{-\frac{1}{3}}$ h $\left(-64x^4y^6\right)^{-\frac{1}{2}}$

What do you think? (A)

1 Decide whether the expressions are **always true**, **sometimes true** or **never true**. Explain your answers.

$a^{\frac{1}{2}} > a^{\frac{1}{4}}$

$\left(a^{\frac{1}{4}}\right)^{\frac{1}{4}} = a^{\frac{1}{8}}$

$a < a^{\frac{1}{3}}$

2 Write each number in the form 16^n

a 256   b 4   c 2

3 Work out each of the following, giving your answers in standard form.

a $\left(4 \times 10^4\right)^{\frac{1}{2}}$ b $\left(1.6 \times 10^5\right)^{\frac{1}{2}}$ c $\left(6.4 \times 10^{10}\right)^{\frac{1}{3}}$

Are you ready? (B) ▰

1 Simplify:

    a $(a^7)^3$
    b $(b^3)^2$
    c $(c^3)^{\frac{1}{3}}$
    d $(d^5)^{\frac{1}{2}}$

2 Work out:

    a $16^{\frac{1}{2}}$
    b $64^{\frac{1}{2}}$
    c $64^{\frac{1}{3}}$
    d $216^{\frac{1}{3}}$

3 Work out:

    a $\left(9^{\frac{1}{2}}\right)^3$
    b $\left(25^{\frac{1}{2}}\right)^3$
    c $\left(8^{\frac{1}{3}}\right)^2$
    d $\left(125^{\frac{1}{3}}\right)^4$

4 Work out:

    a 5^{-1}
    b 4^{-2}
    c 2^{-5}
    d $\left(\dfrac{3}{4}\right)^{-2}$

Now, using the knowledge that $(a^b)^c = a^{bc}$, you can further generalise to give:

$$a^{\frac{m}{n}} = \left(a^{\frac{1}{n}}\right)^m$$
$$= \left(\sqrt[n]{a}\right)^m$$

Alternatively, you could express $a^{\frac{m}{n}}$ as $(a^m)^{\frac{1}{n}} = \sqrt[n]{a^m}$

However, it is usually easier to apply the root first.

Also, since $a^{-p} = \dfrac{1}{a^p}$, you can also have negative fractional indices where $a^{-\frac{m}{n}} = \dfrac{1}{\left(\sqrt[n]{a}\right)^m}$

Example 1

Work out:

a $25^{\frac{3}{2}}$
    b $27^{\frac{2}{3}}$
    c $16^{-\frac{3}{2}}$

Method

Solution	Commentary
a $25^{\frac{3}{2}} = \left(25^{\frac{1}{2}}\right)^3$ $= 5^3 = 125$	You can rewrite $\dfrac{3}{2}$ as $\dfrac{1}{2} \times 3$ Then work out the calculation one step at a time, starting with the brackets.
b $27^{\frac{2}{3}} = \left(27^{\frac{1}{3}}\right)^2$ $3^2 = 9$	Here you can think of $\dfrac{2}{3}$ as $\dfrac{1}{3} \times 2$ Use the fact that $27^{\frac{1}{3}} = \sqrt[3]{27} = 3$ and then square this to get the answer.

c $16^{-\frac{3}{2}} = \left(\frac{1}{16}\right)^{\frac{3}{2}}$

$= \left(\sqrt{\frac{1}{16}}\right)^{3}$

$= \left(\frac{1}{4}\right)^{3} = \frac{1}{64}$

Deal with the negative power first, using the reciprocal.

Then proceed as with parts **a** and **b**.

Example 2

Simplify $\left(9x^{6}\right)^{\frac{3}{2}}$

Method

Solution	Commentary
$\left(9x^{6}\right)^{\frac{3}{2}} = \left(9\right)^{\frac{3}{2}} \times \left(x^{6}\right)^{\frac{3}{2}}$	Consider the coefficient and the variable separately.
$= \left(9^{\frac{1}{2}}\right)^{3} \times \left(x^{6\times\left(\frac{3}{2}\right)}\right)$	For the x part of the expression, use the power of a power index rule.
$= (3)^{3} \times x^{9} = 27x^{9}$	

Practice (B) 📝

1 Work out:

a $9^{\frac{3}{2}}$ **b** $8^{\frac{4}{3}}$ **c** $16^{\frac{3}{2}}$ **d** $9^{\frac{5}{2}}$ **e** $16^{\frac{3}{4}}$

f $32^{\frac{3}{5}}$ **g** $125^{\frac{2}{3}}$ **h** $81^{\frac{3}{2}}$ **i** $16^{\frac{7}{4}}$ **j** $64^{\frac{4}{3}}$

2 Write in the form x^{n}

a $\sqrt{x}$ **b** $\left(\sqrt{x}\right)^{3}$ **c** $\left(\sqrt[3]{x}\right)^{4}$ **d** $\left(\sqrt[6]{x}\right)^{5}$ **e** $\left(\sqrt[m]{x}\right)^{n}$

3 Simplify:

a $\left(16x^{6}\right)^{\frac{3}{2}}$ **b** $\left(27m^{9}\right)^{\frac{2}{3}}$ **c** $\left(64w^{12}\right)^{\frac{5}{6}}$ **d** $\left(125x^{6}y^{9}\right)^{\frac{4}{3}}$

4 Work out:

a $64^{-\frac{2}{3}}$ **b** $16^{-\frac{3}{2}}$ **c** $8^{-1\frac{1}{3}}$ **d** $32^{-\frac{2}{5}}$

e $\left(\frac{4}{9}\right)^{-1.5}$ **f** $\left(1\frac{7}{9}\right)^{-\frac{3}{2}}$ **g** $\left(3\frac{3}{8}\right)^{-\frac{2}{3}}$ **h** $\left(\frac{1}{27}\right)^{-\frac{4}{3}}$

What do you think? (B) 💭

1 Decide whether each statement is **true** or **false**. Give reasons for your answers.

a $8^{\frac{2}{3}} = (-8)^{\frac{2}{3}}$ 　　　**b** $8^{\frac{2}{3}} \times 8^{\frac{2}{3}} = 8^{\frac{4}{3}}$ 　　　**c** $16^{-\frac{3}{2}} \times 4 = 2^{-3}$

2 **a** Write in the form 16^n

　　i 4 　　　**ii** 8 　　　**iii** 32 　　　**iv** 128 　　　**v** 64

b Write in the form 8^n

　　i 4 　　　**ii** 8 　　　**iii** 32 　　　**iv** 128 　　　**v** 64

What do you notice about your answers to parts **a** and **b**?

Consolidate – do you need more?

1 Work out:

a $100^{\frac{1}{2}}$ 　　**b** $8^{\frac{1}{3}}$ 　　**c** $64^{\frac{1}{2}}$ 　　**d** $16^{0.25}$

e $125^{\frac{1}{3}}$ 　　**f** $64^{\frac{1}{3}}$ 　　**g** $\left(\dfrac{4}{9}\right)^{\frac{1}{2}}$ 　　**h** $\left(3\dfrac{3}{8}\right)^{\frac{1}{3}}$

2 Work out:

a $8^{\frac{2}{3}}$ 　　**b** $1000^{\frac{2}{3}}$ 　　**c** $64^{\frac{3}{2}}$ 　　**d** $16^{1.5}$

e $16^{0.75}$ 　　**f** $\left(\dfrac{16}{81}\right)^{\frac{3}{4}}$ 　　**g** $\left(\dfrac{16}{25}\right)^{\frac{3}{2}}$ 　　**h** $\left(5\dfrac{1}{16}\right)^{\frac{5}{4}}$

3 Work out:

a $4^{-\frac{1}{2}}$ 　　**b** $64^{-\frac{2}{3}}$ 　　**c** $25^{-1.5}$ 　　**d** $27^{-\frac{4}{3}}$

e $\left(\dfrac{1}{8}\right)^{-\frac{1}{3}}$ 　　**f** $\left(\dfrac{27}{64}\right)^{-\frac{4}{3}}$ 　　**g** $\left(\dfrac{81}{25}\right)^{-\frac{3}{2}}$ 　　**h** $\left(2\dfrac{10}{27}\right)^{-\frac{5}{3}}$

Stretch – can you deepen your learning?

1 Work out $64^{\frac{1}{2}+\frac{2}{3}}$

2 Work out $16^{\frac{3}{2}} \times \dfrac{1}{32}$, giving your answer:

a as a single power of 2 　　　　**b** as a single power of 4

3 Solve the equations.

a $8 = 32^x$ 　　　**b** $16^5 = 2^x$ 　　　**c** $2^{2x+3} = 32^4$

Indices: exam practice

1 **(a)** Write 1.57×10^{-3} as an ordinary number. [1 mark]

   (b) Write 640 000 in standard form. [1 mark]

   (c) Work out $(8 \times 10^5) \times (9 \times 10^{-2})$

   Give your answer in standard form. [2 marks]

2 Work out $\left(\sqrt{7}\right)^4$ [2 marks]

3 **(a)** Write down the value of 27^0 [1 mark]

   (b) Work out the value of $6^3 \times 6^4 \times 6^{-5}$ [2 marks]

   (c) Work out the value of 2^{-3} [1 mark]

   (d) Work out the value of $64^{\left(\frac{1}{3}\right)}$ [2 marks]

4 Write these numbers in order of size. Start with the smallest number. [2 marks]

   0.000 519 519×10^3 51.9×10^{-3} 5.19×10^4

5 Simplify $(3^{-5} \times 3^9)^2$

   Give your answer as a power of 3 [2 marks]

6 $2^x = 8$ $2^y = \dfrac{1}{4}$ $2^z = 4\sqrt{2}$

   Work out the value of $x + y + z$. [2 marks]

7 Work out the value of $\sqrt[4]{81 \times 10^{12}}$ [2 marks]

8 Simplify $2^5 \times 16 \times 2^7$

   Give your answer as a power of 2 [2 marks]

9 Write $\dfrac{5^x}{25^{x-3}}$ as a power of 5 [2 marks]

10 Express $\sqrt{\dfrac{8^{90}}{8^{100} \div 8^{60}}}$ as a power of 8 [3 marks]

11 Work out the value of $\left(\dfrac{25}{9}\right)^{\frac{3}{2}}$ [2 marks]

12 Given that $16^{\left(-\frac{1}{2}\right)} = 8^{\frac{1}{4}} \div 2^{(x+1)}$, work out the exact value of x. [3 marks]

4–6

7–9

54

In this block, we will cover...

3.1 Simplifying surds

Example

Simplify:

a $\sqrt{5} + 3\sqrt{5}$ **b** $\sqrt{8} - \sqrt{2}$ **c** 5√

Method

Solution	Commenta
a $\sqrt{5} + 3\sqrt{5}$ $= 4\sqrt{5}$	Think of √ three lots
b $\sqrt{8} = \sqrt{4 \times 2} = 2\sqrt{2}$ $\sqrt{8} - \sqrt{2} = 2\sqrt{2} - \sqrt{2} = \sqrt{2}$	As the sur simplest fo

3.2 Multiplying and dividing with surds

Practice (A)

1 Simplify:

     a $\sqrt{5} \times \sqrt{6}$ **b** $\sqrt{7} \times \sqrt{3}$

2 Simplify:

     a $\sqrt{60} \div \sqrt{6}$ **b** $\sqrt{8} \div \sqrt{4}$

3 Write these expressions in their simplest f

     a $\sqrt{6} \times \sqrt{10}$ **b** $\sqrt{6} \times 3\sqrt{3}$

     e $7\sqrt{6} \times 7\sqrt{3}$ **f** $6\sqrt{84} \div 2\sqrt{3}$

3.3 Rationalising the denominator

Consolidate – do you need more

Rationalise the denominator. Give each answer in

1   **a** $\dfrac{5}{\sqrt{5}}$ **b** $\dfrac{6}{\sqrt{3}}$

2   **a** $\dfrac{3 + \sqrt{2}}{\sqrt{6}}$ **b** $\dfrac{\sqrt{3} - 4}{\sqrt{5}}$

Are you ready? (A)

1 List the first 10 square numbers.

2 Write the greatest square factor of each number.

    a 12   **b** 18   **c** 98   **d** 50   **e** 75

3 Work out:

    a $16^{\frac{1}{2}}$ **b** $81^{\frac{1}{2}}$ **c** $144^{\frac{1}{2}}$

A **surd** is a number that cannot be simplified to remove root symbols such as $\sqrt{\ }$ or $\sqrt[3]{\ }$
$\sqrt{2}, \sqrt{5}$ and $\sqrt{30}$ are examples of surds.

$\sqrt{16}$ is **not** a surd as this can be rewritten as just 4

$\sqrt{8}$ is a surd which can be simplified $\sqrt{8} = \sqrt{4 \times 2} = \sqrt{4} \times \sqrt{2} = 2 \times \sqrt{2} = 2\sqrt{2}$

$\sqrt{17}$ cannot be simplified as 17 does not have a square factor other than 1

Surds are **irrational**, which means they cannot be expressed as fractions in the form $\frac{a}{b}$ (where a and b are integers) or as exact decimals.

To simplify surds, you can use the general rules for multiplication and division:

$$\sqrt{a} \times \sqrt{b} = \sqrt{ab}$$

$$\frac{\sqrt{a}}{\sqrt{b}} = \sqrt{\frac{a}{b}}$$

$$\sqrt{a} \times \sqrt{a} = a$$

You will explore multiplication and division of surds in more detail in Chapter 3.2

Example 1

Simplify:

a $\sqrt{18}$ **b** $\sqrt{60}$

Method

Solution	Commentary
a $18 = 9 \times 2$ $\sqrt{18} = \sqrt{9} \times \sqrt{2}$ $\quad\quad = 3 \times \sqrt{2}$ $\sqrt{18} = 3\sqrt{2}$	To write a surd in simplest form, find the greatest factor of 18 that is also a square number, which is 9. Write 18 as a product that includes this factor. Rewrite $\sqrt{9}$ as 3
b $60 = 4 \times 15$ $\sqrt{60} = \sqrt{4} \times \sqrt{15}$ $\quad\quad = 2 \times \sqrt{15}$ $\sqrt{60} = 2\sqrt{15}$	The greatest square factor of 60 is 4

Example 2

Simplify $3\sqrt{18}$

Method

Solution	Commentary
$\sqrt{18} = 3\sqrt{2}$ $3\sqrt{18} = 3 \times 3\sqrt{2}$ $\qquad = 9\sqrt{2}$	$\sqrt{18}$ was simplified in Example 1 Now multiply this result by 3. Notice that the surd part stays the same. You can think of this like $3 \times 3y = 9y$

Practice (A)

1 Decide whether each number is a surd or not.

  a $\sqrt{10}$ **b** $\sqrt{50}$ **c** $\sqrt{36}$ **d** $\sqrt{20}$

  e $\sqrt[3]{8}$ **f** $\sqrt[3]{64}$ **g** $\sqrt{196}$ **h** $\sqrt{8}$

2 Write each surd in its simplest form.

  a $\sqrt{12}$ **b** $\sqrt{8}$ **c** $\sqrt{180}$ **d** $\sqrt{45}$

  e $\sqrt{32}$ **f** $\sqrt{63}$ **g** $\sqrt{250}$ **h** $\sqrt{800}$

3 Simplify:

  a $3\sqrt{48}$ **b** $6\sqrt{24}$ **c** $6\sqrt{50}$ **d** $5\sqrt{75}$

  e $4\sqrt{120}$ **f** $12\sqrt{90}$ **g** $15\sqrt{96}$ **h** $8\sqrt{20}$

4 Write in the form $a\sqrt{b}$.

  a $40^{\frac{1}{2}}$ **b** $128^{\frac{1}{2}}$ **c** $48^{\frac{1}{2}}$ **d** $3 \times 8^{\frac{1}{2}}$

What do you think? (A) 💡

1 Show that $\dfrac{\sqrt{80x}}{\sqrt{5x}}$ can be written as an integer.

2 Write each in the form $\sqrt{a}$.

  a $3\sqrt{5}$ **b** $4\sqrt{6}$ **c** $8\sqrt{21}$ **d** $5\sqrt{42}$

Hint: Think about the result of squaring each number.

Are you ready? (B) 📝

1 Write each surd in its simplest form.

  a $\sqrt{12}$ **b** $\sqrt{27}$ **c** $\sqrt{60}$ **d** $\sqrt{50}$

2 Simplify the expressions.

  a $3x + 7x$ **b** $4x + 7y - 2x$ **c** $3x + 5x^2 + 7x$ **d** $6x + 7y - 2x + 2y$

You can add and subtract surds of the same form in the same way you add and subtract like terms in algebra. For example, $3\sqrt{2} + 5\sqrt{2} = 8\sqrt{2}$ in the same way as $3x + 5x = 8x$

You cannot simplify $6\sqrt{3} - 5\sqrt{7}$, just like you cannot simplify $6x - 5y$, but you may be able to add and subtract different looking surds by writing them in simplest form. See Examples **b** and **c** below.

Example

Simplify:

a $\sqrt{5} + 3\sqrt{5}$ **b** $\sqrt{8} - \sqrt{2}$ **c** $5\sqrt{8} + 3\sqrt{50}$

Method

Solution	Commentary
a $\sqrt{5} + 3\sqrt{5}$ $= 4\sqrt{5}$	Think of $\sqrt{5}$ as $1\sqrt{5}$. You are adding one lot of $\sqrt{5}$ to three lots of $\sqrt{5}$ in the same way as $x + 3x = 4x$
b $\sqrt{8} = \sqrt{4 \times 2} = 2\sqrt{2}$ $\sqrt{8} - \sqrt{2} = 2\sqrt{2} - \sqrt{2} = \sqrt{2}$	As the surd parts aren't the same, write both in their simplest form to simplify. $\sqrt{2}$ is already fully simplified, so write $\sqrt{8}$ as $a\sqrt{2}$
c $5\sqrt{8} = 5 \times 2\sqrt{2} = 10\sqrt{2}$ $3\sqrt{50} = 3 \times 5\sqrt{2} = 15\sqrt{2}$ $5\sqrt{8} + 3\sqrt{50} = 10\sqrt{2} + 15\sqrt{2}$ $\qquad\qquad = 25\sqrt{2}$	As the surds aren't the same, write both in the same form, in this case $a\sqrt{2}$ Now you can simplify.

Practice (B) 📱

1 Simplify:

   a $\sqrt{5} + 3\sqrt{5}$ **b** $3\sqrt{7} - \sqrt{7}$ **c** $8\sqrt{3} - 5\sqrt{3}$ **d** $7\sqrt{11} + 8\sqrt{11}$

2 Simplify:

   a $\sqrt{3} + \sqrt{12}$ **b** $\sqrt{27} - 2\sqrt{3}$ **c** $\sqrt{8} + 3\sqrt{2}$ **d** $9\sqrt{6} + \sqrt{54}$

3 Simplify:

   a $\sqrt{8} + \sqrt{24}$ **b** $\sqrt{50} - 2\sqrt{8}$ **c** $5\sqrt{108} + 2\sqrt{72}$ **d** $15\sqrt{48} - 3\sqrt{27}$

   e $\sqrt{300} - \sqrt{48}$ **f** $3\sqrt{45} - 2\sqrt{20}$ **g** $\sqrt{27} + \sqrt{12} + \sqrt{48}$ **h** $\sqrt{45} - \sqrt{80} + \sqrt{125}$

4 Calculate the perimeter of each shape, giving your answers in their simplest form.

a
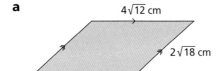

b
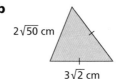

c
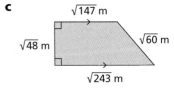

What do you think? (B) 💭

1. List three pairs of single surds $\left(\sqrt{a}\right)$ that add to give $\sqrt{300}$

Consolidate – do you need more?

In questions 1 to 3, write each surd in its simplest form.

1. **a** $\sqrt{12}$ **b** $\sqrt{48}$ **c** $\sqrt{32}$ **d** $\sqrt{40}$
   e $\sqrt{98}$ **f** $\sqrt{80}$ **g** $\sqrt{28}$ **h** $\sqrt{27}$

2. **a** $\sqrt{350}$ **b** $\sqrt{125}$ **c** $\sqrt{1000}$ **d** $\sqrt{196}$
   e $\sqrt{432}$ **f** $\sqrt{120}$ **g** $\sqrt{500}$ **h** $\sqrt{648}$

3. **a** $3\sqrt{50}$ **b** $8\sqrt{24}$ **c** $6\sqrt{8}$ **d** $12\sqrt{240}$

4. Calculate the length of x in each triangle. Give your answer as a surd in its simplest form.

   a
   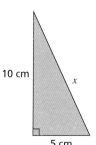
   10 cm, x, 5 cm

   b
   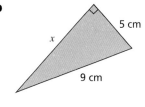
   5 cm, x, 9 cm

5. **a** Write $12\sqrt{12}$ in the form $a\sqrt{3}$ **b** Write $\left(2\sqrt{3}\right)^{3}$ in the form $a\sqrt{3}$
   c Write $\left(3\sqrt{5}\right)^{3} \times \sqrt{30}$ in the form $a\sqrt{6}$ **d** Write $\left(2\sqrt{5}\right)^{4}$ as an integer.

6. Simplify:
   a $\sqrt{3} + 4\sqrt{3}$ **b** $5\sqrt{3} - 2\sqrt{3}$ **c** $\sqrt{7} + 3\sqrt{7}$ **d** $8\sqrt{6} - 2\sqrt{6}$
   e $\sqrt{3} + \sqrt{27}$ **f** $\sqrt{50} - 2\sqrt{8}$ **g** $5\sqrt{8} + 4\sqrt{2}$ **h** $6\sqrt{20} + 4\sqrt{45}$

Stretch – can you deepen your learning?

1. Write $m\sqrt{k}$ in the form $\sqrt{x}$

2. Simplify $\sqrt{x^{3}y^{2}z}$

3. Write $650^{\frac{1}{2}}$ as a surd in its simplest form.

4. Simplify $\sqrt{5a^{2}} + \sqrt{125a^{2}} - \sqrt{45a^{2}b}$

Are you ready? (A) 📝

1 Simplify:

    a $\sqrt{50}$ **b** $\sqrt{60}$ **c** $\sqrt{98}$ **d** $\sqrt{300}$

2 Simplify the expressions.

    a $4x^2 \div x$ **b** $9x \times 3$ **c** $15y \times 3y$ **d** $4mn \div m$

In this chapter you will explore multiplying and dividing surds, using the rules introduced when simplifying surds:

$$\sqrt{a} \times \sqrt{b} = \sqrt{ab} \qquad \frac{\sqrt{a}}{\sqrt{b}} = \sqrt{\frac{a}{b}} \qquad \sqrt{a} \times \sqrt{a} = a$$

Example 1

Simplify: **a** $\sqrt{5} \times \sqrt{7}$ **b** $\sqrt{3} \times \sqrt{6}$ **c** $3\sqrt{5} \times 8\sqrt{3}$

Method

Solution	Commentary
a $\sqrt{5} \times \sqrt{7} = \sqrt{5 \times 7} = \sqrt{35}$	Use the rule $\sqrt{a} \times \sqrt{b} = \sqrt{ab}$
b $\sqrt{3} \times \sqrt{6} = \sqrt{18}$ $\phantom{\sqrt{3} \times \sqrt{6}} = \sqrt{9 \times 2}$ $\phantom{\sqrt{3} \times \sqrt{6}} = 3\sqrt{2}$	Notice that $\sqrt{18}$ is not in simplest form as 18 has a square factor.
c $3\sqrt{5} \times 8\sqrt{3} = 3 \times \sqrt{5} \times 8 \times \sqrt{3}$ $\phantom{3\sqrt{5} \times 8\sqrt{3}} = 3 \times 8 \times \sqrt{5} \times \sqrt{3}$ $\phantom{3\sqrt{5} \times 8\sqrt{3}} = 24\sqrt{15}$	Use the fact that multiplication is commutative to multiply the integer parts and surd parts separately. Note that $\sqrt{15}$ cannot be simplified.

Example 2

Simplify: **a** $\sqrt{50} \div \sqrt{10}$ **b** $\sqrt{54} \div \sqrt{3}$ **c** $9\sqrt{15} \div 3\sqrt{5}$

Method

Solution	Commentary
a $\sqrt{50} \div \sqrt{10} = \sqrt{50 \div 10} = \sqrt{5}$	Use the fact that $\sqrt{a} \div \sqrt{b} = \sqrt{a \div b}$
b $\sqrt{54} \div \sqrt{3} = \sqrt{54 \div 3} = \sqrt{18}$ $\phantom{\sqrt{54} \div \sqrt{3}} = \sqrt{9 \times 2} = 3\sqrt{2}$	Use the same rule and simplify the answer.

c $\quad 9\sqrt{15} \div 3\sqrt{5} = (9 \div 3) \times \sqrt{15 \div 5}$ $\qquad\qquad\qquad = 3\sqrt{3}$	As with multiplication, divide the integer parts and surd parts separately. You might find it helpful to think of this as a fraction you are cancelling down: $\quad 9\sqrt{15} \div 3\sqrt{5} = \dfrac{9\sqrt{15}}{3\sqrt{5}}$

Practice (A)

 **1** Simplify:

     a $\sqrt{5} \times \sqrt{6}$ **b** $\sqrt{7} \times \sqrt{3}$ **c** $2\sqrt{6} \times 3\sqrt{7}$ **d** $9\sqrt{6} \times 3\sqrt{11}$

2 Simplify:

     a $\sqrt{60} \div \sqrt{6}$ **b** $\sqrt{8} \div \sqrt{4}$ **c** $9\sqrt{15} \div 3\sqrt{5}$ **d** $18\sqrt{50} \div 6\sqrt{10}$

3 Write these expressions in their simplest form.

     a $\sqrt{6} \times \sqrt{10}$ **b** $\sqrt{6} \times 3\sqrt{3}$ **c** $6\sqrt{75} \div 3\sqrt{3}$ **d** $\sqrt{50} \times 3\sqrt{8}$

     e $7\sqrt{6} \times 7\sqrt{3}$ **f** $6\sqrt{84} \div 2\sqrt{3}$ **g** $9\sqrt{5} \times \sqrt{8} \times \sqrt{6}$ **h** $6\sqrt{6} \times 4\sqrt{3} \div 3\sqrt{3}$

4 Calculate the area of each shape. Give your answers in their simplest form.

a $4\sqrt{12}$ cm, $6\sqrt{2}$ cm

b $2\sqrt{5}$ m, $2\sqrt{8}$ m, $3\sqrt{6}$ m

c $3\sqrt{5}$ cm, $2\sqrt{17}$ cm, $2\sqrt{12}$ cm, $5\sqrt{5}$ cm

5 A prism is shown.

The area of the cross-section of the prism is $16\sqrt{3}$ cm².

Calculate the volume of the prism, giving your answer as a simplified surd.

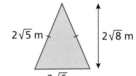

$16\sqrt{3}$ cm² $\sqrt{8}$ cm

6 Calculate the volume of the right-angled triangular prism, giving your answer in its simplest form.

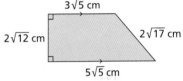

$\sqrt{30}$ cm, $3\sqrt{2}$ cm, $3\sqrt{24}$ cm, $\sqrt{12}$ cm

What do you think? (A)

1 **a** Work out all possible integer values of a and b in which $\sqrt{a} \times \sqrt{b} = 12$

     b Work out all possible integer values of c and d in which $\sqrt{c} \times \sqrt{d} = 20$

     c Compare your answers to parts **a** and **b**.

     d Explore with numbers of your choice.

Are you ready? (B) ✍

1 Simplify:

    a $\sqrt{80}$ **b** $\sqrt{98}$ **c** $\sqrt{63}$ **d** $\sqrt{180}$

2 Simplify:

    a $5 \times \sqrt{6}$ **b** $\sqrt{3} \times \sqrt{5}$ **c** $3\sqrt{6} \times \sqrt{3}$ **d** $3\sqrt{7} \times 2\sqrt{5}$

3 Simplify:

    a $\sqrt{7} + \sqrt{7}$ **b** $7\sqrt{6} - 2\sqrt{6}$ **c** $\sqrt{2} + \sqrt{8}$ **d** $\sqrt{27} - 2\sqrt{3}$

4 Expand and simplify the brackets.

    a $3(2x + 5)$ **b** $5x(2x - 3)$ **c** $(x + 2)(x - 5)$ **d** $(x - 6)^2$

Example 1

a Expand $3(\sqrt{5} + 3)$ **b** Expand and simplify $\sqrt{6}(\sqrt{2} - 3\sqrt{3})$

Method

Solution	Commentary
a $3(\sqrt{5} + 3)$ $= 3\sqrt{5} + 9$	Multiply both terms inside the bracket by 3 $3 \times \sqrt{5} = 3\sqrt{5}$ $3 \times 3 = 9$
b $\sqrt{6}(\sqrt{2} - 3\sqrt{3})$ $= \sqrt{12} - 3\sqrt{18}$ $= \sqrt{4 \times 3} - 3 \times 3\sqrt{2}$ $= 2\sqrt{3} - 9\sqrt{2}$	Multiply both terms inside the bracket by $\sqrt{6}$ $\sqrt{6} \times \sqrt{2} = \sqrt{12}$ and $\sqrt{6} \times 3\sqrt{3} = 3\sqrt{18}$ Simplify the results. The final answer cannot be simplified.

Example 2

Expand and simplify: **a** $(3 + \sqrt{2})(5 + \sqrt{2})$ **b** $(\sqrt{12} - \sqrt{6})(\sqrt{2} + 3\sqrt{3})$

Method

Solution	Commentary
a $(3 + \sqrt{2})(5 + \sqrt{2})$ $= 15 + 5\sqrt{2} + 3\sqrt{2} + 2$ $= 17 + 8\sqrt{2}$	Use the method that you are comfortable with to expand the double brackets. Remember that $\sqrt{2} \times \sqrt{2} = \sqrt{4} = 2$ Then collect integer and surd parts.
b $(\sqrt{12} - \sqrt{6})(\sqrt{2} + 3\sqrt{3})$ $= \sqrt{24} - \sqrt{12} - 3\sqrt{18} + 3\sqrt{36}$ $= 2\sqrt{6} - 2\sqrt{3} - 3 \times 3\sqrt{2} + 3 \times 6$ $= 2\sqrt{6} - 2\sqrt{3} - 9\sqrt{2} + 18$	Expand the brackets. Simplify each term. You may be able to do this mentally by now, or you can show more working.

Practice (B)

1 Expand the brackets, simplifying your answers where possible.

a $3(\sqrt{2} + 4)$ **b** $\sqrt{3}(\sqrt{3} + 4)$ **c** $\sqrt{7}(\sqrt{2} - \sqrt{3})$ **d** $5(\sqrt{7} + \sqrt{3})$

e $\sqrt{6}(\sqrt{3} - \sqrt{2})$ **f** $\sqrt{8}(2\sqrt{3} - 2\sqrt{2})$ **g** $3\sqrt{6}(2\sqrt{3} - 2\sqrt{8})$ **h** $2\sqrt{5}(3\sqrt{2} - \sqrt{5})$

2 Expand and simplify:

a $(3 + \sqrt{5})(2 + \sqrt{5})$ **b** $(\sqrt{2} + \sqrt{6})(\sqrt{5} + \sqrt{6})$ **c** $(\sqrt{3} - \sqrt{5})(\sqrt{2} + \sqrt{5})$

d $(\sqrt{2} + 2)(\sqrt{2} - 3)$ **e** $(\sqrt{7} - \sqrt{2})^2$ **f** $(3 + \sqrt{6})^2$

g $(\sqrt{5} + 4)(\sqrt{5} - 4)$ **h** $(\sqrt{8} - \sqrt{3})(\sqrt{8} + \sqrt{3})$ **i** $(3\sqrt{6} - \sqrt{3})(3\sqrt{6} + \sqrt{3})$

3 Expand and simplify:

a $(2\sqrt{2} - 3)(2\sqrt{6} + 2)$ **b** $(4 + 2\sqrt{3})(6 - 2\sqrt{3})$ **c** $(3\sqrt{5} + 2)(2\sqrt{5} - 1)$

d $(\sqrt{28} - \sqrt{2})(\sqrt{63} + 2)$ **e** $(\sqrt{20} + 3)^2$ **f** $(\sqrt{48} - \sqrt{8})^2$

4 Calculate the area of each shape.

a

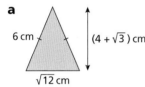

6 cm $(4 + \sqrt{3})$ cm $\sqrt{12}$ cm

b

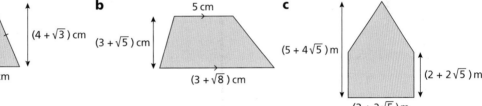

5 cm $(3 + \sqrt{5})$ cm $(3 + \sqrt{8})$ cm

c

$(5 + 4\sqrt{5})$ m $(2 + 2\sqrt{5})$ m $(3 + 2\sqrt{5})$ m

5 Calculate the length of AC.

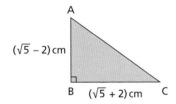

A $(\sqrt{5} - 2)$ cm   B $(\sqrt{5} + 2)$ cm   C

What do you think? (B)

1 Show that $(\sqrt{50} - 3\sqrt{2})^2$ is rational.

2 Beca says that $(3\sqrt{g} + 2g)^2$ is equivalent to $(\sqrt{6g} + 2g)^2$

Do you agree? Prove your answer.

3 Which parts of question 2 in Practice (B) have integer answers? Explain why.

Consolidate – do you need more? ▨

1 Simplify:

   a $\sqrt{50} \div \sqrt{5}$ **b** $\sqrt{7} \times \sqrt{5}$ **c** $\sqrt{8} \times \sqrt{2}$ **d** $\sqrt{80} \div \sqrt{8}$

   e $6\sqrt{2} \times 2\sqrt{6}$ **f** $8\sqrt{6} \div 2\sqrt{2}$ **g** $8\sqrt{6} \div 2\sqrt{2} \times 3\sqrt{8}$ **h** $\dfrac{8\sqrt{6} \times 8\sqrt{8}}{\sqrt{2}}$

2 Expand and simplify, leaving your answers in the form $a + b\sqrt{c}$.

   a $3\left(\sqrt{3} - 2\right)$ **b** $\sqrt{5}\left(2 + \sqrt{3}\right)$ **c** $\sqrt{8}\left(2\sqrt{5} - \sqrt{8}\right)$ **d** $4\sqrt{2}\left(2\sqrt{6} - 2\sqrt{8}\right)$

3 Expand and simplify:

   a $\left(\sqrt{3} + 2\right)\left(\sqrt{3} - 4\right)$ **b** $\left(5 + \sqrt{3}\right)\left(2 + \sqrt{6}\right)$ **c** $\left(3 - \sqrt{5}\right)\left(3 + \sqrt{5}\right)$

   d $\left(5 - \sqrt{2}\right)\left(8 + \sqrt{8}\right)$ **e** $\left(3 - 2\sqrt{5}\right)\left(5 + 3\sqrt{2}\right)$ **f** $\left(4 - 7\sqrt{8}\right)^2$

4 The area of a triangle is $30\sqrt{15}$ cm².

   The base of the triangle is $3\sqrt{5}$ cm.

   Work out the perpendicular height of the triangle, giving your answer in the form $a\sqrt{b}$ where a and b are integers.

5 **a** The area of a square is 40 cm².

      Work out the length of the diagonal of the square, giving your answer in the form $a\sqrt{b}$ where a and b are integers.

   b The diagonal length of a square is $5\sqrt{2}$ cm.

      Calculate the area of the square.

Stretch – can you deepen your learning?

▨ **1** Expand and simplify $\left(\sqrt{10} + \sqrt{8}\right)^2 - \left(\sqrt{10} - \sqrt{8}\right)^2$

2 Given that $\left(5 - \sqrt{x}\right)\left(7 + y\sqrt{3}\right) = 35 - 14\sqrt{2} + 15\sqrt{3} - 6\sqrt{6}$ and x and y are both integers, work out the values of x and y.

3 **a** Write $(12x)^{\frac{1}{2}}$ in the form $a\sqrt{b}$.

   b Write $(18x)^{\frac{3}{2}}$ in the form $a\sqrt{b}$.

   c Write $(27x^3y)^{-\frac{3}{2}}$ in the form $\dfrac{1}{a\sqrt{b}}$

3.3 Rationalising the denominator

Are you ready? ✐

1 Simplify the surds.

  a $\sqrt{50}$ **b** $\sqrt{12}$ **c** $6\sqrt{8}$ **d** $8\sqrt{24}$

2 Simplify:

  a $\sqrt{5} \times \sqrt{3}$ **b** $\sqrt{6} \times \sqrt{12}$ **c** $3\sqrt{6} \times 7\sqrt{3}$ **d** $8\sqrt{6} \times 2\sqrt{2}$

3 Expand the brackets. Write your answers in their simplest form.

  a $3\left(\sqrt{5} - 2\right)$ **b** $\sqrt{2}\left(\sqrt{5} + 3\right)$ **c** $\sqrt{8}\left(\sqrt{3} + \sqrt{6}\right)$ **d** $\sqrt{5}\left(3\sqrt{8} - 2\sqrt{10}\right)$

Surds are **irrational** and so cannot be written in the form $\frac{a}{b}$ where a and b are integers. However, if you have a fraction with a denominator that is a surd, you can simplify it by **rationalising the denominator**.

If the denominator is in the form $\sqrt{n}$, you need to multiply both the numerator and denominator by $\sqrt{n}$. The denominator is then $\left(\sqrt{n}\right)^2 = n$, which is rational.

Example ✐

Write these numbers in their simplest form.

a $\dfrac{4}{\sqrt{6}}$ **b** $\dfrac{5 - \sqrt{6}}{\sqrt{3}}$ **c** $\dfrac{8}{2\sqrt{2}}$

Method

Solution	Commentary
a $\dfrac{4}{\sqrt{6}} \times \dfrac{\sqrt{6}}{\sqrt{6}}$	To rationalise the denominator, make the denominator an integer, not a surd. To make $\sqrt{6}$ into an integer, multiply by $\sqrt{6}$
$= \dfrac{4\sqrt{6}}{6}$	To keep equivalence, multiply the whole fraction by $\dfrac{\sqrt{6}}{\sqrt{6}}$ as this is equal to 1
$= \dfrac{2\sqrt{6}}{3}$	Then write the fraction in its simplest form.
b $\dfrac{5 - \sqrt{6}}{\sqrt{3}} \times \dfrac{\sqrt{3}}{\sqrt{3}}$	To rationalise the denominator, multiply the fraction by $\dfrac{\sqrt{3}}{\sqrt{3}}$
$= \dfrac{\sqrt{3}\left(5 - \sqrt{6}\right)}{3}$	Both terms in the numerator need to be multiplied by $\sqrt{3}$
$= \dfrac{5\sqrt{3} - \sqrt{18}}{3}$	You can then simplify $\sqrt{18}$ to $3\sqrt{2}$
$= \dfrac{5\sqrt{3} - 3\sqrt{2}}{3}$	You could write the answer as $\dfrac{5\sqrt{3}}{3} - \sqrt{2}$

c $\dfrac{8}{2\sqrt{2}} \times \dfrac{\sqrt{2}}{\sqrt{2}}$	$2\sqrt{2} \times \sqrt{2}$ is equivalent to 2×2
$= \dfrac{8\sqrt{2}}{2 \times 2}$	
$= \dfrac{8\sqrt{2}}{4}$	The final step is to simplify the fraction. Both the numerator and denominator can be divided by 4, to give an answer of $2\sqrt{2}$
$= 2\sqrt{2}$	

Practice

In questions 1 to 4, rationalise the denominator of each fraction. Give all answers in their simplest form.

1　**a** $\dfrac{3}{\sqrt{5}}$　　　**b** $\dfrac{5}{\sqrt{11}}$　　　**c** $\dfrac{2}{\sqrt{10}}$　　　**d** $\dfrac{8}{\sqrt{2}}$

　　e $\dfrac{\sqrt{2}}{\sqrt{6}}$　　　**f** $\dfrac{\sqrt{8}}{\sqrt{2}}$　　　**g** $\dfrac{\sqrt{12}}{\sqrt{6}}$　　　**h** $\dfrac{\sqrt{3}}{\sqrt{8}}$

2　**a** $\dfrac{3}{2\sqrt{3}}$　　　**b** $\dfrac{8}{3\sqrt{2}}$　　　**c** $\dfrac{5}{3\sqrt{5}}$　　　**d** $\dfrac{9}{3\sqrt{3}}$

　　e $\dfrac{\sqrt{5}}{2\sqrt{3}}$　　　**f** $\dfrac{\sqrt{8}}{3\sqrt{2}}$　　　**g** $\dfrac{5\sqrt{2}}{\sqrt{10}}$　　　**h** $\dfrac{5\sqrt{3}}{2\sqrt{6}}$

3　**a** $\dfrac{1+\sqrt{3}}{\sqrt{5}}$　　**b** $\dfrac{4-\sqrt{2}}{\sqrt{5}}$　　**c** $\dfrac{5+\sqrt{2}}{\sqrt{5}}$　　**d** $\dfrac{2-2\sqrt{3}}{\sqrt{2}}$

　　e $\dfrac{2\sqrt{6}-3}{\sqrt{3}}$　　**f** $\dfrac{7\sqrt{8}+\sqrt{2}}{\sqrt{2}}$　　**g** $\dfrac{3\sqrt{3}-8}{2\sqrt{3}}$　　**h** $\dfrac{4\sqrt{5}+2\sqrt{8}}{4\sqrt{2}}$

4　**a** $\dfrac{3\left(\sqrt{3}-2\right)}{\sqrt{6}}$　**b** $\dfrac{5\left(2\sqrt{2}+5\right)}{\sqrt{8}}$　**c** $\dfrac{\sqrt{5}\left(\sqrt{8}-3\sqrt{6}\right)}{\sqrt{2}}$　**d** $\dfrac{\left(3+\sqrt{5}\right)\left(3-\sqrt{5}\right)}{\sqrt{2}}$

5　Calculate the heights, h, of the shapes.

　　a　Area = $2\sqrt{15}$ cm²

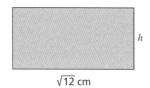

$\sqrt{12}$ cm

　　b　Area = $5\sqrt{18}$ cm²

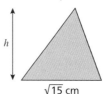

$\sqrt{15}$ cm

6　Work out $\dfrac{\sqrt{24} \times \sqrt{20}}{\sqrt{3}}$

Give your answer in the form $a\sqrt{b}$ where a and b are integers.

What do you think? 💭

1 Jakub thinks that to rationalise $\dfrac{6 + \sqrt{3}}{2\sqrt{6}}$, multiplying by $\dfrac{2\sqrt{6}}{2\sqrt{6}}$ will give the same result as multiplying by $\dfrac{\sqrt{6}}{\sqrt{6}}$

Do you agree with Jakub? Give a reason for your answer.

2 Simplify $\left(3 + \sqrt{5}\right)^2 - \dfrac{\sqrt{500}}{\left(\sqrt{5}\right)^3}$ and decide whether your answer is rational or irrational.

3 Given that $a = \sqrt{2}$, $b = \sqrt{3}$ and $c = \sqrt{6}$, work out in simplest form:

    a $\dfrac{a}{b}$ **b** $\dfrac{b}{a}$ **c** $\dfrac{ab}{c}$ **d** $\dfrac{1}{ab}$ **e** $\dfrac{1}{bc}$ **f** $\dfrac{1}{abc}$

Consolidate – do you need more? 🖊

Rationalise the denominator. Give each answer in its simplest form.

1   **a** $\dfrac{5}{\sqrt{5}}$ **b** $\dfrac{6}{\sqrt{3}}$ **c** $\dfrac{\sqrt{8}}{\sqrt{3}}$ **d** $\dfrac{\sqrt{12}}{3\sqrt{5}}$

2   **a** $\dfrac{3 + \sqrt{2}}{\sqrt{6}}$ **b** $\dfrac{\sqrt{3} - 4}{\sqrt{5}}$ **c** $\dfrac{2\sqrt{5} + \sqrt{3}}{\sqrt{6}}$ **d** $\dfrac{4\sqrt{6} - 3\sqrt{3}}{\sqrt{2}}$

Stretch – can you deepen your learning? 🖊

1 Simplify $\dfrac{a - \sqrt{a}}{\sqrt{a}}$

2 Simplify $\dfrac{\left(6\sqrt{3}\right)^2 - \dfrac{2\sqrt{32}}{\sqrt{2}}}{\sqrt{8}}$

Surds: exam practice

1 Write $\sqrt{48}$ in the form $k\sqrt{3}$, where k is an integer. **[2 marks]** **4–6**

2 Write $\sqrt{200}$ in simplest surd form. **[2 marks]**

3 Write $6\sqrt{20}$ in the form $k\sqrt{5}$, where k is an integer. **[2 marks]**

4 Simplify $\sqrt{75} + \sqrt{27}$, giving your answer in the form $a\sqrt{b}$, where a and b are integers. **[2 marks]**

5 Expand and simplify $\left(3 + \sqrt{2}\right)\left(3 - \sqrt{2}\right)$ **[2 marks]**

6 Expand and simplify $\left(5 + \sqrt{3}\right)\left(2 - \sqrt{3}\right)$ **[2 marks]** **7–9**

7 Simplify $\dfrac{6}{\sqrt{3}}$ **[2 marks]**

8 Write $\left(2 + \sqrt{3}\right)^2$ in the form $a + \sqrt{b}$, where a and b are integers. **[2 marks]**

9 Show that for integer values of a and b, $\left(a + \sqrt{b}\right)\left(a - \sqrt{b}\right)$ is also an integer. **[3 marks]**

10 Express $\left(\dfrac{1}{\sqrt{2}}\right)^5$ in the form $\dfrac{\sqrt{a}}{b}$, where a and b are integers. **[3 marks]**

11 $\sqrt{5}\left(\sqrt{12} + \sqrt{27}\right) = p\sqrt{q}$

Work out the values of p and q. **[3 marks]**

12 Work out the value of $\left(\sqrt{80} - \sqrt{5}\right)^2$ **[3 marks]**

13 Determine whether $\left(4\sqrt{3} - \sqrt{2}\right)\left(4\sqrt{3} + \sqrt{2}\right)$ is rational or irrational. **[3 marks]**

14 Rationalise the denominator of $\dfrac{5 + 6\sqrt{7}}{\sqrt{7}}$ **[3 marks]**

4 Percentages

In this block, we will cover...

4.1 Fraction-decimal-percentage conversion

Example 1

Convert: **a** $0.\dot{4}$ to a fraction **b** $0.2\dot{4}$ to a fr

Method

Solution	Commentary
a Let $x = 0.4444...$ $\quad\quad 10x = 4.4444...$ $-\quad\quad x = 0.4444...$ $\quad\quad\quad 9x = 4$ $\quad\quad\quad\quad x = \dfrac{4}{9}$	Call the number x. A recurs, multiply bot when you subtract y Now subtract x from Solve the resulting e

4.2 Calculating, increasing and decreasing with percentages

Practice (B)

1 Write the decimal multipliers needed to ca
  a 82% of a number   **b** 63% of a n
  c 3% of a number   **d** 0.7% of a r

2 Work out the calculations. Round your ans
  a 32% of 48   **b** 56% of 157
  e 2% of 456 g   **f** 85.2% of 846 g

3 Write down the decimal multipliers for cal

4.3 Repeat percentages

Consolidate – do you need more

1 Write down the single calculation needed
  a increase 75 by 10%, then by 20%
  c decrease 220 by 12%, then by 15%

2 There are 200 cells in a Petri dish.
The number of cells increases by 6% each
How many cells are in the Petri dish after:
  a 1 hour   **b** 5 hours

3 Chloe bought a house for £250 000

4.4 Reverse percentages

Stretch – can you deepen your le

1 A strawberry milkshake is made from strav
80% of the strawberries are frozen.
60 g of frozen strawberries are used.

How many grams of milk are needed for th

2 Rhys and Lida share some money in the rat
Rhys then saves 30% of his money.
Rhys saves £45

How much money did they share?

4.1 Fraction-decimal-percentage conversion

Are you ready? (A)

1 Write the fractions as decimals.

    a $\dfrac{1}{2}$ **b** $\dfrac{2}{5}$ **c** $\dfrac{3}{4}$ **d** $\dfrac{1}{3}$ **e** $\dfrac{5}{8}$ **f** $\dfrac{3}{10}$

2 Write the decimals as percentages.

    a 0.8   **b** 0.4   **c** 0.03   **d** 0.84   **e** 1.07   **f** 0.015

3 Write the percentages as fractions in their simplest form.

    a 32%   **b** 45%   **c** 23%   **d** 3%   **e** 0.2%   **f** 105%

Fractions, decimals and percentages are equivalent and can all be converted to the other forms.

You will probably be familiar with some common equivalences such as $\dfrac{1}{10} = 0.1 = 10\%$, $\dfrac{1}{4} = 0.25 = 25\%$, $\dfrac{1}{100} = 0.01 = 1\%$. You can deduce other facts from these and their multiples. For example, $0.17 = 17 \times 0.01 = \dfrac{17}{100} = 17\%$

Other useful facts include $\dfrac{1}{3} = 0.\dot{3} = 33.\dot{3}\%$ (or $33\frac{1}{3}\%$) and $\dfrac{1}{5} = 0.2 = 20\%$

Some fractions are equivalent to **terminating** decimals, for example $\dfrac{9}{20} = 0.45$

Others result in **recurring** decimals with patterns that repeat forever, for example $\dfrac{5}{6} = 0.833\,333...$, written as $0.8\dot{3}$

Fractions with denominators that are factors of 100 are also easily converted using equivalence, for example, $\dfrac{8}{25} = \dfrac{32}{100} = 32\%$

Example 1

a Write $\dfrac{5}{8}$ as a decimal.   **b** Write $\dfrac{1}{6}$ as a decimal.   **c** Write $\dfrac{7}{40}$ as a percentage.

d Write 28% as: **i** a decimal   **ii** a fraction.

Method

Solution	Commentary
a $\dfrac{5}{8} = 5 \div 8$ $\begin{array}{r} 0.625 \\ 8\,\overline{)5\,.^50\,^20\,^40} \end{array}$ $\dfrac{5}{8} = 0.625$	A fraction can be written as a division. Using the formal method to complete the division, work out the equivalent decimal.

---

b $\frac{1}{6} = 1 \div 6$ $\quad\quad 0 \, . \, 1 \; 6 \; 6 \; 6...$ $6 \, \overline{\smash{\big)}\, 1 \, . \, {}^{1}0 \, {}^{4}0 \, {}^{4}0 \, {}^{4}0}$ $\frac{1}{6} = 0.1\dot{6}$	Write the fraction as a division and use short division to convert it to a decimal. The 6 is going to repeat forever; it is a recurring decimal. To show a recurring decimal, write a dot above the first and last recurring digits. Here, the dot is only written above the 6
c $\frac{7}{40} = 7 \div 40 = 0.175$ $0.175 \times 100\% = 17.5\%$	First, complete the division to convert the fraction to a decimal. You could do this by working out $7 \div 4 = 1.75$ and dividing the answer by 10 using place value knowledge. Then multiply by 100 to convert the decimal to a percentage.
d i $28\% = \frac{28}{100}$ $\quad\quad\quad = 28 \div 100 = 0.28$	Use the fact that percent means out of 100, and your knowledge of dividing by powers of 10
ii $28\% = \frac{28}{100} = \frac{7}{25}$	Simplify the fraction by dividing the numerator and denominator by their highest common factor.

Example 2

Write the numbers in order of size, starting with the smallest.　　　$0.4, \frac{3}{8}, 38\%$

Method

Solution	Commentary
$\frac{3}{8} = 0.375$ $38\% = 0.38$	To compare the numbers, they need to be in the same form. Either decimals or percentages are the easiest to compare.
$0.375, 0.38, 0.4$	Once all the numbers are written in the same form, they can be written in order.
$\frac{3}{8}, 38\%, 0.4$	In the answer, write the numbers in their original form.

Practice (A)

1 Copy and complete the table with equivalent values.

Fraction	Decimal	Percentage
$\frac{1}{4}$		
	0.7	
		63%
	0.2	
$\frac{3}{8}$		
		2%
$\frac{2}{25}$		
		0.5%
	0.00025	

2 Write each fraction as: **i**   a decimal   **ii**   a percentage.

   a $\frac{5}{8}$ **b** $\frac{9}{20}$ **c** $\frac{9}{25}$ **d** $\frac{23}{40}$

   e $\frac{145}{200}$ **f** $\frac{47}{50}$ **g** $\frac{63}{80}$ **h** $\frac{17}{40}$

3 Write each number as an equivalent fraction in its simplest form.

   a 70%   **b** 0.24   **c** 32%   **d** 95%

   e 0.04   **f** 0.3%   **g** 0.875   **h** 63.8%

4 Write each set of numbers in order of size, starting with the smallest.

   a   0.4 $\frac{1}{4}$ 35% $\frac{1}{3}$ **b** $\frac{2}{3}$ 60%   0.66 $\frac{13}{20}$

   c $\frac{9}{10}$ $\frac{19}{20}$ 92%   0.89   **d**   105%   1.01 $\frac{11}{10}$ $\frac{10}{9}$

What do you think? (A) 💡

1 Given that $\frac{1}{7} = 0.\dot{1}4285\dot{7}$, convert the fractions to decimals.

   a $\frac{1}{70}$ **b** $\frac{2}{7}$ **c** $\frac{13}{7}$ **d** $\frac{1}{14}$

Compare your strategies with a partner's.

Are you ready? (B)

1 Work out the following, giving your answers in recurring decimal notation.

   a   3 ÷ 7   **b**   2 ÷ 3   **c**   5 ÷ 9   **d**   8 ÷ 11

2 Write the first six decimal places of each number.

   a $0.\dot{5}$ **b** $0.\dot{3}\dot{6}$ **c** $0.5\dot{2}$ **d** $0.\dot{2}4\dot{6}$

3 Write each decimal as a fraction in its simplest form.

   a   0.5   **b**   0.7   **c**   0.84   **d**   0.06

4 Which of these fractions are equivalent to a recurring decimal?

   $\frac{2}{5}$ $\frac{3}{7}$ $\frac{5}{6}$ $\frac{5}{11}$ $\frac{3}{6}$ $\frac{8}{15}$

All recurring decimals are **rational** as they can be written in the form $\frac{a}{b}$ where a and b are integers and $b \neq 0$. To convert a recurring decimal to a fraction, you need to use algebra and manipulate the number until the decimal parts are equal, as shown in the examples below.

Example 1

Convert: **a** $0.\dot{4}$ to a fraction   **b** $0.\dot{2}\dot{4}$ to a fraction   **c** $0.4\dot{6}$ to a fraction.

Method

Solution	Commentary
a Let $x = 0.4444...$ $10x = 4.4444...$ $-\underline{\quad x = 0.4444...}$ $9x = 4$ $x = \dfrac{4}{9}$	Call the number x. As this number has only one digit that recurs, multiply both sides of the equation by 10, so that when you subtract you can eliminate the recurring part. Now subtract x from $10x$ to be left with $9x = 4$ Solve the resulting equation to find x as a fraction.
b Let $x = 0.242\,424...$ $100x = 24.242\,424...$ $-\underline{\quad x = \;0.242\,424...}$ $99x = 24$ $x = \dfrac{24}{99} = \dfrac{8}{33}$	As there are two recurring digits, multiply by 100 Subtract to get $99x = 24$ After dividing both sides of the equation by 99, the fraction needs simplifying to give $\dfrac{8}{33}$
c Let $x = 0.466\,66...$ $100x = 46.6666...$ $-\underline{\;\; 10x = \;\;4.6666...}$ $90x = 42$ $x = \dfrac{42}{90} = \dfrac{7}{15}$	There is a non-recurring digit before the 6 that recurs. To eliminate the decimal part, multiply x by two different powers of 10. Here, use 100 and 10 as there is just one digit that doesn't recur. Then proceed as before.

Example 2

Write $1.3\dot{8}\dot{7}$ as a mixed number.

Method

Solution	Commentary
Let $x = 1.387\,8787...$ $1000x = 1387.878\,787...$ $-\underline{\;\; 10x = \;\;\;\;13.878\,787...}$ $990x = 1374$ $x = \dfrac{1374}{990} = \dfrac{229}{165} = 1\dfrac{64}{165}$	Multiply by two powers of 10 so you can eliminate the decimal part. Rewrite the improper fraction as a mixed number.

Practice (B)

1 Write each fraction as a recurring decimal.

    a $\dfrac{2}{3}$ **b** $\dfrac{5}{6}$ **c** $\dfrac{6}{7}$ **d** $\dfrac{8}{11}$

2 Convert each recurring decimal to a fraction.

    a 0.5555...   **b** 0.8888...   **c** 0.4848...   **d** 0.723 723...

3 Convert each recurring decimal to a fraction in its simplest form.

    a $0.\dot{3}$ **b** $0.\dot{2}\dot{7}$ **c** $0.\dot{8}\dot{1}$ **d** $0.\dot{2}\dot{6}$

    e $0.\dot{7}\dot{4}$ **f** $0.\dot{3}0\dot{4}$ **g** $0.\dot{6}3\dot{3}$ **h** $0.\dot{8}4\dot{6}$

4 Convert each recurring decimal to a fraction in its simplest form.

    a $0.4\dot{6}$ **b** $0.2\dot{5}$ **c** $0.4\dot{1}$ **d** $0.03\dot{6}$

    e $0.38\dot{4}$ **f** $0.4\dot{2}\dot{4}$ **g** $0.82\dot{5}\dot{3}$ **h** $0.2\dot{3}0\dot{6}$

5 Convert each recurring decimal to a mixed number in its simplest form.

    a $1.\dot{5}$ **b** $2.3\dot{8}$ **c** $4.0\dot{7}\dot{3}$ **d** $3.08\dot{4}$

6 Work out $66 \times 0.6\dot{5}$, giving your answer as a mixed number.

7 Work out $\sqrt{0.69\overline{4}}$

8 Write the numbers in order of size, starting with the smallest.

    $\dfrac{4}{7}$ $0.\dot{5}\dot{7}$ $\dfrac{3}{5}$ $\dfrac{31}{55}$

What do you think? (B) 🌐

1 Given that $0.\dot{3}\dot{9} = \dfrac{13}{33}$, write $0.8\dot{3}\dot{9}$ as a fraction.

2 Work out $\dfrac{2}{3} + 0.0\dot{4}$, giving your answer as a fraction.

    Compare your strategy with a partner's.

3 How would the decimal equivalent of $\dfrac{n}{99}$ be different if n was a one-digit or a two-digit number?

Consolidate – do you need more?

1 Copy and complete the table with equivalent values.

Fraction	Decimal	Percentage
$\frac{1}{5}$		
	0.9	
		54%
	0.13	
$\frac{3}{20}$		
		8%
$\frac{9}{25}$		
		0.74%
	0.0084	

2 Write each set of numbers in order of size, starting with the smallest.

a $\frac{1}{2}$ 0.55 47% $\frac{13}{25}$

b $\frac{3}{20}$ $\frac{1}{6}$ 14% 0.17

c 0.09 $\frac{1}{10}$ $\frac{3}{15}$ 0.099

d 0.84 86% $\frac{17}{20}$ $\frac{4}{5}$

3 Write each decimal as a fraction in its simplest form.

a $0.\dot{7}$ **b** $0.\dot{2}$ **c** $0.6\dot{5}$ **d** $0.8\dot{4}$

e $0.5\dot{7}$ **f** $0.80\dot{4}$ **g** $0.43\dot{8}$ **h** $0.18\dot{5}$

4 Write each decimal as a fraction in its simplest form.

a $0.0\dot{4}$ **b** $0.3\dot{8}$ **c** $0.03\dot{9}$ **d** $0.46\dot{4}$

e $0.264\dot{8}$ **f** $2.\dot{3}$ **g** $1.3\dot{8}$ **h** $7.23\dot{7}$

Stretch – can you deepen your learning?

1 Work out the calculations, giving your answers as fractions in their simplest form.

a $0.4\dot{5} \times 0.\dot{6}$ **b** $2 - 0.3\dot{2}$ **c** $0.1\dot{8} + 0.2\dot{8} \times 0.3\dot{4}$

2 If $x = 0.\dot{n}$, where n is a digit, determine the value of $0.4\dot{n}$ in terms of x.

3 If $y = 0.\dot{m}$, where m is a digit, determine the value of $0.3\dot{m}$ in terms of y.

Are you ready? (A)

1 Work out:

a 10% of 80 **b** 50% of 72 **c** 1% of 650 **d** 5% of 91

2 Work out:

a 650 ÷ 100 **b** 34 ÷ 100 **c** 48 ÷ 10 **d** 3.8 ÷ 10

There are many ways to **calculate percentages of amounts**.

For example, to find 45% of 60 you could:

- work out 50% (60 ÷ 2 = 30), work out 5% (50% ÷ 10 = 30 ÷ 10 = 3) and subtract (45% = 50% − 5% = 30 − 3 = 27)
- work out 1% (60 ÷ 100) and multiply the answer by 45
- use the fraction multiplication $\frac{45}{100} \times 60$
- use a **decimal multiplier** by working out 0.45 × 60; this is the most efficient way if you have a calculator, especially when working with more challenging percentages such as 37% or 1.6%

Section A will explore methods that do not use a calculator and section B will explore using multipliers.

Example 1

Calculate:

a 40% of 60 **b** 35% of 340 **c** 12% of 260

Method

Solution	Commentary
a **Method A** 100% is 60 20% is 60 ÷ 5 = 12 40% is 24	There are many different ways to find 40% of an amount. One method is to find 20%, and then multiply by 2 To find 20% of an amount, divide the amount by 5
Method B 100% is 60 10% is 60 ÷ 10 = 6 40% is 24	Another method is to find 10%, then multiply by 4 To find 10% of an amount, divide the amount by 10
	There are many other possibilites, for example finding 50% (half) of the amount and subtracting 10% of the amount from this.

b Method A	
100% is 340	One method is a 'build-up' method.
10% is 340 ÷ 10 = 34	Find 10% by dividing the amount by 10
30% is 34 × 3 = 102	
5% is 34 ÷ 2 = 17	5% is half of 10%.
35% is 30% + 5%	To find 35%, simply add together the 30% and 5%.
35% is 102 + 17 = 119	
Method B	
100% is 340	An alternative method is to find 1% by dividing by 100, then multiplying by the percentage (35).
1% is 3.4	
35% is 119	This method can be adapted for every question, but also requires the added skill of multiplying decimals.
c 100% is 260	When finding a percentage that is not a multiple of 5, it is often easiest to find 1%, then multiply by the percentage that is needed.
1% is 2.6	
12% is 31.2	

Example 2

a   Increase 58 by 30%.   b   Decrease 128 by 8%.

Method

Solution	Commentary
a 100% is 58	
10% is 58 ÷ 10 = 5.8	
30% is 5.8 × 3 = 17.4	The first step is to find 30% of 58
Final amount is 58 + 17.4 = 75.4	Then add this on to the original amount.
b 100% is 128	
1% is 128 ÷ 100 = 1.28	
8% is 10.24	Start by calculating 8% of 128
Subtract 8% from 128	This time, subtract it from 100% as we are decreasing.
128 − 10.24 = 117.76	

Practice (A)

1 Work out:

   a   30% of 52   b   85% of 30   c   55% of 120   d   65% of 20

   e   75% of 570   f   15% of 374   g   95% of 18   h   70% of 456

2 Work out:

   a   13% of 90   b   28% of 520   c   62% of 340   d   87% of 170

3 Work out:

    a 50% of 58   **b** 25% of 58   **c** 12.5% of 58   **d** 0.25% of 58

4 **a** Increase 32 by 20%.   **b** Decrease 84 by 55%.

    c Decrease 62 by 12%.   **d** Increase 260 by 3%.

5 Mario gets paid £12 an hour. He gets a pay rise of 20%.

What is his new hourly wage?

6 There are 220 students in Year 10.

55% of them get the bus to school.

20% of them walk to school.

15 cycle to school.

The rest get a lift.

How many Year 10 students get a lift to school?

7 A computer costs £650

In a sale, the price of the computer is reduced by 15%.

What is the sale price of the computer?

8 A standard box of tea bags contains 120 tea bags.

A special offer box contains 40% more tea bags.

How many tea bags are in the special offer box?

9 Emily owns a café. In August she sold 3250 cupcakes.

She sold 8% more cupcakes in September than in August.

How many cupcakes did Emily sell in September?

10 Abdullah puts £260 into a savings account that pays 3% interest per year.

How much money does Abdullah have in the savings account a year later?

What do you think? (A)

1 In a sale, all prices are reduced by 20%.

Faith has a discount card which gets her a further 10% off.

Faith says that this means she has got 30% off in total. Is she correct?

2 **a** Show that x% of y is the same as y% of x.

    b Use the result in part **a** to work out 64% of 25

    c Make up similar questions to part **b** and challenge a partner.

Are you ready? (B)

 1 Write each percentage as a decimal.

    a 30%   **b** 12%   **c** 68%   **d** 3%

2 Round each number to the nearest integer.

    a 248.6   **b** 305.48   **c** 19.98   **d** 168.314

3 Round each amount to the nearest penny.

    a £354.328   **b** £24.234   **c** £1.284   **d** £5684.387

Using your calculator

There is a percentage button on most calculators that can be used to find percentages of amounts. To use this button, enter the amount multiplied by the percentage. For example, to find 62% of 80, type into your calculator

Example 1

Calculate:

a 28% of 56   **b** 3.5% of 524

Method

Solution	Commentary
a 28% is 0.28	When using a calculator to find percentages, it is easiest to use the multiplier method. To work out the multiplier, convert the percentage to the equivalent decimal by dividing by 100
28% of 56 = 0.28 × 56 = 15.68	Then multiply the amount by the multiplier.
b 3.5% is 0.035 524 × 0.035 = 18.34	The multiplier for 3.5% is 0.035 as 3.5 ÷ 100 = 0.035

Example 2

a In March, the amount of profit that Sven made was £154

    In April, this increased by 13%.

    What was Sven's profit in April?

b Amina buys a computer for £564

    The value of the computer depreciates by 5.2% per year.

    What is the value of the computer after 1 year?

> 'Depreciates' means the value decreases.

Method

Solution	Commentary
a 100% + 13% = 113%	In April, Sven's profit was higher than March. This means the question is to increase 154 by 13%. As it is increasing by 13%, add this to 100%.
113% is 1.13	Then convert this into a decimal by dividing 113 by 100 to find the decimal multiplier.
£154 × 1.13 = £174.02	Then, multiply the amount by the multiplier.
b 100% − 5.2% = 94.8% 94.8% is 0.948	You need to decrease 564 by 5.2%. For a decrease, subtract the percentage reduction from 100% and convert this to a decimal multiplier.
£564 × 0.948 = £534.672	Multiply the value by the multiplier.
The value of the computer is £534.67	As this is money, the answer should be rounded to 2 decimal places (the nearest penny).

Practice (B)

1 Write the decimal multipliers needed to calculate these percentages.

a 82% of a number **b** 63% of a number

c 3% of a number **d** 0.7% of a number

2 Work out the calculations. Round your answer to 2 decimal places where appropriate.

a 32% of 48 **b** 56% of 157 **c** 68% of £231 **d** 7% of 4566

e 2% of 456 g **f** 85.2% of 846 g **g** 0.4% of £924 **h** 0.35% of £6589

3 Write down the decimal multipliers for calculating these percentage changes.

a 23% increase **b** 14% decrease **c** 28% increase **d** 68% increase

e 52% decrease **f** 3% decrease **g** 6.8% decrease **h** 9.2% increase

4 Use percentage multipliers to:

a increase 82 by 42% **b** decrease 468 by 13% **c** increase 246 by 28%

d decrease 146 by 84% **e** decrease 45 by 12% **f** increase 63 by 3%

g decrease 76 by 0.5% **h** increase 12 by 102%

5 Ed is booking a holiday. He pays a 12% deposit.

The holiday costs £2350

a What percentage of the total cost does Ed still have to pay?

b How much does Ed have left to pay?

6 Marta's salary is £38 250 per annum.

She pays 20% tax on her income.

What is her monthly salary after tax?

7 Beca buys a car for £12 580

She pays 6% interest on the payment.

How much does the car cost in total?

8 Ten years ago, the population of a town was 85 960

The population has increased by 2.5% since then.

What is the population of the town now?

What do you think? (B)

1 Explain why 32% of 84 = 84% of 32 but increasing 32 by 84% is not the same as increasing 84 by 32%

2 Beth thinks of a number. She increases it by 10% and then decreases the result by 10%. Is her final answer **greater than, less than** or **equal to** the number she started with? Justify your answer.

Are you ready? (C)

1 Write each statement as a fraction in its simplest form.

   a 15 out of 100   **b** 30 out of 40   **c** 37 out of 50   **d** 39 out of 60

2 Convert each fraction to a percentage.

   a $\dfrac{67}{100}$ **b** $\dfrac{27}{50}$ **c** $\dfrac{17}{20}$ **d** $\dfrac{19}{25}$

3 Round each value to 2 decimal places.

   a 198.391   **b** £6.8434   **c** £652.498   **d** 3574.304

To express one amount as a percentage of another, you can:

- express as a fraction and convert
  For example, a mark of 24 out of 60 as a percentage is $\dfrac{24}{60} = \dfrac{2}{5} = \dfrac{4}{10} = 40\%$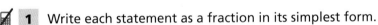

- divide and convert the decimal if you have a calculator
  For example, if 6 out of 32 students are absent then $6 \div 32 = 0.1875 = 18.75\% \approx 19\%$

Example 1

Express 28 as a percentage of 40

Method

Solution	Commentary
$\dfrac{28}{40} = \dfrac{14}{20} = \dfrac{70}{100} = 70\%$	Write 28 out of 40 as a fraction then use equivalent fractions to make the denominator 100 (as percentages are out of 100).

Example 2 ▦

Express 18 out of 32 as a percentage.

Method

Solution	Commentary
$\frac{18}{32}$ = 18 ÷ 32 = 0.5625	First convert the fraction into a decimal.
0.5625 × 100% = 56.25%	Then multiply this by 100 to give a percentage.

Example 3 ▦

Jackson buys a vase for £32. He then sells it for £25.60

Work out his percentage loss.

Method

Solution	Commentary
Method A	
Loss = £32 – £25.60 = £6.40	Calculate the amount of money that Jackson lost on the vase.
$\frac{6.40}{32}$ = 0.2 = 20%	Express this amount as a percentage of the original cost.
Jackson made a 20% loss.	
Method B	
$\frac{25.60}{32}$ = 0.8 = 80%	Write the new amount as a percentage of the original amount. This shows that the new amount is 80% of the original amount.
20% loss	It has been reduced by 20%, as 100% – 80% = 20%

Practice (C)

1 Express each of these as a percentage.

    a 18 out 25   **b** 37 out of 50   **c** 56 out of 80   **d** 13 out of 20

    e 124 out of 200   **f** 180 out of 250   **g** 37 cm out of 1 m   **h** 18p out of £3

2 Express each of these as a percentage. Give your answers to an appropriate degree of accuracy.

    a 45 out of 80   **b** 32 out of 65   **c** 3.8 out of 5   **d** 32.8 out of 63

    e 214 out of 280   **f** 38.4 out of 125   **g** 45 cm out of 2.5 m   **h** 65p out of £2.60

3 Emily has £45 in savings.

    She spends £12 on a book and £18 on a new top.

    What percentage of her savings does she have left?

4 Seb scores 26 out of 30 on a Maths test and 35 out of 40 on an English test.

By converting his scores to percentages, in which test did he do better?

5 Mario buys a car for £12 500
He later sells the car for £10 500

What is his percentage loss?

6 Tiff buys supplies to make 20 frames for a total of £250
She sells the frames for £19.50 each.

What is her percentage profit?

7 Flo invests £350 into a savings account.
After 1 year, she has £369.25

What is the interest rate of the savings account?

8 Zach buys 10 kg of biscuits for £45 and puts them into packs of 200 g each.

If he sells each pack for £1.20, what will his percentage profit be? Give your answer to 1 decimal place.

9 Huda starts a new business making cakes. It costs 75p to make each cake.
During a single day, she sells 215 cakes and takes £154.80

Work out her percentage profit or loss.

10 Ed invests £3500 into a company.
After 1 year, he gets £3200 back.

What is his percentage loss over the year?

11 The population of a city is 3.58×10^7
The number of people in the city who are retired is 5.871×10^6

What percentage of the population are retired? Give your answer to 1 decimal place.

Consolidate – do you need more?

1 Work out:

    a 40% of 182   **b** 20% of 384   **c** 35% of 92   **d** 65% of 36

    e 72% of 40   **f** 37% of 830   **g** 3% of 53   **h** 12% of 80

2 **a** Increase 180 by 35%.   **b** Decrease 350 by 20%.   **c** Decrease 96 by 85%.

    d Increase 32 by 40%.   **e** Decrease 65 by 12%.   **f** Increase 120 by 28%.

    g Increase 654 by 2%.   **h** Decrease 40 by 3.5%.

🖩 **3** Work out:

    a   12% of 380   **b**   8% of 65   **c**   30% of 87   **d**   0.7% of 680

🖩 **4**   **a**   Increase 35 by 27%.   **b**   Increase 114 by 2%.   **c**   Decrease 16 by 4%.

    d   Increase 68 by 15%.   **e**   Increase 30 by 74%.   **f**   Decrease 110 by 85%.

    g   Decrease 8.5 by 18%.   **h**   Decrease 289 by 0.4%.

5 Express each as a percentage.

    a   3 out of 20   **b**   18 out of 25   **c**   23 out of 40

    d   123 out of 200   **e**   85p out of £3.50   **f**   37 cm out of 3 m

    g   275 g out of 1.5 kg   **h**   2 weeks out of a year

6 Samira buys a new car that costs £15 500

    She pays a 4% deposit, and pays the rest in monthly payments for 4 years.

    How much is her monthly payment?

Stretch – can you deepen your learning?

1 All of the prices in a shop are reduced by 25%.

    Two weeks later, all the prices are increased by 30%.

    Are the prices after two weeks higher or lower than the original prices?
Justify your answer.

2 Bobbie invests £N in a bank.

    Write an expression to show the value of her investment if the investment:

    a   grows in value by a%   **b**   decreases in value by a%.

3 Faith is investigating the areas of various rectangles.

    a   The length of a rectangle is increased by 25% but its area remains the same.

        By what percentage is the width decreased?

    b   The length of another rectangle is decreased by 36% and its area remains the same.

        By what percentage is the width increased?

4 Ali is making 480 cookies to sell.

    The cookies are either white chocolate, plain, milk chocolate, or raisin.

    $\frac{1}{6}$ of the cookies are white chocolate.

    45% of the cookies are plain.

    The ratio of milk chocolate to raisin cookies is 3 : 5

    What percentage of the cookies are raisin cookies?

4.3 Repeat percentages

Are you ready?

1. Work out:

   a 45% of 60   b 32% of 230   c 80 increased by 15%   d 38 decreased by 12%.

2. Write the decimal multiplier for a:

   a 20% increase   b 30% decrease   c 12% increase   d 62% decrease.

3. Work out:

   a 1.2×1.3 b 1.05×0.95 c 0.85×1.15 d 0.9×0.75

In this chapter you will explore **repeated percentage change**, where an amount is increased or decreased by a percentage more than once.

It is easiest to calculate the effect of repeated percentage changes using decimal multipliers. For example, if a population of 60 000 decreases by 12% a year for three successive years, the final population can be found by working out

$$60\,000 \times 0.88 \times 0.88 \times 0.88 \quad \text{or} \quad 60\,000 \times 0.88^3$$

Example 1

There are 200 bacteria cells in a dish. Each hour, the number of cells increases by 5%.

Work out the number of cells in the dish after:

a 1 hour   b 3 hours.

Method

Solution	Commentary
a Multiplier for 5% increase = 1.05	100% + 5% = 105%, 105 ÷ 100 = 1.05
$200 \times 1.05 = 210$	Multiply the original number by the multiplier.
b **Method A**	
After: 1 hour = 210 cells 2 hours = 210 × 1.05 = 220.5 3 hours = 220.5 × 1.05 = 231.525	As the number of cells increases by 5% each hour, you need to multiply the previous value by 1.05 each time.
After 3 hours, there will be 232 cells in the dish.	Round the answer to the nearest integer as you cannot have decimal parts of a cell. The model of a 5% increase per hour cannot be exact.
Method B Number of cells after 3 hours = 200 × 1.05 × 1.05 × 1.05 = 200×1.05^3 = 232	Instead of writing the calculations in full, one hour at a time, you can use indices to reduce the working to a single calculation.

Example 2

A car costs £25 000

The value of the car depreciates at a rate of 6% per year.

What is the value of the car after 4 years?

> Remember, 'depreciate' means to reduce in value.

Method

Solution	Commentary
Method A	
Multiplier for a 6% decrease = 0.94	The first step is to find the multiplier for a 6% decrease:
	100% − 6% = 94% so the multiplier is 0.94
After 1 year: 25 000 × 0.94 = 23 500	Use the multiplier to calculate the value of the car after each year.
After 2 years: 23 500 × 0.94 = 22 090	
After 3 years: 22 090 × 0.94 = 20 764.6	
After 4 years: 20 764.6 × 0.94 = 19 518.724	
Value of the car after 4 years = £19 518.72	Round the answer to 2 decimal places (the nearest penny).
Method B	
25 000 × 0.94 × 0.94 × 0.94 × 0.94	Rewriting the repeated multiplication as a power is far more efficent.
25 000 × 0.94^4 = 19 518.724	
After 4 years, £19 518.72	

Example 3

Marta invests £2500 into a savings account that gives 3.2% compound interest.

Work out the money that is in Marta's account after 5 years.

> 'Compound interest' means the interest is based on the balance at the end of each year.

Method

Solution	Commentary
100% + 3.2% = 103.2% = 1.032	The first step is to find the multiplier for a 3.2% increase.
2500 × 1.032 × 1.032 × 1.032 × 1.032 × 1.032	Using Method B from Example 2, each year the balance is multiplied by the multiplier. Writing the repeated multiplication as a power is far more efficent.
= 2500 × 1.032^5	
= 2926.432 391	
After 5 years, Marta has £2926.43	As this is a money caculation, the answer needs to be rounded to 2 decimal places.

Practice

1 Write down the calculations needed to perform these percentage changes.

    a   Increase 20 by 10%, then by 10%.   **b**   Decrease 120 by 10%, then by 10%.

    c   Decrease 60 by 20%, then by 25%.   **d**   Increase 160 by 30%, then by 15%.

    e   Increase 40 by 15%, then by 30%.   **f**   Increase 85 by 12%, then by 6%.

    g   Decrease 163 by 23%, then by 5%.   **h**   Decrease 82 by 5%, then increase by 3.5%.

2 In a sale, all prices are reduced by 10%.

Jackson is buying a jumper that costs £45

He also has a discount card for a further 5% off.

Work out the amount Jackson pays for the jumper.

3 The value of a car depreciates by 4% per year.

Lida buys a car for £12 500

What is the value of the car after:

    a   1 year   **b**   2 years   **c**   5 years?

4 Tiff invests £500 into a savings account that gives 3% compound interest per year.

    a   How much money does Tiff have in the account after 3 years?

    b   How much interest does Tiff gain after 8 years?

5 A ball is dropped from a height of 5 m.

After each bounce, it loses 50% of its height.

What height, in centimetres, will the ball reach after:

    a   1 bounce   **b**   6 bounces?

6 Abdullah has invested £800 into a savings account.

The savings account pays 5% interest for the first year, then 1.6% interest per year after that.

How much will be in Abdullah's account after 3 years?

7 A square has a length of 6 cm.

Each side of the square is enlarged by 10%.

What is the new area of the square?

8 A cube has a volume of 216 cm³.

Each side of the cube is increased by 20%.

What is the volume of the enlarged cube?

9 A savings account offers a compound interest rate of 6% for the first year, and 2.7% for any following years.

Junaid is going to invest £3000 into this savings account.

How much money will he have after 4 years?

10 The world population is increasing at a rate of approximately 0.9% per year.

In 2021 the population was approximately 7.88 billion.

a Estimate the world population in 2030. Give your answer to 3 significant figures.

b Assuming this rate is constant, in which year will the population first be greater than 10 billion?

What do you think?

1 The height of a tree is 1.2 m.

The height of the tree increases by 12% every year.

a After how many years will the height of the tree be over 3 m?

b How realistic do you think this model is?

2 a Calculate the overall percentage change if:

i a 30% decrease is followed by a 12% decrease

ii a 20% increase is followed by an 8.5% increase.

b The value of a new car decreases by 5% in the first year, then 2% in the second year.

What is the overall percentage change after 2 years?

Consolidate – do you need more?

1 Write down the single calculation needed to:

a increase 75 by 10%, then by 20%   **b** increase 150 by 15%, then by 8%

c decrease 220 by 12%, then by 15%   **d** increase 342 by 20%, then decrease by 3%.

2 There are 200 cells in a Petri dish.

The number of cells increases by 6% each hour.

How many cells are in the Petri dish after:

a 1 hour   **b** 5 hours   **c** 1 day?

3 Chloe bought a house for £250 000

In the first year, the value of the house increased by 3%.

In the second year, the value of the house decreased by 2%.

In the third year, the value of the house increased by 2%.

Work out the value of the house after 3 years.

4 Marta buys a television for £600

She sells it to Sven and makes a 10% loss.

Sven then sells the television to Huda for a 15% loss.

How much does Huda pay for the television?

5 Samira invests £1500 into a savings account that pays 2.6% compound interest.

How much is in the savings account after:

a 1 year **b** 3 years **c** 10 years?

6 A scientist believes that the number of otters in a lake is increasing by 12.5% each year.
In the first year of study, there are 2500 otters in the lake.

How many otters does the scientist's model suggest will be in the lake in each of the three following years?

Stretch – can you deepen your learning?

1 Flo invests money into a savings account that pays 3% compound interest per year.
After 4 years, Flo has £877.90 in her account.

How much money did Flo invest into the savings account?

2 There are 800 birds in a forest.
The number of birds in the forest decreases by 2% each year.

After how many years are there fewer than 600 birds in the forest?

3 The population of a town increased by n% per year.
In 2014, the population was 32 000
In 2024, the population was 40 841

Calculate the value of n. Give your answer to 3 decimal places.

4 Benji invests £15 000 into a savings account that offers x% compound interest per year.
After 5 years, there is £18 425.95 in the account.

Work out the value of x.

White Rose
M▲THS

Are you ready?

1 Convert each percentage to a decimal.

    a 80%   **b** 65%   **c** 28%   **d** 3.8%

2 Write the decimal multiplier for a:

    a 20% increase   **b** 40% decrease   **c** 32% increase   **d** 2% decrease.

3 Solve the equations.

    a $1.2x = 48$ **b** $0.8y = 64$ **c** $1.05a = 47.25$ **d** $0.65w = 0.455$

Finding a **reverse percentage** is the process of finding an original amount given the value after a percentage change. This can be done using a non-calculator or a calculator method, depending on the numbers.

When a value has been increased by 20%, the new value is 120% of the **original value**.

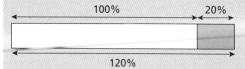

When a value has been decreased by 20%, the new value is 80% of the original value.

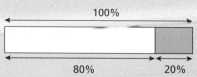

You can often work out percentages of a number using proportional reasoning. For example, if you know 20% of a number, you can halve this amount to find 10% of the number or multiply by 3 to find 60% of the number.

Example 1

The price of a jumper is reduced by 20% to £13.60

Calculate the original price of the jumper.

Method

Solution	Commentary
Method A	
100% − 20% = 80%	Work out what percentage of the original price is left after the reduction.
80% of the original price is £13.60	
10% would be £13.60 ÷ 8 = £1.70	80% is £13.60 so 10% is 80% ÷ 8
The original price would be £1.70 × 10 = £17	The original price is 100%, which is 10% × 10

Method B	Work out the percentage that is remaining after the reduction and convert to the decimal multiplier. For a 20% decrease, the multiplier is 0.8
$100\% - 20\% = 80\% = 0.8$ Let x be the original cost of the jumper. $$0.8x = £13.60$$ $\div\,0.8$ $\div\,0.8$ $$x = £17$$ The original price of the jumper was £17	Set up an equation to show the change that has taken place. Solve the equation by balancing.

Example 2

Chloe's salary is increased by 20% to £39 000

What was Chloe's salary before the increase?

Method

Solution	Commentary
Method A	
$100\% + 20\% = 120\%$	Work out what percentage of the original salary the increased salary is.
120% 100% 20%	
120% of original salary is £39 000	
So 10% would be £39 000 ÷ 12 = £3250	120% is £39 000 so 10% is 120% ÷ 12
The original salary was 10 × £3250 = £32 500	The original salary is 100%, which is 10% × 10
Method B	
120% is £39 000	You could have worked this out by finding 1%.
1% is £39 000 ÷ 120 = £325	
100% is £32 500	You could also have set up an equation as in Example 1 ($1.2x = £39\,000$).
Chloe's salary before the increase was £32 500	

Example 3

Kath earns 3% interest on her savings each year. At the end of the first year, Kath has £2575 in her savings account.

Calculate the amount that was in her savings account before the interest was added.

Method

Solution	Commentary
100% + 3% = 103% = 1.03 Let x be the original amount saved. $1.03x = £2575$ $\div 1.03 \left(\quad \right) \div 1.03$ $x = £2500$ Original amount Kath saved was £2500	Work out the percentage multiplier for a 3% increase. Form an equation to represent the change. Solve the equation by balancing.

Practice ▦

1. 40% of a number is 200

   Work out:

   a 10% of the number   **b** 20% of the number

   c 80% of the number   **d** 100% of the number.

2. 120% of a number is 288

   Work out:

   a 20% of the number   **b** 60% of the number

   c 10% of the number   **d** 100% of the number.

3. 40% of the students in a class have blonde hair.

   10 students in the class have blonde hair.

   How many students are in the class?

4. In a sale, all prices are reduced by 15%.
   A T-shirt costs £19.55 in the sale.

   Work out the price of the T-shirt before the sale.

5. After a 15% pay rise, Zach earns £14.72 per hour.

   Work out Zach's wage before his pay rise.

6. The price of a computer is £450, including 20% VAT.

   Work out the price before VAT.

7. A large bag of chocolates contains 30% more than the standard bag.
   A large bag of chocolates contains 845 g.

   What is the mass of the standard bag of chocolates?

8 Ali sells a sofa for £570, making a 14% profit.

How much profit does Ali make?

9 Jakub earns £42 354 per year after a 17% pay rise.

How much did Jakub earn before his pay rise?

10 In a sale, all phone prices are reduced by 20%.

After a further discount of 26%, Lida pays £402.56 for a phone.

What was the price of the phone before the discounts were applied?

What do you think? 💭

1 The lengths of the sides of a cube are increased by 10%.
The volume of the cube is now 681.472 cm³.

Work out the volume of the cube before the sides were increased.

2 Junaid sells his computer to Mario and makes a 20% loss.
Mario then sells the computer to Rob for £132 and makes a 10% profit.

How much did Junaid originally buy the computer for?

3 Investigate which would be better value for a customer:
- Increasing the size of a chocolate bar by 10%   *or*
- Decreasing the cost of the chocolate bar by 10%

Consolidate – do you need more? 🖩

1 60% of a number is 45
Work out:
- **a** 20% of the number
- **b** 10% of the number
- **c** 5% of the number
- **d** 100% of the number.

2 A mirror costs £216, including 20% VAT.

What is the cost of the mirror before VAT is added?

3 Bev earns £17 480 after a 15% pay rise.

What was Bev's salary before the increase?

4 The average rainfall for May was 36 mm.
This was a decrease of 55% from April.

What was the average rainfall for April?

5 A pair of headphones costs £125.40 in a 20% off sale.

What is the normal cost of the headphones?

6 Junaid sells painted baskets. He makes 65% profit on each basket and sells them for £46.20 each.

If Junaid wants to make 75% profit, how much does he need to sell the baskets for?

7 Kath has a savings account that pays 4% interest.

After 1 year, she withdraws £150 and has £604.52 left in the account.

How much money did Kath invest to begin with?

8 A bag contains four types of shapes: red circles, red triangles, red squares, and blue squares.

40% of the squares are red.

30% of the red shapes are squares.

There are 12 red squares in the bag.

Work out the total number of shapes in the bag.

Stretch – can you deepen your learning? 🖩

1 A strawberry milkshake is made from strawberries and milk in the ratio 2 : 5

80% of the strawberries are frozen.

60 g of frozen strawberries are used.

How many grams of milk are needed for the milkshake?

2 Rhys and Lida share some money in the ratio 3 : 2

Rhys then saves 30% of his money.

Rhys saves £45

How much money did they share?

3 a After an increase of x%, a car costs £2700

Write an expression for the value of the car before the increase.

b After a decrease of y%, a calculator costs £5.60

Write an expression for the cost of the calculator before the decrease.

c The price of a T-shirt is reduced by a%, then by a further b%, to a sale price of t.

Write an expression for the price of the T-shirt before the reductions.

4 Amina invests money into a savings account that offers 2.4% compound interest.

After 5 years, there is £1400.52 in the account (to the nearest penny).

Calculate the amount that Amina originally invested into the savings account.

Percentages: exam practice

1 Seb pays £8 for 24 bags of sweets. He sells all the bags of sweets for 75p each.

Work out Seb's percentage profit. **[3 marks]**

4–6

2 Amira buys a car for £16 000

The car depreciates at a rate of 18% in the first year and 15% a year in following years.

Work out the value of the car after 3 years. **[3 marks]**

3 When a number is reduced by 15%, the answer is 7344

What was the original number? **[3 marks]**

4 The sides of cube A are 5 cm long.

Cuboid B measures 4 cm by 4 cm by 5 cm.

Express the volume of cuboid B as a percentage of the volume of cube A. **[3 marks]**

5 Work out the percentage increase from 40 to 240 **[3 marks]**

6 Rhys gets a quote of £532 to build a shed.

The quote includes 8 hours of labour at £27.50 an hour and a 20% discount off the cost of the materials needed.

Work out the cost of the materials before the discount. **[3 marks]**

7 The population of a city was 800 000 this year.

The population is expected to grow by 5% a year.

Calculate the expected population of the city in 3 years' time. **[3 marks]**

8 Convert the fractions shown on the cards into decimals and state whether or not each decimal is terminating or recurring.

$\frac{3}{8}$ $\frac{5}{9}$ $\frac{8}{11}$

[4 marks]

7–9

9 Write $0.1\dot{2}$ as a fraction in its simplest form. **[3 marks]**

10 Write $0.3\dot{7}$ as a fraction in its simplest form. **[3 marks]**

11 Work out $0.5\dot{6} + \frac{1}{4}$

Give your answer as both a fraction and a recurring decimal. **[4 marks]**

5 Accuracy and estimation

White Rose
MATHS

In this block, we will cover...

5.1 Rounding and estimating

Example

Estimate the value of:

a $76.4 \times 87.3 + 23.7$ b $\dfrac{207 \times 21.8}{0.43}$ c

Method

Solution	Commentary
a $76.4 \times 87.3 + 23.7$	
$\approx 80 \times 90 + 20$	Round each numb
$\approx 7200 + 20$	Follow the order o
≈ 7220	

5.2 Upper and lower bounds

Practice (B)

1 A rectangular field is 230 m long and 70 m
   a   Calculate the lower bound for the peri
   b   Calculate the upper bound for the area

2 A pack of fries has 250 fries, to the nearest
   Calculate the error interval for the number

3 A bucket has a capacity of 50 litres, to 1 sig
   A carton of water has a capacity of 200 ml,
   Write the error interval for the number of

Are you ready?

1 Round each number to 1 significant figure.

    a 566   **b** 23874   **c** 35.74   **d** 0.3184   **e** 0.000684

2 Which of the numbers below round to 1000 to 1 significant figure?

| 500 | 600 | 870.1 | 1020.846 | 984 | 949 |

| 1250 | 499.99 | 1499.999 |

3 Work out:

    a $\dfrac{6}{0.5}$ **b** $18 \div 0.4$ **c** $\dfrac{90}{2.5}$ **d** $\dfrac{3^2}{0.3}$ **e** $\dfrac{33.6}{1.2}$

> You can **estimate** the answer to a calculation by rounding all numbers in the calculation to 1 significant figure.

Example

Estimate the value of:

a $76.4 \times 87.3 + 23.7$ **b** $\dfrac{207 \times 21.8}{0.43}$ **c** $\sqrt{52}$

Method

Solution	Commentary
a $76.4 \times 87.3 + 23.7$	
$\approx 80 \times 90 + 20$	Round each number to 1 significant figure.
$\approx 7200 + 20$	Follow the order of operations to work out the estimate.
≈ 7220	
b $\dfrac{207 \times 21.8}{0.43} \approx \dfrac{200 \times 20}{0.4}$	Round each number to 1 significant figure.
$\approx \dfrac{4000}{0.4}$	
$\approx \dfrac{40\,000}{4}$	Work out the numerator. Multiply both the numerator and denominator by 10 to make the denominator an integer.
$\approx 10\,000$	Perform the division to get the final answer.
c $\sqrt{49} < \sqrt{52} < \sqrt{64}$	$\sqrt{52}$ lies between 7 and 8. It is closer to 7, as 52 is closer to 49 than 64
$7 < \sqrt{52} < 8$	
$\sqrt{52} \approx 7.2$	The square root of 52 is approximately 7.2. Answers in this region are acceptable.

For more on estimating roots, see Chapter 2.1

Practice

1 By rounding each number to 1 significant figure, estimate the answer to each calculation.

a 7.6×9.6

b $32.7 + 48 \times 2.4$

c $\dfrac{684 - 215}{23.8}$

d $\dfrac{5.84^2}{3.12}$

e $\dfrac{83.84 + 77.2}{18 - 4.1}$

f $\dfrac{12.84 \times 8.6}{12}$

g $9 + 8.4 \times 6.3 - 2$

h $(-3.7)^2 - 2.67$

2 Estimate the answer to each calculation.

a $\dfrac{32.8 + 12.94}{0.47}$

b $\dfrac{98.124 \times 32.4}{0.42}$

c $\sqrt{70}$

d $\sqrt{2.7 \times 6.8}$

e $\dfrac{(7.4 + 3.05)^2}{0.234}$

f $\dfrac{\sqrt{8.4 \times 4.95}}{0.35}$

g $\dfrac{95.4 \times 3.8^2}{3.68 \times 0.529}$

h $\dfrac{\sqrt{200} + 6.4}{0.481}$

3 Estimate the area of each shape.

a

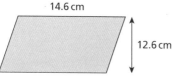

14.6 cm

12.6 cm

b

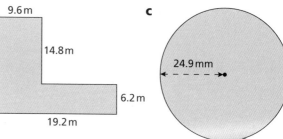

9.6 m

14.8 m

6.2 m

19.2 m

c

24.9 mm

4 **a** Estimate the area of the trapezium.

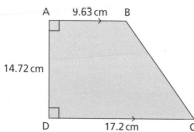

A   9.63 cm   B

14.72 cm

D   17.2 cm   C

b Estimate the length of BC. Give your answer in centimetres to 1 decimal place.

5 The shape is made up from a rectangle and two identical quarter circles.

Estimate the area of the shaded region.

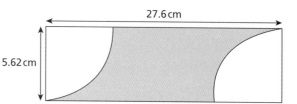

27.6 cm

5.62 cm

6 Express the error in the estimate of these calculations, when numbers are rounded to 1 significant figure, as a percentage of the actual answer. Give your answers to 2 decimal places.

a $\dfrac{32.7 - 16.9}{3.48}$

b $\dfrac{4.6 + 9.7 \times 3.28}{0.48}$

c $(9.42 + 13.8 \times 2.94)^2$

> You can work out the percentage error by expressing the difference between the estimate and the actual value as a percentage of the actual value.

What do you think? 💡

1. A cycle race is 380.5 miles.

   A cyclist can ride a maximum of 96 miles per day.

   a Estimate the minimum number of days that the race will take.

   b Can you tell if your answer is an overestimate or an underestimate? How do you know?

2. Estimate each calculation, and decide if your estimate is an **overestimate**, an **underestimate** or you **cannot tell**.

   a $\dfrac{32.6 + 21.8}{5.2}$

   b $(98.4 - 38.6 \times 2.8)^2$

   c $\dfrac{6.45 - 2.48}{0.472}$

   d $\dfrac{3.877 \times 32.8 - 8.975}{2.483}$

Consolidate – do you need more?

1. Estimate the answer to each calculation.

   a $\dfrac{38.4 \times 3.84}{13.8}$

   b $\dfrac{68.4 - 2.4 \times 9.6}{4.92}$

   c $0.12^2 + 0.032$

   d $64.7^2 - 32.4 \times 64.8$

   e $846 + 6.43 \times 0.48$

   f $\dfrac{72.4 - 18}{8.4}$

   g $\dfrac{6.84^2}{2.41}$

   h $32.84 \div 64.128$

2. Estimate the answer to each calculation.

   a $\dfrac{38.4 - 21.6}{0.524}$

   b $\dfrac{16.1 + 5.4^2}{0.42}$

   c $\sqrt{84}$

   d $\sqrt{150} - 11.4$

   e $\dfrac{68.7 \times 128.12}{0.818 - 0.29}$

   f $\dfrac{\sqrt{90} \times 7.2}{2.1}$

   g $\dfrac{\sqrt{6.2^2 - 12.8}}{0.42}$

   h $\sqrt{\dfrac{38.4 \times 2.9}{0.48}}$

3. Emily earns £13.24 per hour.

   She works 37.5 hours per week.

   Estimate Emily's wage for 1 week.

4. A theatre has 62 rows.

   Each row has 128 seats.

   75% of the seats are taken.

   Estimate the number of seats that are empty in the theatre.

5. The distance between Land's End and John O'Groats by road is 837.4 miles.

   Rob is driving from Land's End to John O'Groats at an average speed of 58 mph.

   Estimate how long it will take Rob to drive from Land's End to John O'Groats.

Stretch – can you deepen your learning?

1 A cycle race is 3492 miles long.

A cyclist rides at an average speed of 26 mph for 6 hours per day.

a Estimate how many days it will take the cyclist to complete the race.

b If the cyclist increased their average speed to 32 mph, how would it affect your answer to part **a**?

2 A sphere has a volume of 3256 cm³.

Given that the formula for the volume of a sphere of radius r is given by $V = \dfrac{4}{3}\pi r^3$, calculate an estimate for the radius of the sphere.

3 A pool is in the shape of a trapezoidal prism, as shown.

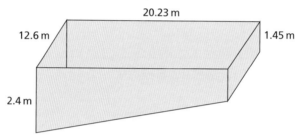

The pool fills with water at a rate of 120 litres per minute.

Given that 1 m³ = 1000 litres, estimate the amount of time it takes for the pool to be filled.

4 Think of a question in which there would be a different answer when the numbers are rounded to 1 significant figure, compared to being truncated to 1 significant figure.

What causes the difference?

Remember that when a number is truncated, digits are missed off after a certain point without rounding. For example, 0.683 truncated to 1 decimal place is 0.6

Are you ready? (A)

1 Which of these numbers will round to 200, when rounded to the nearest 10?

192   197   201.58   205.4   206.9   194.8

2 Round each number to 1 decimal place.

  a 84.354   **b** 3.4856   **c** 0.314   **d** 98.97

3 Round each number to 1 significant figure.

  a 324   **b** 0.0846   **c** 38851   **d** 384975.345

4 Truncate each number to the specified degree of accuracy.

  a 0.0842, to 2 decimal places   **b** 6.8495, to 3 decimal places

  c 0.00318, to 1 significant figure   **d** 3.6845, to 2 decimal places

When you use a measurement, it is never totally accurate. For example, if your height was given as 155 cm, you would be very unlikely to be exactly 155 cm tall. You could be 154.7 cm or 155.34 cm, for example.

You can use **bounds** to consider what the least and greatest values are when given a degree of accuracy. For example, if the length of a train carriage is 22 m to the nearest metre, it could be as short as 21.5 m and as long as any length up to but not including 22.5 m. This can be shown on a number line.

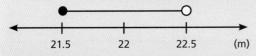

If the length of the train carriage was 22.5 m, this would round to 23 m. For this reason, we often use inequalities to write upper and lower bounds. These are often called **error intervals**. For this example, the error interval for the length would be 21.5 m $\leqslant l <$ 22.5 m.

The lower bound is 21.5 and the upper bound is 22.5. This error interval is said as 'l is greater than or equal to 21.5 m, and less than 22.5 m'.

An error interval is the representation of the upper and lower bound for a number. It is usually written as an inequality in the form _____ $\leqslant x <$ _____

Example 1

A box has a mass of 145 g to the nearest gram.

a Write the lower bound for the mass, m, of the box.

b Write the upper bound for the mass, m, of the box.

c Write the error interval for the mass, m, of the box.

Method

Solution	Commentary
a 144.5 g	144.5 g is the least possible value that rounds to 145 g.
b 145.5 g	Values that are 145.5 g or greater will round to 146 g.
c 144.5 g $\leqslant m <$ 145.5 g	

Example 2

The price of a house is £140 000, rounded to 2 significant figures.

a Write the lower bound for the price of the house.

b Write the upper bound for the price of the house.

c Write the error interval for the price of the house.

Method

Solution	Commentary
a £135 000	This time the number has been rounded to 2 significant figures. Notice that this means the bounds are going to contain 3 significant figures.
b £145 000	
c £135 000 $\leqslant x <$ £145 000	£145 000 would round to £150 000 when rounded to 2 significant figures, so it is not included in the error interval.

Example 3

The mass of a grain of sand is 0.004 grams, truncated to 3 decimal places.

a Write the lower bound for the mass.

b Write the upper bound for the mass.

c Write the error interval for the mass.

Method

Solution	Commentary
a 0.004	Truncated means that digits have been removed, in this case the digits after the thousandths column. Any number of the form 0.003$abcd$…will truncate to 0.003 whatever the values of a, b, c, d and so on.
b 0.005	In the same way, 0.004 999… will truncate to 0.004, so the bound is 0.005
c 0.004 g $\leqslant x <$ 0.005 g	

Practice (A)

1 For each rounded number, write:

i the lower bound **ii** the upper bound **iii** the error interval.

a $a = 300$, to the nearest 100 **b** $b = 240$, to the nearest 10

c $c = 45\,700$, to the nearest 100 **d** $d = 32\,500$, to the nearest 10

e $e = 19\,000$, to the nearest 1000 **f** $f = 350$, to the nearest integer

g $g = 2000$, to the nearest 10 **h** $h = 456$, to the nearest integer

2 Write the error interval for each rounded number.

    a $a = 500$, to 1 significant figure   **b** $b = 80\,000$, to 1 significant figure

    c $c = 520$, to 2 significant figures   **d** $d = 1850$, to 3 significant figures

3 Write the error interval for each rounded number.

    a $a = 3.5$, to 1 decimal place   **b** $b = 2.36$, to 2 decimal places

    c $c = 0.584$, to 3 decimal places   **d** $d = 320.3$, to 1 decimal place

    e $e = 84.35$, to 2 decimal places   **f** $f = 3.584$, to 3 decimal places

    g $g = 3.25$, to 3 significant figures   **h** $h = 0.084$, to 2 significant figures

4 Write the error interval for the **truncated** numbers.

    a $a = 3.5$, to 1 decimal place   **b** $b = 8.4$, to 2 significant figures

    c $c = 9534$, to the nearest integer   **d** $d = 0.08$, to 1 significant figure

5 A square has an estimated area of $400\,\text{cm}^2$ when rounded to 1 significant figure.

    Write the error interval for the length of the square.

6 The perimeter of a regular pentagon is $32.7\,\text{cm}$, to 3 significant figures.

    Calculate the error interval for the length of one side of the pentagon.

7 Lida writes an answer of 8.45, but she has only copied down the first three digits from her calculator.

    What is the error interval for Lida's answer?

8 The perimeter of the rectangle is $95\,\text{mm}$, to 2 significant figures.

    Calculate the error interval for x.

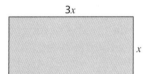

What do you think? (A) 💭

1 A number, n, when rounded is 100

    Write the error interval for the number if it was rounded to:

    a 1 significant figure   **b** 2 significant figures   **c** 3 significant figures.

2 The area of a circle is $30\,\text{cm}^2$.

    Write the error interval for the radius if the area was rounded to:

    a 1 significant figure   **b** 2 significant figures.

3 The number $p = 25.8$ has been rounded up to the next tenth.

    Write the error interval for p.

Are you ready? (B)

1 Write the error interval for each **rounded** number.

  a 18, to the nearest integer

  b 248, to 3 significant figures

  c 350, to the nearest 10

  d 6500, to the nearest 100

  e 84 000, to the nearest 1000

  f 3500, to 2 significant figures

2 Write the error interval for each **truncated** number.

  a 8.4, to 2 significant figures

  b 6.49, to 3 significant figures

  c 0.84, to 2 decimal places

  d 0.09, to 1 significant figure

  e 84.67, to 2 decimal places

  f 945.324, to 3 decimal places

In this section, the error intervals from section A will be applied to calculations. You will consider the upper and lower bounds for each value in the calculation, and how to combine them to give the bounds of the answer.

Example 1

A rectangle measures 8 cm by 3 cm, both correct to the nearest centimetre.

a Calculate the error interval for the perimeter of the rectangle.

b Calculate the error interval for the area of the rectangle.

Method

Solution	Commentary
a Length $7.5 \leqslant l < 8.5$ Width $2.5 \leqslant w < 3.5$	Start by writing the error intervals for both the length and width of the rectangle.
$P_{lower} = 7.5 + 2.5 + 7.5 + 2.5$ $= 20$ cm	To work out the lower bound for the perimeter, add the least possible values for the sides.
$P_{upper} = 8.5 + 3.5 + 8.5 + 3.5$ $= 24$ cm	To work out the upper bound for the perimeter, add the greatest possible values for the sides.
20 cm $\leqslant P < 24$ cm	Once both the lower and upper bounds have been calculated, write them in an error interval.
b $A_{lower} = 7.5 \times 2.5$ $= 18.75$ cm²	To calculate the lower bound for the area, find the product of the least possible values.
$A_{upper} = 8.5 \times 3.5$ $= 29.75$ cm²	For the upper bound, find the product of the greatest possible values.
18.75 cm² $\leqslant A < 29.75$ cm²	Write the lower and upper bounds in an error interval.

Example 2

$a = 7.8$, $b = 3.2$, both rounded to 1 decimal place.

a Calculate the error interval for $a - b$

b Calculate the error interval for $\frac{a}{b}$

Method

Solution	Commentary
a $7.75 \leqslant a < 7.85$ $3.15 \leqslant b < 3.25$	Write the error intervals for the values of a and b.
lower $= 7.75 - 3.25$ $= 4.5$	The lower bound will be the least possible value for a subtract the greatest possible value for b. This will give the least possible result.
upper $= 7.85 - 3.15$ $= 4.7$	The upper bound will be the greatest possible value for a subtract the least possible value for b. This will give the greatest possible result.
$4.5 \leqslant x < 4.7$	Once both the lower and upper bounds have been calculated, write them in an error interval.
b	Again, find the least possible values and greatest possible values for the calculation.
lower $= \dfrac{7.75}{3.25}$ $= 2.3846$	To calculate the lower bound of $\frac{a}{b}$, calculate the lower bound for a divided by the upper bound for b.
upper $= \dfrac{7.85}{3.15}$ $= 2.4921$	To calculate the upper bound, do the upper bound for a divided by the lower bound for b. This gives the greatest possible value.
$2.3846 \leqslant x < 2.4921$	Write the lower and upper bounds in an error interval.

Example 3

$a = bc$

$b = 5.8$ to 1 decimal place

$c = 32.48$ to 2 decimal places

Calculate a to a suitable degree of accuracy.

Method

Solution	Commentary
$5.75 \leqslant b < 5.85$ $32.475 \leqslant c < 32.485$	Again, begin with writing the lower and upper bounds for both values.
$a_{\text{lower}} = 5.75 \times 32.475$ $= 186.73125$ $a_{\text{upper}} = 5.85 \times 32.485$ $= 190.03725$	Then work out the lower and upper bounds for a.

$186.73125 \leqslant a < 190.03725$ Both of these numbers round to 190 to the nearest 10 $a = 190$, to the nearest 10	This time, instead of writing the error interval for a, find the most accurate number that both the lower and upper bounds would round to. Here, both values round to 190 when rounded to the nearest 10, so this is the value for a. It could also be rounded to 2 significant figures.

Practice (B) 🖩

1 A rectangular field is 230 m long and 70 m wide, both rounded to the nearest 10 m.

   a Calculate the lower bound for the perimeter of the field.

   b Calculate the upper bound for the area of the field.

2 A pack of fries has 250 fries, to the nearest 10

Calculate the error interval for the number of fries in five packs.

3 A bucket has a capacity of 50 litres, to 1 significant figure.

A carton of water has a capacity of 200 ml, to 1 significant figure.

Write the error interval for the number of cartons that can be emptied into the bucket.

4 $a = 119$, rounded to the nearest integer

$b = 320$, rounded to the nearest 10

$c = 84.2$, rounded to 1 decimal place

$d = 680$, rounded to the nearest 10

Calculate the error interval for each expression. Give your answers to 2 decimal places.

   a $a + b$ **b** $b + c$ **c** $c + d$ **d** $a + d$

   e ab **f** bc **g** cd **h** ad

5 $e = 84.9$, rounded to 1 decimal place

$f = 60$, rounded to the nearest 10

$g = 8.84$, rounded to 3 significant figures

$h = 20$, rounded to the nearest 10

Calculate the error interval for each expression. Give your answers to 2 decimal places.

   a $e - f$ **b** $f - g$ **c** $h - g$ **d** $e - h$

6 $i = 1300$, rounded to the nearest 100

$j = 30$, rounded to the nearest 10

$k = 19.2$, rounded to 1 decimal place

$l = 80$, rounded to 1 significant figure

Calculate the error interval for each expression. Give your answers to 2 decimal places.

   a $\dfrac{i}{j}$ **b** $\dfrac{j}{k}$ **c** $\dfrac{l}{k}$ **d** $\dfrac{i}{l}$

7 $m = 35$, rounded to 2 significant figures

$n = 0.8$, rounded to 1 significant figure

$o = 80.2$, rounded to 1 decimal place

$p = 620$, rounded to the nearest 10

Calculate the error interval for each expression. Give your answers to 2 decimal places.

a $\dfrac{\sqrt{m}}{n}$ **b** $o^2 - p$ **c** $\dfrac{n + p}{o}$ **d** $n(m + p)$

8 a The length of a rectangle is 8 cm, to the nearest centimetre.

The width of the rectangle is 6.4 cm, to the nearest millimetre.

Calculate the upper bound for the area of the rectangle.

b The area of a different rectangle is 96 cm², to the nearest integer.

The length of the rectangle is 25 cm, to the nearest 5 centimetres.

Calculate the lower bound for the width of the rectangle, giving your answer to 1 decimal place.

9 Ed is driving 84 miles, rounded to the nearest mile.

This trip takes 2 hours, to the nearest hour.

Calculate the upper bound for Ed's average speed.

10 a ABC is a right-angled triangle.

The lengths of AB and BC are given correct to the nearest centimetre.

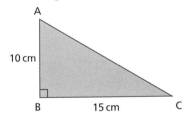

Calculate the upper and lower bounds for AC, giving your answers to 2 decimal places.

b ABC is a right-angled triangle.

The lengths of AB and AC are given correct to the nearest millimetre.

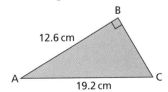

Calculate the upper and lower bounds for BC, giving your answers to 2 decimal places.

What do you think? (B) 🌐

1. A cube has an estimated volume of 8000 cm³.

   What is the upper bound for the surface area of the cube?

2. Investigate whether there is a link between the size of the error interval for the length of a side of a shape, compared to the error interval for the perimeter and area of the shape.

Consolidate – do you need more?

1. Write the error interval for each **rounded** number.

   a 32, to the nearest integer
   b 620, to the nearest 10
   c 800, to the nearest 100
   d 25 000, to the nearest 1000
   e 9000, to the nearest 100
   f 9000, to 1 significant figure
   g 3500, to 2 significant figures
   h 3500, to the nearest 10

2. Write the error interval for each **rounded** number.

   a 9.2, to 1 decimal place
   b 95.48, to 2 decimal places
   c 0.0008, to 1 significant figure
   d 984.6, to 1 decimal place
   e 3.6, to 2 significant figures
   f 8.568, to 3 decimal places
   g 0.0648, to 3 significant figures
   h 3.84, to 3 significant figures

3. Write the error interval for each **truncated** number.

   a 352, to the nearest integer
   b 9.50, to 2 decimal places
   c 6.802, to 3 decimal places
   d 0.002, to 1 significant figure
   e 32.84, to 2 decimal places
   f 962.4, to 1 decimal place
   g 94.6, to 3 significant figures
   h 0.008, to 1 significant figure

4. $a = 15.8$, rounded to 1 decimal place

   $b = 70$, rounded to 1 significant figure

   $c = 94$, rounded to the nearest integer

   Write the error interval for:

   a $a + b$ **b** $\dfrac{b}{c}$ **c** a^2 **d** bc **e** $c - a$

5. $x = 13.28$, rounded to 2 decimal places

   $y = 23.8$, rounded to 3 significant figures

   $z = 500$, rounded to 2 significant figures

   Write the error interval for:

   a $x + y$ **b** $\dfrac{z}{y}$ **c** xy^2 **d** $(xy)^2$ **e** $z - (x + y)$

Stretch – can you deepen your learning?

1 The area of a circle is 500 cm², to 1 significant figure.

Calculate the error interval for the circumference of the circle.

2 $x = 3.75$, correct to 3 significant figures

$y = 3.4$, correct to 2 significant figures

$z = 0.4$, correct to 1 significant figure

a $a = \dfrac{\sqrt{x}}{y}$

Work out the error interval for a, giving your answer to 3 significant figures.

b $b = \dfrac{y - x}{z}$

Work out the error interval for b, giving your answer to 2 decimal places.

3 A sphere has a radius of 0.85 m, truncated to 2 decimal places.

Surface area of a sphere = $4\pi r^2$

Volume of a sphere = $\dfrac{4}{3}\pi r^3$

a Write the error interval for the surface area of the sphere.

b Write the error interval for the volume of the sphere.

4 A building yard has 150 tonnes of gravel, measured to the nearest tonne.

The gravel is to be stored in bags that can hold 450 kg, measured to the nearest 10 kg.

Calculate the minimum number of bags that may be needed.

Accuracy and estimation: exam practice

1 Use your calculator to work out $\dfrac{18.3 + \sqrt{42.6}}{3.9 \times 6.05}$

   (a) Write down all the figures on your calculator display. **[2 marks]**

   (b) Write your answer to part **(a)** correct to 2 significant figures. **[1 mark]**

---

2 Rhys runs an average of 31.2 miles a week.

   (a) Work out an estimate for the number of miles Rhys runs in a year. **[2 marks]**

   (b) Is your answer to part **(a)** an underestimate or an overestimate?

      Give a reason for your answer. **[1 mark]**

---

3 Estimate the answer to $\dfrac{2.97^3 + \sqrt{102}}{0.48}$

   You must show all your working. **[3 marks]**

---

4 Amina uses her calculator to work out the value of a number x.

   She writes down the first two digits of the answer on her calculator display.

   She writes 4.7

   Complete the error interval for x.   .... $\leqslant x <$ .... **[2 marks]**

---

5 Rob writes down the value of y correct to 2 decimal places.

   He writes 0.64

   Complete the error interval for y.   .... $\leqslant y <$ .... **[2 marks]**

---

6 A circle has a radius of 21 cm.

   (a) Work out an estimate for the area of the circle. **[2 marks]**

   (b) Is your answer to part **(a)** an underestimate or an overestimate?

      Give a reason for your answer. **[1 mark]**

---

7 Estimate the value of $\sqrt{24.5 \times 103.7}$ **[2 marks]**

---

8 The sides of a square are measured as 7.6 cm correct to the nearest millimetre.

   Work out the lower bound of the area of the square, giving your answer correct to 2 significant figures. **[2 marks]**

7–9

9 $a = 8$ and $b = 5$, both correct to the nearest integer.

Benji thinks the maximum possible value of $a - b$ is $8.5 - 5.5 = 3$

Explain why Benji is wrong. **[2 marks]**

10 In 2021, Jackson bought a bike for £1400

In 2022, the bike went down in value by 16% to the nearest 1%.

In 2023, the bike went down in value by a further 10% to the nearest 1%.

What was the greatest possible value of the bike by the end of 2023?

Give your answer to the nearest pound. **[3 marks]**

11 Textbooks each have mass 1.2 kg correct to 1 decimal place.

A shelf can hold 15 kg correct to the nearest kilogram.

What is the greatest number of these textbooks that can be safely put on the shelf? **[3 marks]**

12 $a = 350$ to 2 significant figures.

$b = 80$ to 1 significant figure.

Work out the minimum value of $\frac{a}{b}$, giving your answer correct to 3 significant figures. **[3 marks]**

13 $q = \sqrt{\dfrac{p}{5t}}$

$p = 2.8$ correct to 1 decimal place and $t = 0.09$ correct to 1 significant figure.

Work out the lower bound for the value of q, giving your answer correct to 3 significant figures. **[3 marks]**

1 Work out $\dfrac{1.6 \times 10^3}{2 \times 10^{-2}}$

Give your answer as an ordinary number. **[2 marks]**

4–6

2 Work out:

$\left(\dfrac{3}{4} - \dfrac{2}{5}\right) \div 1\dfrac{1}{2}$ **[3 marks]**

3 Work out the value of

$\dfrac{3^8 \times 3^{-4}}{3^2}$ **[2 marks]**

4 Express $\sqrt{2} + \sqrt{32}$ in the form $a\sqrt{2}$, where a is an integer. **[2 marks]**

5 At a bus stop, four people get on a bus and nobody gets off the bus.

This increases the number of people on the bus by 20%.

In total, how many people are on the bus now? **[2 marks]**

6 An antique has decreased in value by 30% from last year.

The antique now costs £560

Calculate how much the antique cost last year. **[2 marks]**

7 Work out an estimate for the value of $\sqrt{80.5 \times 99.6}$ **[3 marks]**

8 $a = 2^4 \times 3^2 \times 5^4$ and $b = 2^3 \times 3^5 \times 5^2$

Work out the lowest common multiple of a and b, giving your answer in index form. **[2 marks]**

9 Work out the interest received on an investment of £5000 over 3 years at a compound interest rate of 2.6%. Give your answer to the nearest penny. **[3 marks]**

10 Convert $0.7\dot{3}$ to a fraction. **[3 marks]**

11 Work out the value of

$$\left(\frac{8}{27}\right)^{-\frac{4}{3}}$$ **[3 marks]**

12 Work out:

$$\frac{\left(3\frac{1}{16}\right)^{-\frac{1}{2}} \times 2\frac{1}{3}}{6^{-2}}$$ **[3 marks]**

13 Beach huts are 280 cm long, correct to the nearest 10 cm.

A wall on a beach is 40 m long, correct to the nearest metre.

Show that it may **not** be possible to fit 14 of these beach huts along the wall. **[3 marks]**

14 Solve the equation $4^{x+5} = 8^{x-1}$ **[2 marks]**

6 Understanding algebra

White Rose MATHS

In this block, we will cover...

6.1 Algebraic structures

Example 2

Work out the value of each expression when $a =$

a $4a - b$ **b** $2a^2$

Method

Solution	Commentary
a $4a - b$	
$= 4 \times 4 - {-5}$	Substitute the values of a a
$= 16 - {-5}$	
$= 16 + 5$	

6.2 Working with brackets

Practice (A)

1. Expand:

   a $3(2x + 4)$ **b** $4(4 - 2m)$

   d $-6(c + 4)$ **e** $-9(q - 4)$

2. Multiply out:

   a $b(b - 6)$ **b** $3t(4 - 3t)$

   d $2h(h^2 + g)$ **e** $w^2(2y - 3w)$

3. Expand and simplify:

   a $3(c + 4) + 2(c + 6)$ **b**

6.3 Factorising quadratic expressions

Consolidate – do you need more

1. Factorise:

   a $x^2 + 8x + 7$ **b** $x^2 - 4x - 5$

2. Factorise:

   a $2x^2 + 7x + 5$ **b** $2x^2 + 2x - 1$

3. Simplify:

   a $\dfrac{20a}{12}$ **b** $\dfrac{16b}{14}$

Are you ready?

 1 Calculate:

    a $-4 + 3$ **b** $6 - -12$ **c** $-12 - 4$

2 Simplify the expressions.

    a $2x + 3y - 4x + 4y$ **b** $6 + 3t - 4 + 5t$

    c $3w^2 + 6w - w^2 + 6w$ **d** $3b + 2ab - a + b - 4ab - 3b$

3 Decide whether each statement is **true** or **false**.

    a $y \times y \times y = 3y$ **b** $2y + y = 3y$ **c** $y \times y \times z = yz^2$ **d** $5y^2 = 5 \times y \times y$

4 Write down a simplified algebraic expression for each description.

    a 2 more than k **b** 7 less than k **c** One half of k

    d k divided by 4   **e** k multiplied by j

You need to know these key words.

Variable: A numerical quantity that might change, often denoted by a letter, such as x or t.

Term: A single number or variable, or a number and variable combined by multiplication or division.

Coefficient: A number placed before and multiplying the variable in an algebraic expression. For example, for $4x$ the coefficient of x is 4

Like terms: Terms whose variables are the same; for example, $7x$ and $12x$.

Expression: A collection of terms involving mathematical operations, such as $3c + 4$

Equation: A statement with an equals sign that can be solved to find the value (or values) of the variable in the equation. For example, $2x + 5 = 22$ is an equation that can be solved to find $x = 8.5$

Equations will be explored in more detail throughout the Algebra part of the book (especially in Blocks 7, 11 and 13).

Formula: A rule connecting variables written with mathematical symbols. For example, the area, A, of a trapezium is given by the formula $A = \frac{1}{2}(a + b)h$

Identity: A statement that is true no matter what the values of the variables are. The symbol $\equiv$ is used to show an identity. $\equiv$ means 'equivalent to'. For example, $2(a + b) \equiv 2a + 2b$ is true for all values of a and b.

Identities will be looked at in depth in Block 8.

Index: An index (or power) tells you how many times to multiply a number by itself.
(pl. Indices)

Remember that you can **simplify** an expression by adding or subtracting **like terms**.

For example, $8a + 4a = 12a$, because $8 + 4 = 12$

Similarly, $6mn - 2mn = 4mn$, because $6 - 4 = 2$

You can add or subtract the coefficients of the like terms. You cannot simplify unlike terms by adding or subtracting.

There are three **laws of indices**.

The **addition law**:	$a^m \times a^n = a^{m+n}$	For example, $a^7 \times a^3 = a^{7+3} = a^{10}$
The **subtraction law**:	$a^m \div a^n = a^{m-n}$	For example, $a^7 \div a^3 = a^{7-3} = a^4$
The **multiplication law**:	$(a^m)^n = a^{mn}$	For example, $(a^7)^3 = a^{7 \times 3} = a^{21}$

Example 1

Write down a simplified expression for the area of each shape.

a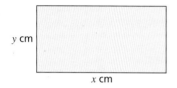

b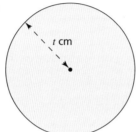

c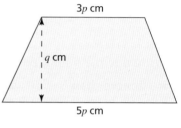

Method

Solution	Commentary
a Area = length × width	Recall the formula for the area of a rectangle.
$= x \times y$	Here, the length is x and the width is y.
$= xy$	Remember that you don't use the multiplication symbol when writing expressions in their simplest form. Note that xy is an expression for the area of the rectangle and $A = xy$ is a formula for the area of the rectangle.
b Area = πr^2	Recall the formula for the area of a circle.
$= \pi \times t^2$	Here the radius is t cm.
$= \pi t^2$	This is the expression in its simplest form.
c Area = $\frac{1}{2}(a + b)h$	Recall the formula for the area of a trapezium.
$= \frac{1}{2} \times (3p + 5p) \times q$	Here $a = 3p$, $b = 5p$ and $h = q$, so you can substitute these values.
$= \frac{1}{2} \times (8p) \times q$	Now simplify by adding the like terms.
$= 4pq$	Simplify further by calculating half of $8p$, which equals $4p$, and omitting the × symbol.

Example 2

Work out the value of each expression when $a = 4$ and $b = -5$

a $4a - b$ **b** $2a^2$ **c** $\dfrac{10a}{b^2}$

Method

Solution	Commentary
a $4a - b$ $= 4 \times 4 - -5$ $= 16 - -5$ $= 16 + 5$ $= 21$	Substitute the values of a and b into the expression.
b $2 \times a^2$ $= 2 \times 4^2$ $= 2 \times 16$ $= 32$	Remember the order of operations: $2a^2$ means $2 \times a^2$, **not** $2a \times 2a$, which would be written $(2a)^2$ Substitute the value of a into the expression and evaluate.
c $\dfrac{10a}{b^2}$ $= \dfrac{10 \times 4}{(-5)^2}$ $= \dfrac{40}{25}$ $= \dfrac{8}{5}$	Remember that $(-5)^2$ means -5×-5, which is positive 25 Simplify your answer.

Example 3

Simplify:

a $6a^2b \times 4a^3b^4$ **b** $(4x^2y)^3$

Method

Solution	Commentary
a $6a^2b \times 4a^3b^4$	Consider each part of the product separately. $6 \times 4 = 24$ $a^2 \times a^3 = a^5$ (using the addition law of indices) $b \times b^4 = b^5$
$= 24a^5b^5$	
b $(4x^2y)^3$	The power of 3 applies to each term in the bracket.
$= 4^3(x^2)^3y^3$	$4^3 = 64$ $(x^2)^3 = x^6$ (using the multiplication law of indices)
$= 64x^6y^3$	

Practice

1. Decide if each of the following is an **expression, identity, equation** or **formula**.

   a $3y + 2x^2 + 5$

   b $v = u + at$

   c $3x + 12 = x - 6$

   d $3(x + 4) \equiv 3x + 12$

2. Write a simplified expression for the perimeter of the rectangle.

   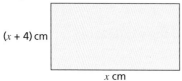

   $(x + 4)$ cm

   x cm

3. Simplify:

   a $2t \times t^2$

   b $3x^2 \times 4x^3$

   c $3ab^2 \times 2b$

   d $\dfrac{6cd^2}{2c}$

   e $\dfrac{12f^2g^3}{4fg}$

   f $(2gh^2)^3$

4. One can of cola costs t pence.

   One bar of chocolate costs s pence.

   Write down an expression for the cost of 4 cans of cola and 3 bars of chocolate.

5. Seb is x years old.

   Flo is 4 years older than Seb.

   Jackson is three years younger than Flo.

   a Write an expression for Flo's age.

   b Write an expression for Jackson's age.

   c Write an expression for the sum of their three ages.

6. A car rental company charges £y per day, plus £x per mile.

   Benji hires a car for 2 days and travels 45 miles.

   Write down an expression for the total cost, C, in terms of x and y.

7. Given that $a = 5$, $b = -6$ and $c = -1$, work out the value of each expression.

   a $3a + b$

   b $3b - c$

   c $a^2 + 2b$

   d $c - 3b^2 + a$

8. The formula used to work out a monthly gas bill is $C = £12.50 + £0.33u$, where u is the number of units of gas used.

   Faith's family use 335 units of gas in one month.

   Use the formula to work out the cost of the bill.

What do you think? 💭

1 Given that r is an even number and t is an odd number, state whether each expression is **odd**, **even**, or could be **either**.

a $2r + t$

b $t^2 r$

c $r - t$

d $3t + r$

2 p and q are both odd numbers.

a Write expressions in terms of both p and q that are:

  i definitely odd

  ii definitely even.

b Is it possible to write an expression in terms of both p and q that could be odd or even?

Consolidate – do you need more?

1 Given $x = 4$ and $y = -2$, work out the value of:

a $2x + 3y$　　　　**b** $2x - 3y$　　　　**c** $2xy$　　　　**d** $2x^2 y$

2 A taxi company uses the following formula to work out fares.

$F = £4 + £0.75$ per mile, where F is the fare in pounds sterling.

a Beca travels 4 miles.

Work out the cost of her taxi journey.

b Jackson travels 5.5 miles.

Work out the cost of his journey.

c Faith has £10 in her purse and lives 7.5 miles away.

Does she have enough money to get home?

3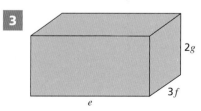

> You will revisit volume and surface area of shapes in *Collins White Rose Maths AQA GCSE 9–1 Higher Student Book 2*, Block 5.

a Write an expression for the volume of the cuboid.

b Write an expression for the surface area of the cuboid.

4 Decide if each of the following is an **expression**, **identity**, **equation** or **formula**.

a $a^2 + b^2 = c^2$　　　　　　**b** $6(x - 4) \equiv 6x - 24$

c $2g + 14 = 12 - 5g$　　　　**d** $3t + 2t^2 - 2$

Stretch – can you deepen your learning?

1 Which expression is equivalent to $2p - p \times 5p + 4p$?

A $5p^2 + 4p$ **B** $9p^2$ **C** $6p - 5p^2$

For the two incorrect expressions, make up your own expression involving four terms and a minimum of one multiplication (×) and one addition (+) which would simplify to that expression.

2 Ali has p marbles.

Benji has twice as many marbles as Ali.

Flo has 3 more marbles than Benji.

Seb has three times as many marbles as Flo.

Write an expression in terms of p for the number of marbles that Seb has.

3 $\dfrac{a + b}{5a - c} = \dfrac{1}{2}$

Given that $c = 2b$, work out the ratio $a : b$.

4 Given the identity $6x + 15 + ax - 2b \equiv bx + 11$, work out the values of a and b.

5 Beca is playing a number trick.

The trick says.

> *Choose a number.*
>
> *Add 3*
>
> *Multiply by 2*
>
> *Add 4*
>
> *Subtract twice the number you originally thought of.*

Use algebra to show that the trick always gives the same answer, whatever number you start with.

Are you ready? (A)

1 Simplify the expressions.

    a $2a \times 3$ **b** $4y \times -4$ **c** $2b \times b$

    d $-3j \times 2k$ **e** $-5t \times 6t^2$

2 Work out the highest common factor of each set of numbers.

    a   6 and 9   **b**   12 and 42   **c**   28 and 70   **d**   12, 24 and 40

You need to know these key words:

Expand or **Multiply out**: Multiply to remove brackets from an expression.

Factorise: Write an expression as a product of its factors.

If you are asked to **factorise fully**, it means you are looking for more than one factor. For example, the expression $6ab + 9bc$ could be factorised as $3(2ab + 3bc)$ or $b(6a + 9c)$, but the full factorisation would be $3b(2a + 3c)$.

Example 1

Expand and simplify $3(m + 2n) - 2(4n - 3m)$

Method

Solution	Commentary
$3(m + 2n) - 2(4n - 3m)$	First, **expand** both sets of brackets.
	Be careful with the second bracket:
$\equiv 3m + 6n - 8n + 6m$	$-2 \times 4n$ gives $-8n$ but $-2 \times -3m$ gives $+6m$.
$\equiv 9m - 2n$	Then you collect the like terms, being careful with any negative terms.

Example 2

Factorise fully $5ab - 10b^2$

Method

Solution	Commentary
$5ab - 10b^2$	The highest common factor of 5 and 10 is 5
	$5ab$ and $10b^2$ also have a factor of b in common, hence the highest common factor of both terms is $5b$.
	This means you can write $5ab - 10b^2$ in the form $5b\,(\boxed{} - \boxed{})$.

$5ab - 10b^2 \equiv 5b(a - 2b)$	For the first term, think "What do I need to multiply $5b$ by to get $5ab$?" The answer is a.
	For the second term, think "What do I need to multiply $5b$ by to get $10b^2$?" The answer is $2b$.

Practice (A)

1. Expand:

   a $3(2x + 4)$ **b** $4(4 - 2m)$ **c** $3(w + 10)$

   d $-6(c + 4)$ **e** $-9(q - 4)$ **f** $12(x - 2y)$

2. Multiply out:

   a $b(b - 6)$ **b** $3t(4 - 3t)$ **c** $5g(3g - 1)$

   d $2h(h^2 + g)$ **e** $w^2(2y - 3w)$ **f** $xy(2x - 3y^2)$

3. Expand and simplify:

   a $3(c + 4) + 2(c + 6)$ **b** $4(x - 3) + 3(x + 4)$

   c $3(4e + 3) - 5(e + 6)$ **d** $5(5t + 6) - 3(4t - 1)$

   e $g(2g + 4) + 5g(g - 3)$ **f** $k(2 - 7j) - 2k(3j - 4)$

4. Factorise:

   a $4x + 8$ **b** $6x + 24$ **c** $12x - 24$

   d $8x^2 + 16x$ **e** $15x^2 + 35x$ **f** $7x + 14x^2$

5. Factorise fully:

   a $4t^2 + 16ts$ **b** $12ab^2 + 28a^2b$ **c** $15cd^2 + 10c^2d^2 + 25cd$

6. Write a simplified expression for the area of the hexagon.

   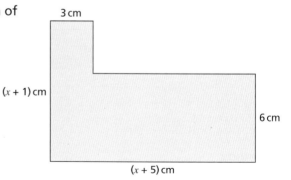

7. Write a simplified expression for the shaded area.

   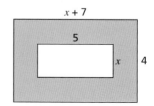

8 Write a simplified expression for the area of the trapezium.

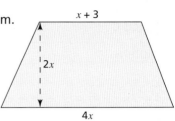

Are you ready? (B)

1 Simplify:

a $2y \times 3y$ b $3t \times -4t$ c $-5g \times -2g$

2 Factorise:

a $2x^2 + 4x$ b $16x^2 - 24x$ c $8x^2 + 20x$

3 Simplify:

a $2x + 2y - 3x + 5y$ b $24 + 5x - 7 - 8x$ c $16 + 2x - 11 + 8x$

Example 1

Multiply out $(a + 3)(b + 4)$

Method A

Solution	Commentary
$(a + 3)(b + 4) \equiv a(b + 4) + 3(b + 4)$	$(a + 3)(b + 4)$ means $(a + 3)$ lots of $(b + 4)$.
	Split this into a lots of $(b + 4)$ and 3 lots of $(b + 4)$, and then multiply out a single bracket.
$\equiv ab + 4a + 3b + 12$	Notice the four terms are unlike, so the expression cannot be simplified.

Method B

Solution	Commentary
$(a + 3)(b + 4)$ $\equiv ab + 4a + 3b + 12$	Think of $(a + 3) \times (b + 4)$ as finding the area of a rectangle with dimensions $a + 3$ and $b + 4$ Divide the sides into sections of a and 3, and b and 4 Find the area of each section. Then add to find the total area and so the total value of the expanded expression.

Method C

Solution	Commentary
$(a + 3)(b + 4) \equiv ab + 4a + 3b + 12$	This is the same as the first method, but shown in a different way.
	The arrows above the expression indicate multiplying a by $b + 4$ and the arrows below indicate multiplying 3 by $b + 4$

Example 2

Expand $(x - 1)(x + 2)(2x + 4)$

Method

Solution	Commentary
$(x - 1)(x + 2) \equiv x^2 - x + 2x - 2$ $\equiv x^2 + x - 2$	Expand the first pair of brackets using one of the methods from above.
$(x^2 + x - 2)(2x + 4)$	Now you need to multiply this answer by the expression in the third bracket.
$\begin{array}{c\|c\|c\|c} \times & x^2 & x & -2 \\ \hline 2x & 2x^3 & 2x^2 & -4x \\ \hline 4 & 4x^2 & 4x & -8 \end{array}$ $2x^3 + 2x^2 - 4x + 4x^2 + 4x - 8$	Multiply each term in the first bracket by each term in the second bracket.
$2x^3 + 6x^2 - 8$	Finally, simplify by collecting like terms.

Practice (B)

1. Multiply out and simplify:

   a $(x + 3)(x + 4)$　　**b** $(x + 3)(x + 1)$　　**c** $(x - 6)(x + 2)$

   d $(x + 4)(x - 6)$　　**e** $(x - 4)(x - 5)$　　**f** $(x - 1)(x - 1)$

2. Expand and simplify:

   a $(x + 4)(2x - 9)$　　**b** $(2x + 4)(2x - 1)$　　**c** $(3x + 4)(5x - 9)$

   d $(2x + 3)(4x + 1)$　　**e** $(x - 4)^2$　　**f** $(x + 7)^2$

3. Multiply out:

   a $(x - 3)(x + 3)$　　**b** $(x - 6)(x + 6)$　　**c** $(3x + 1)(3x - 1)$

   What do you notice?

④ Write a simplified expression for the area of each shape.

a

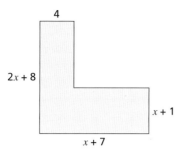

b

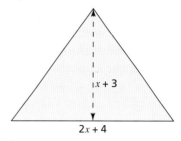

⑤ **a** Write a simplified expression for the area of the trapezium.

b Given that the trapezium has area 100 square units, show that $2x^2 + 3x - 25 = 0$

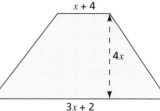

You will learn to solve equations of this form in Chapter 6.3

⑥ **a** Expand and simplify $(x + 2)(x + 4)(x - 5)$

b Expand and simplify $(x - 1)(x + 3)(2x + 1)$

What do you think? 💡

❶ $(x + p)(x + q) \equiv x^2 + ax - 24$ where p, q and a are integers.

What is the smallest possible value of a?

❷ Investigate expanding expressions of the form $(x + a)^3$ and $(x - a)^3$

What is the same and what is different?

❸ Investigate expanding expressions of the form $(x \pm a)(x \pm b)(x \pm c)$.

What happens when you change the signs of the constants?

Are you ready? (C) ▨

1 Work out:

a $\dfrac{1}{2} + \dfrac{2}{3}$ **b** $\dfrac{2}{5} - \dfrac{1}{7}$ **c** $\dfrac{4}{3} + \dfrac{8}{9}$

2 Multiply out:

a $3(x - 3)$ **b** $4(x + 5)$ **c** $2x(x - 5)$

3 Expand and simplify:

a $3(x + 3) + 2(x - 2)$ **b** $4(x - 1) - 2(x + 6)$ **c** $6(2x - 1) - 3(5 - 2x)$

In this section you will explore working with **algebraic fractions**; these have at least one algebraic term in the numerator, the denominator, or both. The processes are the same as for working with numerical fractions. You will often need to use your understanding of expanding and factorising with brackets when working out and simplifying answers. You will learn more about algebraic fractions in Chapter 6.3

Example

Simplify:

a $\dfrac{x}{3} + \dfrac{2x}{5}$

b $\dfrac{2}{x + 3} - \dfrac{5}{x - 4}$

Method

Solution	Commentary
a $\dfrac{x}{3} + \dfrac{2x}{5}$	To add fractions, you need a common denominator. 3 and 5 are both factors of 15
$\times 5 \left(\dfrac{x}{3} + \dfrac{2x}{5} \right) \times 3$ $= \dfrac{5x}{15} + \dfrac{6x}{15}$	3 has been multiplied by 5 to give 15, so you need to multiply the numerator, x, by 5 as well. 5 has been multiplied by 3 to give 15, so you need to multiply the numerator, $2x$, by 3 as well.
$= \dfrac{11x}{15}$	Now you can add the numerators.
b $\dfrac{2}{x + 3} - \dfrac{5}{x - 4}$	You also need a common denominator to subtract fractions. In part **a**, multiplying the denominators (3 and 5) would give 15, which is a common multiple of 3 and 5. You can use the same idea here.
$\times \dfrac{(x - 4)}{(x - 4)} \left(\dfrac{2}{x + 3} - \dfrac{5}{x - 4} \right) \times \dfrac{(x + 3)}{(x + 3)}$ $= \dfrac{2(x - 4)}{(x + 3)(x - 4)} - \dfrac{5(x + 3)}{(x + 3)(x - 4)}$	To keep the fractions equivalent, multiply the numerator by the same expression as the denominator. Keep everything in brackets so you don't make any sign errors.
$= \dfrac{2(x - 4) - 5(x + 3)}{(x + 3)(x - 4)}$	Now write as one fraction.
$= \dfrac{2x - 8 - 5x - 15}{(x + 3)(x - 4)}$	Then expand the numerator, taking care with signs.
$= \dfrac{-3x - 23}{(x + 3)(x - 4)}$	Finally simplify. You don't need to expand the denominator, but you can if you want to.

Practice (C)

1 Write as a single fraction and simplify.

a $\dfrac{x}{8} + \dfrac{2x}{7}$

b $\dfrac{5k}{6} - \dfrac{k}{30}$

c $\dfrac{t}{5} + \dfrac{t+2}{4}$

d $\dfrac{g+6}{4} - \dfrac{g}{9}$

e $\dfrac{h-3}{6} + \dfrac{h+2}{7}$

2 Write as a single fraction.

a $\dfrac{4}{p} + \dfrac{3}{p+1}$

b $\dfrac{2}{c} - \dfrac{4}{c-2}$

c $\dfrac{2}{v+1} + \dfrac{4}{v-1}$

d $\dfrac{5}{t+1} + \dfrac{6}{t+2}$

e $\dfrac{2}{3e+1} - \dfrac{4}{2e+4}$

3 Simplify:

a $\dfrac{2x}{3x+1} + \dfrac{4x}{x+1}$

b $\dfrac{5x}{x+10} - \dfrac{2x}{x-5}$

c $\dfrac{7x}{x+2} + \dfrac{6x}{2x-1}$

Consolidate – do you need more?

1 Expand:

a $5(x-4)$

b $7(x+5)$

c $12(x+3)$

d $2a(a+5)$

e $4b(6-b)$

f $12x(y+x)$

2 Multiply out and simplify:

a $2(x+3) + 3(4x-5)$

b $5(x-3) - 6(3-x)$

c $4(2x+1) + 6(x-4)$

3 Factorise fully:

a $4g + 18$

b $6h + 48$

c $3t^2 + 27t$

d $5c^2 + 25c$

e $4ab + 16a^2b - 32ab^2$

4 Multiply out and simplify:

a $(x+3)(x-8)$

b $(x+7)(x-7)$

c $(x+1)^2$

d $(2x+3)(x-5)$

e $(x+3)(2x-4)$

5 Expand and simplify $(x+2)(x-1)(x+3)$

6 Simplify:

a $\dfrac{x}{3} + \dfrac{2x}{7}$

b $\dfrac{3x}{5} - \dfrac{2x}{9}$

c $\dfrac{2x+1}{4} - \dfrac{x}{9}$

d $\dfrac{4}{x+2} + \dfrac{2}{x-1}$

Stretch – can you deepen your learning?

1 ABC is a right-angled triangle.

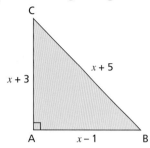

Show that $x^2 - 6x - 15 = 0$

2 Simplify:

a $\dfrac{h-5}{h+3} + \dfrac{h+1}{h+4}$ **b** $\dfrac{x+1}{x+2} - \dfrac{x-4}{x-3}$

3 Show that $2 - \dfrac{x+2}{x+4} - \dfrac{x-3}{x-4}$ can be written as a single fraction of the form

$\dfrac{ax+b}{x^2-16}$ where a and b are constants to be found.

4 Faith walks 500 m at a speed of $(x + 2)$ m/s and then a further 600 m at a speed of $(x + 1)$ m/s.

The total walk takes Faith 30 minutes.

Show that $18x^2 + 43x + 19 = 0$

6.3 Factorising quadratic expressions

Are you ready? (A)

1 Factorise fully:

   a $3k + 12$ **b** $f^2 - 5f$ **c** $6t^2 - 12t$

2 Simplify:

   a $3x - 2 + x - 5$ **b** $3x^2 + 2x - 5 - 4x$ **c** $4x + 12 - 6x + 4x^2 - x$

3 Write down a pair of numbers that have:

   a a sum of 6 and a product of 5   **b** a sum of –4 and a product of –32

   c a sum of –10 and a product of 21

In general, to factorise a quadratic of the form $x^2 + bx + c$ you need to find two values that have product c and sum b. You can see this by using an area model with algebra tiles.

When the **coefficient** of x^2 is not 1, you need to think more carefully.

The value of a in the first term of a quadratic expression $ax^2 + bx + c$ gives you a clue about the first term in each bracket, and c gives you a clue about the last term in each bracket. It can be helpful to list the factors of c and see which ones work. Again, using algebra tiles or other area models can be useful.

The area is $x^2 + 5x + 6$

So $x^2 + 5x + 6 \equiv (x + 2)(x + 3)$

Example 1

Factorise:

a $x^2 + 7x + 10$ **b** $x^2 + 2x - 15$

Method

Solution	Commentary
a $x^2 + 7x + 10$ <table><tr><td></td><td>x</td></tr><tr><td>x</td><td>x^2</td></tr><tr><td></td><td>10</td></tr></table>	Use an area model, filling in the x^2 and constant term (10). The side lengths can be partly filled in with x, as you know $x \times x = x^2$
$2 \times 5 = 10$ and $2 + 5 = 7$ <table><tr><td></td><td>x</td><td>5</td></tr><tr><td>x</td><td>x^2</td><td>$5x$</td></tr><tr><td>2</td><td>$2x$</td><td>10</td></tr></table>	Now find a pair of values which multiply to give 10 and sum to make 7 The only possible values are 2 and 5

So $x^2 + 7x + 10 \equiv (x + 2)(x + 5)$	Write the quadratic as the product of the two 'side lengths'.
b $x^2 + 2x - 15$ 	Use an area model, filling in the x^2 and the constant term. Look for a pair of values with multiply to make −15 and sum to give 2 The only possible values are −3 and 5
$x^2 + 2x - 15 \equiv (x - 3)(x + 5)$	Write the quadratic as the product of the two 'side lengths'.

The area model shows:

	x	-3
x	x^2	$-3x$
$+5$	$5x$	-15

Example 2

Factorise:

a $2x^2 + 13x + 15$ **b** $x^2 - 25$

Method

Solution	Commentary
a $2x^2 + 13x + 15 \equiv$ $2x^2$ $(2x\ \ \ )(x\ \ \ )$	The factors of 2 are 1 and 2, so you know one bracket contains a $2x$ term and the other bracket contains an x term. Remember that $1x$ is written as x.
Factor pairs of 15: 1 and 15 3 and 5	List the factor pairs of 15
x $(2x + 1)(x + 15)$ $30x$	Try the first factor pairs in the brackets, remembering that the sum of the x terms needs to be $13x$. The total of the terms is $31x$ so try a different combination.
$3x$ $(2x + 3)(x + 5)$ $10x$	Try the different factor pairs in the brackets: $10x + 3x = 13x$, so the correct solution is shown. With practice you will be able to 'spot' which factor pair to use, and which bracket they go in, more easily.
$2x^2 + 13x + 15 \equiv (2x + 3)(x + 5)$	Remember that you can check your solution by expanding $(2x + 3)(x + 5)$.
b $x^2 - 25 \equiv x^2 + 0x - 25$	$x^2 - 25$ is a special case known as the **difference of two squares**. You can think of it as $x^2 + 0x - 25$
$5 \times -5 = -25$ $5 + -5 = 0$	You now need two values with a product of −25 and a sum of 0 The pair of values 5 and −5 is the only possibility.
$x^2 - 25 \equiv (x + 5)(x - 5)$	

In general, when factorising expressions that are the difference of two squares,
$x^2 - a^2 \equiv (x + a)(x - a)$

Practice (A)

1 Factorise:

a $p^2 + 14p + 24$ **b** $q^2 + 18q + 32$ **c** $r^2 - 8r + 15$ **d** $s^2 - s - 12$

e $u^2 - 2u - 15$ **f** $v^2 - 6v + 9$ **g** $w^2 - 8w - 20$ **h** $x^2 + 3x - 10$

i $y^2 + 5y - 14$ **j** $z^2 + 12z + 36$

2 Factorise:

a $x^2 - 9$ **b** $z^2 - 16$ **c** $w^2 - 100$ **d** $4x^2 - 25$

e $4u^2 - 9$ **f** $9t^2 - 49$ **g** $4k^2 - p^2$ **h** $9w^2 - 4u^2$

i $16x^2 - 4y^2$ **j** $4x^4 - 9$

3 Factorise:

a $2a^2 + 5a + 2$ **b** $3b^2 - 16b - 12$ **c** $5c^2 + 19c + 12$ **d** $4d^2 + 16d + 7$

e $3e^2 - e - 24$ **f** $2f^2 - 7f - 15$ **g** $3g^2 - 16g - 35$ **h** $2h^2 + 15h - 8$

i $4i^2 - 8i + 3$ **j** $4j^2 - 9j + 5$

4 Factorise:

a $2x^2 - 11x + 15$ **b** $x^2 + x - 42$ **c** $4x^2 + 4x - 3$

d $x^2 - 169$ **e** $x^2 - 14x + 40$ **f** $4x^2 - 81$

5 Use factorisation to show that $x^2 + 2xy + y^2$ is always a square number.

6 The expression for the area of this rectangle is $x^2 - 5x - 50$

$x - 10$

Work out an expression for the unknown side length.

7 Factorise $81g^2 - 64h^2$

8 Factorise fully $20x^2 - 5$

9 Jackson and Beca each factorise $6x^2 + 24x + 18$

Jackson writes, $(3x + 9)(2x + 2)$

Beca writes, $(x + 3)(6x + 6)$

a By expanding, show that they are both correct.

b Show that both answers can be factorised further to a more simplified answer.

What do you think? (A) 💭

1. Factorise the expression $4x^2 - 1$

   Use your answer to help find the prime factors of 399

2. How could you factorise $x^4 - 4x^2 - 32$?

Are you ready? (B)

1. Simplify the fractions.

   a $\dfrac{4}{12}$ b $\dfrac{16}{56}$ c $\dfrac{49}{70}$

2. Factorise the expressions.

   a $6x + 24$ b $x^2 - 9$ c $x^2 + 5x + 6$ d $x^2 + x$

Example

Simplify:

a $\dfrac{2xy}{4y^2}$ b $\dfrac{3x + 12}{x^2 + x - 12}$

Method

Solution	Commentary
a $\dfrac{2xy}{4y^2}$	To simplify fractions, divide the numerator and denominator by a common factor. Here the common factor is $2y$.
$\dfrac{2xy}{4y^2} = \dfrac{x}{2y}$	Dividing the numerator by $2y$ gives x. Dividing the denominator by $2y$ gives $2y$.
b $\dfrac{3x + 12}{x^2 + x - 12}$	Before you can cancel common factors, factorise so the numerator and denominator are written as products.
$3x + 12 \equiv 3(x + 4)$	$3x + 12$ factorises into a single bracket using the common factor 3
$x^2 + x - 12 \equiv (x + 4)(x - 3)$	$x^2 + x - 12$ is a quadratic expression that factorises into two brackets.
$\dfrac{3\cancel{(x + 4)}}{\cancel{(x + 4)}(x - 3)}$	Now you can cancel the common factor of $x + 4$
$\dfrac{3}{x - 3}$	

Practice (B)

1 Simplify the fractions.

a $\dfrac{5a}{15}$　　b $\dfrac{12b}{30}$　　c $\dfrac{5c}{c^2}$　　d $\dfrac{2xy}{4x}$　　e $\dfrac{12y^2z}{3yz}$

2 Simplify fully:

a $\dfrac{24d + 12}{18}$　　b $\dfrac{60e - 40}{12}$　　c $\dfrac{10f + 5}{5f - 15}$

d $\dfrac{g^2 + 3g}{2g^2 - g}$　　e $\dfrac{4h^2 - 6h}{h^2 + 3h}$

3 Simplify fully:

a $\dfrac{x^2 - 6x + 8}{x - 2}$　　b $\dfrac{x^2 - 4}{x + 2}$　　c $\dfrac{x^2 + 7x + 12}{x^2 + 2x - 8}$　　d $\dfrac{x^2 + x - 20}{x^2 - 16}$

What do you think? (B)

1 Write as a single fraction.

a $\dfrac{a}{2b} \times \dfrac{b}{a^2}$

b $\dfrac{a^2}{b} \times \dfrac{b^2}{a}$

c $\dfrac{ab}{c^2} \times \dfrac{c}{2a}$

Compare your method with a partner. Did you cancel terms before or after multiplying?

2 Investigate multiplying and dividing expressions like those in question 1, for example $\dfrac{a}{2b} \times \dfrac{b}{a^2}$

Consolidate – do you need more?

1 Factorise:

    a $x^2 + 8x + 7$ **b** $x^2 - 4x - 5$ **c** $x^2 - x - 20$ **d** $x^2 - 10x + 25$

2 Factorise:

    a $2x^2 + 7x + 5$ **b** $2x^2 + 2x - 12$ **c** $3x^2 - 11x + 6$

3 Simplify:

    a $\dfrac{20a}{12}$ **b** $\dfrac{16b}{14}$ **c** $\dfrac{5k}{10k^2}$

4 Simplify fully:

    a $\dfrac{6g + 4}{2g + 6}$ **b** $\dfrac{24 - 6m}{6 + 18m}$ **c** $\dfrac{15b^2 + 30b}{20b^2 - 20b}$

5 Simplify fully:

    a $\dfrac{c^2 - 4}{c^2 - 2c}$ **b** $\dfrac{d^2 + 2d - 15}{d^2 - 5d + 6}$ **c** $\dfrac{g^2 - 9}{g^2 + 2g - 15}$

6 **a** The rectangle shown has area $x^2 + 10x + 21$

    $x + 7$

    Work out an expression for the unknown length.

    b The triangle shown has area $x^2 + 4x - 5$

    $2x - 2$

    Work out an expression for the perpendicular height of the triangle.

Stretch – can you deepen your learning?

1 Show that the value of $\dfrac{10a + 30 + 2a - 6}{5a + 12 - a - 4}$ is always the same whatever the value of a.

2 **a** Expand and simplify $(n + 4)^2 - (n - 2)^2$

    b Use your answer to show that when n is an integer, $(n + 4)^2 - (n - 2)^2$ is always an even number.

3 n is an integer greater than 1

    By factorising $n^2 - n$, explain why $n^2 - n$ can never be an odd number.

Understanding algebra: exam practice

1 Work out the value of $5a^2 + b^2$ when $a = 3$ and $b = -4$ [2 marks]

2 Simplify:

    (a) $6a^2b^3 \times 2ab^{-2}$ [2 marks]

    (b) $6a^2b^3 \div 2ab^{-2}$ [2 marks]

3 Decide if each of the following is an **expression**, **identity**, **equation** or **formula**.

    (a) $P = 2(l + w)$ [1 mark]

    (b) $3(2a + b) \equiv 6a + 3b$ [1 mark]

    (c) $\dfrac{3t - 5}{2} = 8$ [1 mark]

    (d) $b^2 - 4ac$ [1 mark]

4 A regular polygon has n sides, each of length x cm.

    Write down an expression for the perimeter of the polygon.   [2 marks]

5 Multiply out and simplify $(x + 7)(x - 2)$ [2 marks]

6 Factorise fully:

    (a) $8a + 24b$ [1 mark]

    (b) $6pq^2 - 15q^2t$ [2 marks]

7 Factorise $y^2 - 7y + 12$ [2 marks]

8 Factorise:

    (a) $x^2 - y^2$ [1 mark]

    (b) $16p^2 - 25q^2$ [2 marks]

9 Given that n is an integer, state whether each expression is odd, even, or could be either.

     n^2 $2n + 1$ $n + 7$

     [2 marks]

10 Expand and simplify $(x + 2)(x - 3)(x + 4)$ [3 marks]

11 Write as a single fraction $\dfrac{3}{x + 1} + \dfrac{4}{x - 2}$ [3 marks]

12 Simplify fully $\dfrac{4x^2 + 8x}{x^2 - 4}$ [3 marks]

7 Functions and linear equations

In this block, we will cover...

7.1 Linear equations

Example

Solve:

a $\frac{x}{3} + 8 = 15$ **b** $\frac{x - 12}{5} = 8$

Method

Solution	Commentary
a $\frac{x}{3} + 8 = 15$ $-8 \quad\quad -8$ $\frac{x}{3} = 7$ $\times 3 \quad\quad \times 3$	You need to isolate is the 'balance meth First you can subtra Then you can multi

7.2 Functions and inverses

Practice (A)

1. Given that $f(x) = 3x + 9$, work out:

   a $f(3)$ **b** $f(-1)$

   c $f\left(\frac{1}{2}\right)$ **d** the value of k if

2. Given that $g(x) = x^2 - 9$, work out:

   a $g(1)$ **b** $g(-2)$

   c $g(1.5)$ **d** the value of m if

3. Given that $h(x) = \frac{2x - 1}{3}$, work out:

   a $h(5)$ **b** $h(-3)$

7.3 Composite functions

Consolidate – do you need more

1. Given $f(x) = 5x - 8$ and $g(x) = x^2$, evaluate:

   a $ff(4)$ **b** $gf(-1)$

2. Given $f(x) = 6x - 7$ and $g(x) = 2x^2 + 1$, work

   a $fg(x)$ **b** $gf(x)$

3. $f(x) = 3(x + 4)$ and $g(x) = x^3 + 5$

   Work out an expression for:

   a $f^{-1}(x)$ **b** $g^{-1}(x)$

Are you ready? (A)

1 Solve:

   a $2x = 16$ **b** $x + 5 = 12$ **c** $x - 8 = 22$ **d** $\frac{x}{4} = 13$

2 Solve:

   a $14 - x = 9$ **b** $3x = -18$ **c** $x + 4 = 1$ **d** $\frac{x}{6} = -3$

You can use different representations to support you when solving linear equations, such as algebra tiles or bar models.

$2x + 3 = 5x + 1$

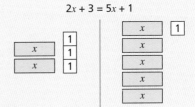

Here is the same equation shown as a bar model:

| x | x | 1 | 1 | 1 |

| x | x | x | x | x | 1 |

Sometimes you need to **form** an equation yourself before solving it.

For example, the problem "Work out my number if 5 more than double my number is 79" can be solved by writing the equation $2x + 5 = 79$ and then solving for x.

Example

Solve:

a $\frac{x}{3} + 8 = 15$ **b** $\frac{x - 12}{5} = 8$

Method

Solution	Commentary
a $\frac{x}{3} + 8 = 15$ $-8 \quad\quad\quad\quad -8$ $\frac{x}{3} = 7$ $\times 3 \quad\quad\quad\quad \times 3$ $x = 21$	You need to isolate the x term to work out its value. One way is the 'balance method'. First you can subtract 8 from both sides. Then you can multiply both sides by 3 Check your answer by substituting it back into the original equation. When $x = 21$, $\frac{x}{3} + 8 = \frac{21}{3} + 8 = 7 + 8 = 15$ as required.
b $\frac{x - 12}{5} = 8$ $\times 5 \quad\quad\quad\quad \times 5$ $x - 12 = 40$ $+12 \quad\quad\quad\quad +12$ $x = 52$	Start by multiplying both sides by 5 to eliminate the fraction. Then add 12 to both sides to find the value of x.

Practice (A)

1 Solve:

    a $3x + 9 = 24$ **b** $2x - 9 = 33$ **c** $5x + 19 = 49$

    d $16 - 2x = 4$ **e** $3(3x + 6) = 63$ **f** $4(3x - 10) = 56$

> You could start by expanding the brackets or by dividing both sides by 3.
> Compare these methods and choose your preferred strategy for part **f**.

2 Solve:

    a $3x - 5 = -11$ **b** $2x + 16 = 6$ **c** $4 - 2x = 18$ **d** $3(4 - x) = 12$

3 Solve:

    a $\dfrac{a}{2} - 4 = 6$ **b** $\dfrac{b}{10} + 4 = 12$ **c** $\dfrac{c}{3} - 9 = -8$ **d** $\dfrac{d}{3} + 6 = 1$

4 Solve:

    a $\dfrac{x + 1}{2} = 3$ **b** $\dfrac{3x - 1}{4} = 5$ **c** $\dfrac{5x}{6} = 10$ **d** $\dfrac{7x + 3}{2} = 12$

5 Beca thinks of a number.

    She multiplies this number by 7 and adds 5

    The answer is 68

    a   Write down an equation to show this information.

    b   Solve your equation to find the number Beca was thinking of.

6 Work out the value of x.

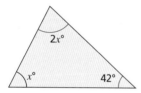

7 The rectangle has perimeter 48 cm.

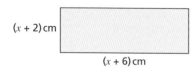

    a   Work out the value of x.

    b   Calculate the area of the rectangle.

8 Work out the size of the largest angle in this triangle.

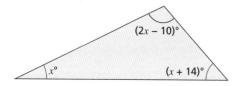

9 Work out the mean of these expressions.

$7x + 2$ $6x - 9$ $5x + 14$ $2x + 1$

10 The trapezium has area 24 cm².

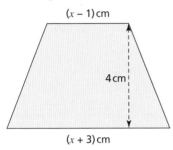

$(x - 1)$ cm

4 cm

$(x + 3)$ cm

Work out the value of x.

11 Three consecutive integers have a sum of 39
What is the smallest of these numbers?

Hint: Assign a letter to the value of the smallest of the three integers.

Are you ready? (B)

 1 Solve:

    a $3x + 9 = 18$ **b** $2x + 15 = 34$ **c** $4 - 2x = 12$ **d** $\frac{x}{5} + 3 = 21$

2 Simplify:

    a $2x + 4 - 3x - 7$ **b** $2 + 6x - 14 + x$ **c** $5x - 3 - x - 8$ **d** $3 - 5x + 2 - 9x$

In this section you will explore equations with unknowns on both sides.

Example

Solve:

a $2x + 3 = 6x - 5$ **b** $\frac{x + 7}{5} = 2x - 4$

Method

Solution	Commentary

a $2x + 3 = 6x - 5$
$-2x$ $-2x$
      $3 = 4x - 5$
$+5$ $+5$
      $8 = 4x$
$\div 4$ $\div 4$
      $2 = x$
 	
So $x = 2$
	To balance the equation, you could choose to subtract $2x$ or $6x$ from each side. Subtracting $6x$ would give $-4x$ on one side and it is best to avoid negatives, so subtract $2x$ from each side.
	
	Now add 5 to both sides.
	
	Finally, divide both sides by 4

b

$$\frac{x+7}{5} = 2x - 4$$

×5 $\Big($ $\Big)$ ×5

$$x + 7 = 5(2x - 4)$$

First multiply both sides by 5

Then expand the bracket on the right-hand side.

$$x + 7 = 10x - 20$$

−x $\Big($ $\Big)$ −x

$$7 = 9x - 20$$

+20 $\Big($ $\Big)$ +20

$$27 = 9x$$

÷9 $\Big($ $\Big)$ ÷9

$$3 = x$$

So $x = 3$

Now subtract x from both sides.

Then add 20 to both sides.

And finally divide both sides by 9

Practice (B)

1 Solve:

a $4x + 2 = 2x + 8$ **b** $x + 5 = 5x + 1$ **c** $6x - 10 = 4x - 2$ **d** $2x + 19 = 4x + 3$

e $5x + 3 = 17 - 2x$ **f** $3x - 2 = 22 - x$ **g** $7x + 8 = 38 - 3x$ **h** $-11x = -3x + 48$

2 Solve:

a $5(x + 3) = 3(x + 8)$ **b** $8(x - 2) = 4(x + 1)$

c $4(x + 2) = 10(x - 1)$ **d** $4(2x + 1) - 3(x + 4) = 3x + 1$

3 Solve:

a $\dfrac{5x - 8}{6} = 2x + 1$ **b** $\dfrac{4(4x - 11)}{3x} = 20$ **c** $\dfrac{9 - x}{2} = 2x + 12$ **d** $4 - 6x = \dfrac{4 - 12x}{4}$

4 A rectangle is shown.

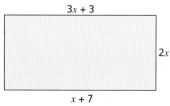

3x + 3

2x

x + 7

a Work out the value of x.

b Hence work out the area of the rectangle.

5 Work out the perimeter of the triangle.

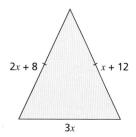

2x + 8

x + 12

3x

6 Solve $(x + 3)(x - 4) = (x + 1)^2$, giving your answer as a mixed number.

Consolidate – do you need more?

1 Solve:

a $2x - 9 = 8$ **b** $3x - 4 = 17$

c $3 - 2x = 12$ **d** $12 - 5x = 2$

2 Solve:

a $2(x + 3) = 18$ **b** $5(2x - 1) = 60$

c $3(x + 3) = -12$ **d** $4(2x + 1) = -28$

3 Solve:

a $\dfrac{a}{3} + 2 = 10$ **b** $\dfrac{b}{7} - 3 = 12$

c $\dfrac{c + 2}{12} = 10$ **d** $\dfrac{2d + 1}{4} = 0.5$

4 Solve:

a $2x + 11 = 3x - 2$ **b** $4x + 3 = 7x - 9$

c $2(3x + 1) = 7x - 11$ **d** $3(x + 7) = 11 - 2x$

5 Work out the value of x in each diagram.

a 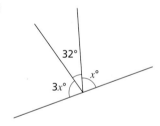   **b**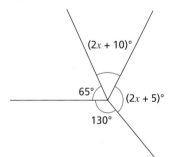

6 The sum of three consecutive odd numbers is 69

Work out the largest of these numbers.

Stretch – can you deepen your learning?

1 Solve:

a $\dfrac{x + 3}{2} + \dfrac{x + 1}{4} = 10$ **b** $\dfrac{2x - 5}{7} - \dfrac{2x - 1}{2} = 3$

2 Ed runs $(2x + 3)$ metres at a speed of 4 m/s and a further $(x - 4)$ metres at a speed of 5 m/s.

He runs for a total of 5 minutes.

a Set up an equation to show this information.

b Solve the equation and work out the total distance Ed ran, giving your answer to the nearest metre.

Are you ready? (A)

1 Work out the value of the expressions when $x = 3$ and $y = -2$

    a $3x + y$ **b** $x^2 - 2y$ **c** xy^2

    d $y(x + 5)$ **e** $y^3 + 4x$

2 Solve the equations.

    a $4x + 1 = 19$ **b** $6x - 3 = 21$ **c** $\dfrac{2x + 1}{4} = 21$

    d $\dfrac{2}{x + 1} = 4$ **e** $\dfrac{x + 2}{x - 2} = 9$

A **function** describes a relationship between a set of input values and a set of output values. You may have already seen this in the form of function machines.

Here, the function is 'multiply by 3 and add 4'.

If the input is x, then you can write the output as $3x + 4$

This is written as $f(x) = 3x + 4$ and is read as 'f of x is equal to $3x + 4$'.

Example

$f(x) = 3x^2 + 2$

a Work out $f(3)$

b Work out $f(-2)$

c Given that $f(k) = 50$, work out two possible values for k.

Method

Solution	Commentary
a $f(3) = 3 \times 3^2 + 2$ $ = 27 + 2$ $ = 29$	Substitute 3 into the expression for $f(x)$.
b $f(-2) = 3 \times (-2)^2 + 2$ $ = 3 \times 4 + 2$ $ = 14$	Substitute -2 into the expression for $f(x)$. Remember that $(-2)^2 = 4$, not -4
c $-2 \left(\;\; 3k^2 + 2 = 50 \;\; \right) -2$ $\div 3 \left(\;\;\;\;\;\; 3k^2 = 48 \;\;\;\;\;\; \right) \div 3$ $\sqrt{\;} \left(\;\;\;\;\;\; k^2 = 16 \;\;\;\;\;\; \right) \sqrt{\;}$ $k = 4 \text{ or } k = -4$	Substitute k into the expression for $f(x)$ and set up an equation. Solve the equation. Remember the order of operations – you need to divide by 3 before you square root. Remember that both 4^2 and $(-4)^2$ are equal to 16

Practice (A)

1 Given that $f(x) = 3x + 9$, work out:

    a $f(3)$ **b** $f(-1)$

    c $f\left(\dfrac{1}{2}\right)$ **d** the value of k if $f(k) = 12$

2 Given that $g(x) = x^2 - 9$, work out:

    a $g(1)$ **b** $g(-2)$

    c $g(1.5)$ **d** the value of m if $g(m) = 16$

3 Given that $h(x) = \dfrac{2x - 1}{3}$, work out:

    a $h(5)$ **b** $h(-3)$

    c $h(0)$ **d** the value of n if $h(n) = 12$

4 $f(x) = 2x + 5$ and $g(x) = 5x - 4$

    Work out the value of a for which $f(a) = g(a)$

5 $f(x) = \dfrac{4 + x}{2 - x}$

    a Work out $f(4)$

    b What value cannot be in the domain (input) of $f(x)$?

    c Solve $f(x) = 8$

6 Given $g(x) = 3x - 12$, solve $g(x) = 2x$

7 Given $h(x) = \dfrac{2x + 6}{3}$, solve $h(x) = 4x$

8 Given $f(x) = 2x + 4$, write expressions for:

   a $f(2x)$

   b $f(x + 2)$

Are you ready? (B)

1 Rearrange each formula to make x the subject.

   a $y = 2x + 1$ **b** $y + 4x = 3$ **c** $y + x^2 - 3 = 0$

   d $y = 5x^2$ **e** $y = \dfrac{2x}{x + 4}$

> Rearranging formulae is reviewed in Chapter 8.2 if you need a reminder.

2 Given $a = 5$ and $b = -1$, work out the value of each expression.

   a $a + 2b$ **b** $2a - 3b$ **c** $5ab$

   d $2a^2b$ **e** $b(6a - 4)$

The **inverse** of a function $f(x)$ is written as $f^{-1}(x)$. This works out the input for given outputs. For the function $f(x) = 3x + 4$, $f^{-1}(x) = \dfrac{x - 4}{3}$, because to find the outputs from the inputs you perform the operations in reverse: subtract 4 first and then divide by 3

Example

Work out the inverse of each function.

a $f(x) = 3x - 1$ **b** $g(x) = 2x^2 + 3$

Method

Solution	Commentary
a $+1 \left(\begin{array}{c} 3x - 1 = y \\ 3x = y + 1 \end{array} \right) +1$ $\div 3 \left(\begin{array}{c} \\ x = \dfrac{y + 1}{3} \end{array} \right) \div 3$	To find the inverse function, form an equation by labelling the output with another letter, such as y. Then rearrange to make x the subject.
$f^{-1}(x) = \dfrac{x + 1}{3}$	Then rewrite the expression using inverse function notation and the letter x in place of y.
b $-3 \left(\begin{array}{c} 2x^2 + 3 = y \\ 2x^2 = y - 3 \end{array} \right) -3$ $\div 2 \left(\begin{array}{c} \\ x^2 = \dfrac{y - 3}{2} \end{array} \right) \div 2$ $\sqrt{} \left(\begin{array}{c} \\ x = \sqrt{\dfrac{y - 3}{2}} \end{array} \right) \sqrt{}$	Form an equation as before. Rearrange to make x the subject, remembering to divide by 2 before you square root.
$g^{-1}(x) = \sqrt{\dfrac{x - 3}{2}}$	Now rewrite the expression using the inverse function notation and the letter x in place of y.

Practice (B)

1 Work out the inverse of each function.

   a $f(x) = 4x + 3$ **b** $f(x) = x^2 - 7$ **c** $f(x) = \dfrac{x + 1}{6}$

   d $f(x) = 12 - 3x$ **e** $f(x) = \dfrac{4}{x + 2}$ **f** $f(x) = \dfrac{5}{x + 3}$

2 Given $g(x) = 3x^3 - 1$

   a Write an expression for $g^{-1}(x)$

   b Solve the equation $g^{-1}(x) = 2$

3 Work out the inverse of each of the functions $g(x) = 6 - x$ and $h(x) = \dfrac{3}{x}$

   What do you notice?

4 Given $h(x) = 3x + 19$ and $k(x) = 2x - 10$, work out the value of x for which $h^{-1}(x) = k^{-1}(x)$

5 Work out the inverse of $h(x) = \dfrac{3x + 4}{5x - 6}$

6 **a** Show that $\dfrac{x^2 + x}{x^2 + 4x + 3}$ can be written as $\dfrac{x}{x + a}$ where a is a positive integer.

   b Hence, given $f(x) = \dfrac{x^2 + x}{x^2 + 4x + 3}$, work out an expression for $f^{-1}(x)$

Consolidate – do you need more?

1 Given $f(x) = 7x + 12$

   a Work out $f(-4)$

   b Work out k such that $f(k) = 61$

   c Work out $f^{-1}(x)$

2 Given $g(x) = 5x^2 + 9$

   a Work out $g(6)$

   b Work out k such that $g(k) = 54$

   c Work out $g^{-1}(x)$

3 Given $h(x) = \dfrac{3x + 1}{4}$

   a Work out $h(0.5)$

   b Work out k such that $h(k) = 0$

   c Work out $h^{-1}(x)$

4 $f(x) = 3x - 7$ and $g(x) = \dfrac{3x - 1}{3}$

   Work out a such that $f(a) = g(a)$

5 $h(x) = \dfrac{3x + 4}{5}$

   Solve $h(x) = 2x$

6 $f(x) = x^2 + 2x$

Write an expression for $f(3x)$

7 $h(x) = \dfrac{2x + 1}{x - 1}$

Write an expression for $h^{-1}(x)$

Stretch – can you deepen your learning?

1 $f(x) = 2x^2 + 3x - 2$

Write a simplified expression for $f(x + 2)$

2 **a** Write an expression for $f^{-1}(x)$ when $f(x) = \dfrac{3x + 7}{2x - 8}$

b Write a general expression for $g^{-1}(x)$ when $g(x) = \dfrac{ax + b}{cx + d}$

3 $f(x) = 4x + 2$ and $g(x) = px + q$

$g(4) = 7$ and $f^{-1}(18) = g(3)$

Work out the values of p and q.

White Rose MATHS

Are you ready?

1 Given $f(x) = 4x - 9$ and $g(x) = x^2 + x + 1$, evaluate:

    a $f(3)$ **b** $g(2)$ **c** $f(-5)$ **d** $g(-1)$

2 Given $f(x) = 2x + 1$ and $g(x) = x^2 - 9$, write expressions for:

    a $f(2x)$ **b** $f(x + 1)$ **c** $g(3x)$ **d** $g(x - 1)$

3 Expand and simplify:

    a $3(x + 1) + 4(x - 1)$ **b** $4x(x + 2) - 2x(7 - x)$

    c $(x + 1)(x - 3)$ **d** $(x - 2)^2$

Composite functions are made when two functions are applied consecutively so that the output from the first function becomes the input to the second function.

For two functions $f(x)$ and $g(x)$, you write $fg(x)$ to mean 'apply g first and then f', or write $gf(x)$ to mean 'apply f first and then g'.

$x \xrightarrow{\text{function g}} g(x) \xrightarrow{\text{function f}} fg(x)$

$x \xrightarrow{\text{function f}} f(x) \xrightarrow{\text{function g}} gf(x)$

Example

$f(x) = 2x + 1$ and $g(x) = x^2 + 3$

a Evaluate $fg(3)$

b Evaluate $ff(4)$

c Work out an expression for $fg(x)$

Method

Solution	Commentary
a $g(3) = 3^2 + 3 = 12$	$fg(3)$ means apply the function f to $g(3)$, so work out $g(3)$ first.
$f(12) = 2 \times 12 + 1 = 25$	$g(3) = 12$, so now work out $f(g(3))$, which is the same as $f(12)$.
So, $fg(3) = 25$	
b $f(4) = 2 \times 4 + 1 = 9$	$ff(4)$ means you need to work out $f(4)$ first.
$f(9) = 2 \times 9 + 1 = 19$	$f(4) = 9$, so now work out $f(f(4))$, which is the same as $f(9)$.
So $ff(4) = 19$	

c $\quad g(x) = x^2 + 3$	You need to find an algebraic expression for fg(x). This means apply the function g first and then substitute the result into f(x).
$f(g(x)) = 2 \times g(x) + 1$ $f(g(x)) = 2 \times (x^2 + 3) + 1$ $\quad = 2x^2 + 6 + 1$	You replace x in f(x) with g(x), i.e. g(x) is the input. Replace g(x) with $x^2 + 3$
$\quad = 2x^2 + 7$	Finally, simplify the expression.

Practice

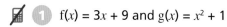

1 $f(x) = 3x + 9$ and $g(x) = x^2 + 1$

Work out:

   a ff(1)   **b** fg(4)   **c** gf(–1)   **d** gg(3)

2 $g(x) = \dfrac{2x + 1}{3}$ and $h(x) = 2x^3$

Work out:

   a gg(1)   **b** hg(–1)   **c** gh(3)   **d** hh(0.5)

3 Given $f(x) = 5x - 7$, $g(x) = 12 - 2x$ and $h(x) = x^2 + 4$, work out simplified expressions for the composite functions.

   a fh(x)   **b** hf(x)   **c** ff(x)   **d** gf(x)   **e** fg(x)

4 $f(x) = x^2 - 3$ and $g(x) = x - 12$

   a Work out an expression for gf(x)

   b Work out an expression for fg(x)

   c Solve the equation gf(x) = fg(x)

5 $f(x) = 3x + 4$ and $g(x) = ax^2$

   a Work out an expression for $f^{-1}(x)$

   b Given that fg(3) = 85, work out the value of a.

6 $f(x) = ax + 8$ and $g(x) = 2x - 3$

   a Work out an expression for $f^{-1}(x)$

   b Given fg(2) = 13, find the value of a.

7 $f(x) = x^2 - 1$ and $g(x) = 4x + 5$

   a Work out an expression for $f^{-1}(x)$

   b Work out an expression for $g^{-1}(x)$

   c Given that fg(x) = 3gf(x), show that $4x^2 + 40x + 21 = 0$

What do you think? 💭

1. Investigate $f^{-1}f(x)$ for some functions of your own choice.

   What do you notice?

2. $f(x) = 2x + 4$ and $g(x) = 3x - 8$

   a Work out $f^{-1}(x)$ and $g^{-1}(x)$

   b Work out an expression for $fg(x)$

   c Work out an expression for the inverse of $fg(x)$

   d Is $(fg)^{-1}(x)$ the same as $f^{-1}g^{-1}(x)$?

   e Is $(fg)^{-1}(x)$ the same as $g^{-1}f^{-1}(x)$?

Consolidate – do you need more?

1. Given $f(x) = 5x - 8$ and $g(x) = x^2$, evaluate:

   a $ff(4)$ **b** $gf(-1)$ **c** $fg(10)$ **d** $gg(6)$

2. Given $f(x) = 6x - 7$ and $g(x) = 2x^2 + 1$, work out simplified expressions for:

   a $fg(x)$ **b** $gf(x)$ **c** $ff(x)$ **d** $gg(x)$

3. $f(x) = 3(x + 4)$ and $g(x) = x^3 + 5$

   Work out an expression for:

   a $f^{-1}(x)$ **b** $g^{-1}(x)$ **c** $fg(x)$ **d** $ff(x)$

4. $g(x) = x - 12$ and $h(x) = x^2 - 10$

   Solve the equation $gh(x) = hg(x)$

5. $f(x) = 2x + k$ and $g(x) = kx + 3$

   $fg(x) = 10x + m$

   Work out the values of k and m.

6. $g(x) = 3x + 2$ and $h(x) = 6x + 5$

   Show that $gh(x) = hg(x)$

Stretch – can you deepen your learning?

1. $f(x) = x + 2$ and $g(x) = \dfrac{1}{x - 1}$

   a Work out $g^{-1}(x)$

   b Work out $fg(x)$ and give your answer in its simplest form.

   c Work out the inverse of $fg(x)$, i.e. $(fg)^{-1}(x)$

2. $g(x) = \dfrac{2x + 1}{x - 1}$ and $h(x) = \dfrac{x + 3}{x - 5}$

   Work out an expression for $gh(x)$

Functions and linear equations: exam practice

1 Solve $4x - 2 = 40$ [2 marks] **4–6**

2 Solve $5(y - 3) = 65$ [2 marks]

3 Solve $\dfrac{3p}{2} = 12$ [2 marks]

4 Solve $23 = 2a + a + a - 5$ [2 marks]

5 Solve $\dfrac{t - 4}{3} = 30$ [2 marks]

6 Solve $7m = 3m + 10$ [2 marks]

7 Given that $g(x) = 2x^2$, work out $g(3)$ [2 marks]

8 Given that $f(x) = 3x - 5$
    (a) Work out $f(2)$ [1 mark]
    (b) Solve $f(x) = -11$ [2 marks]

9 Solve $12 - 3w = w - 8$ [3 marks]

10 $f(x) = x^2 + 3x - 1$
    Write down an expression for $f(2x)$ [2 marks] **7–9**

11 $f(x) = 2x + 1$ and $g(x) = x^2$
    (a) Work out the value of $fg(3)$ [2 marks]
    (b) Work out an expression for $f^{-1}(x)$ [2 marks]

12 $f(x) = x^2 - 1$
    Show that $f(x + 2)$ can be written in the form $(x + a)(x + b)$
    where a and b are integers. [3 marks]

13 $f(x) = 3x - 2$ and $g(x) = \dfrac{x - 4}{2}$
    Solve $f^{-1}(x) = g^{-1}(x)$ [3 marks]

14 $f(x) = \dfrac{x}{3} + 2$ and $g(x) = 6x^2 - 9$
    Work out an expression for $fg(x)$ [3 marks]

8 Identities, formulae and proof

White Rose MATHS

In this block, we will cover...

8.1 Identities

Example 1

a Show that $2(8x + 4) \equiv 8(1 + 2x)$

b Work out the values of a and b in the identity

Method

Solution	Comme
a LHS: $2(8x + 4) \equiv 16x + 8$	Starting the ider could c (RHS) if
$16x + 8 \equiv 8 + 16x \equiv 8(1 + 2x)$	Now fa

8.2 Rearranging formulae

Practice (A)

1. Rearrange to make x the subject of each fo
   a $c = ax + b$ **b** $d = \frac{x}{3} + b$

2. Rewrite these to express x in terms of a an
   a $\frac{x}{a} + 2 = b$ **b** $\frac{1}{2}ax = b$

3. Rearrange to make y the subject of each fo
   a $x = y^2$ **b** $x = y^2 + 4$
   d $x = \sqrt{y + a}$ **e** $x = \sqrt{y} + 2a$

8.3 Proof

Consolidate – do you need more

1. p is an even integer and q is an odd intege
   Decide whether each statement is **true** or
   a $p + q$ is even   **b** $p - q$ is even

2. Prove that $(5n + 2)^2 - (5n - 1)^2$ is always a m

3. Prove algebraically that the sum of any thr
   multiple of 6

4. Prove algebraically that the sum of four co

Are you ready?

1 Multiply out and simplify:

a $3(t + 3) + 2(t - 8)$ b $4(y - 5) - 12(y - 1)$ c $(x + 3)(x - 7)$ d $(2p + 1)(p - 5)$

2 Solve:

a $3a + 7 = 34$ b $6k - 5 = 31$ c $3d + 3 = 8d - 12$ d $2(n + 3) = 3n + 17$

3 Simplify:

a $\dfrac{9}{12}$ b $\dfrac{44}{121}$ c $\dfrac{6}{42}$ d $\dfrac{15}{45}$

4 Evaluate the following, giving your answer as a simplified fraction.

a $\dfrac{3}{4} \times \dfrac{5}{6}$ b $\dfrac{3}{5} \div \dfrac{4}{9}$ c $\dfrac{9}{10} \times \dfrac{4}{7}$ d $\dfrac{1}{3} \div \dfrac{9}{2}$

An **identity** is true for all values of the variable(s) it features. The sign $\equiv$ means 'equivalent to' or 'identically equal to'.

For example, $2(x + y) \equiv 2x + 2y$ as it is true for whatever values of x and y you choose.

Example 1

a Show that $2(8x + 4) \equiv 8(1 + 2x)$

b Work out the values of a and b in the identity $4x + 12 + ax - b \equiv 6x + 10$

Method

Solution	Commentary
a LHS: $2(8x + 4) \equiv 16x + 8$	Starting with the left-hand side (LHS for short) of the identity, first expand the brackets. Note you could choose to start with the right-hand side (RHS) if you prefer.
$16x + 8 \equiv 8 + 16x \equiv 8(1 + 2x)$ This is the same as the RHS.	Now factorise this expression to show it is the same as the right-hand side.
So $2(8x + 4) \equiv 8(1 + 2x)$	

An alternative method is to multiply out both sets of brackets.

LHS $2(8x + 4) \equiv 16x + 8$

RHS $8(1 + 2x) \equiv 16x + 8$

Both sides are equivalent to $16x + 8$, so they are identically equal.

b $4x + 12 + ax - b \equiv (4 + a)x + 12 - b$	Start by simplifying the left-hand side of the identity.
$4 + a = 6$	Now you can compare coefficients.
	On the left-hand side, the coefficient of x is $4 + a$.
	On the right-hand side, the coefficient of x is 6
So $a = 2$	You can form and solve an equation to find the value of a.
$12 - b = 10$	Next you can equate the constant terms.
	On the left-hand side the constant term is $12 - b$.
So $b = 2$	On the right-hand side the constant term is 10
$a = 2$ and $b = 2$	

Example 2

Show that:

a $\quad \dfrac{7}{4x} \times \dfrac{4}{5x} \equiv \dfrac{7}{5x^2}$

b $\quad \dfrac{5n}{18n - 27} \div \dfrac{4}{2n - 3} \equiv \dfrac{5n}{36}$

Method

Solution	Commentary	
a $\quad \dfrac{7}{\cancel{4}x_{1}} \times \dfrac{\cancel{4}^{1}}{5x}$	You can cancel common factors before multiplying fractions. In this case, you can cancel the common factor 4	See Chapter 1.2, section B, for a reminder of how to do this.
$\dfrac{7}{x} \times \dfrac{1}{5x} \equiv \dfrac{7}{5x^2}$	You can now multiply the numerators and denominators.	
So $\dfrac{7}{4x} \times \dfrac{4}{5x} \equiv \dfrac{7}{5x^2}$		
b $\quad \dfrac{5n}{18n - 27} \times \dfrac{2n - 3}{4}$	To divide by a fraction, you multiply by its reciprocal.	
$\dfrac{5n}{9(2n - 3)} \times \dfrac{2n - 3}{4}$	$18n - 27$ can be factorised to give $9(2n - 3)$.	
$\dfrac{5n}{9(2n-3)_{1}} \times \dfrac{{}^{1}\cancel{2n - 3}}{4}$	Cancel the common factors before multiplying. In this case, you can cancel the common factor $2n - 3$	
$\dfrac{5n}{9} \times \dfrac{1}{4} \equiv \dfrac{5n}{36}$	You can now multiply the numerators and denominators.	
So $\dfrac{5n}{18n - 27} \div \dfrac{4}{2n - 3} \equiv \dfrac{5n}{36}$		

Practice

1. Show that $7(2p + 6) \equiv 2(21 + 7p)$

2. Show that $7(4x + 2y + 10) \equiv 14(5 + x + y) + 14x$

3. Show that $(a - b)(a + b) \equiv a^2 - b^2$

4. Work out the values of p and q in the following identities.

   a $3(3x - p) \equiv qx - 15$

   b $(x + 3)(x - 4) \equiv x^2 + px + q$

   c $(x - 4)(x + p) \equiv x^2 + 3x + q$

   d $2(x + 4) + p(3x - 2) \equiv 14x + q$

   e $x^2 + 8x - 5 \equiv (x + p)^2 + q$

5. Show that $(a + b)^3 \equiv a^3 + 3a^2b + 3ab^2 + b^3$

6. Show that:

   a $\dfrac{7x}{4} \times \dfrac{3}{2x} \equiv \dfrac{21}{8}$

   b $\dfrac{4x}{9} \times \dfrac{2x}{6} \equiv \dfrac{4x^2}{27}$

   c $\dfrac{6x}{21x + 7} \times \dfrac{3x + 1}{5x} \equiv \dfrac{6}{35}$

   d $\dfrac{5y}{7x - 7} \times \dfrac{6x - 6}{7} \equiv \dfrac{30y}{49}$

7. Show that:

   a $\dfrac{7}{10x} \div \dfrac{5x}{9} \equiv \dfrac{63}{50x^2}$

   b $\dfrac{5x}{7} \div \dfrac{9x}{5} \equiv \dfrac{25}{63}$

   c $\dfrac{5x + 20}{6} \div \dfrac{9x + 27}{4} \equiv \dfrac{10(x + 4)}{27(x + 3)}$

   d $\dfrac{4x}{3x - 12} \div \dfrac{7}{2x - 8} \equiv \dfrac{8x}{21}$

8. Show that:

   a $\dfrac{x^2 - 2x - 3}{4} \times \dfrac{3}{x^2 + 2x + 1} \equiv \dfrac{3x - 9}{4x - 4}$

   b $\dfrac{x^2 - 2x - 8}{2} \div \dfrac{x^2 - 7x + 12}{3} \equiv \dfrac{3(x + 2)}{2(x - 3)}$

Consolidate – do you need more?

1 Show that $6(3x + 10) \equiv 9(2x + 6) + 6$

2 Show that $(x + 2y)^2 \equiv x^2 + 4xy + 4y^2$

3 Work out the values of a and b in these identities.

    a $4x + 10 + ax + b \equiv 5x + 22$ **b** $2(x + 4) + b(x - 5) \equiv 6x + a$

    c $(x + 4)(x - 9) + 3 \equiv x^2 + ax + b$ **d** $(x + 5)(x - a) \equiv x^2 - 3x + b$

4 Show that:

    a $\dfrac{9}{10x} \times \dfrac{5x}{3} \equiv \dfrac{3}{2}$ **b** $\dfrac{7}{6x} \times \dfrac{5}{3x} \equiv \dfrac{35}{18x^2}$

    c $\dfrac{3}{5x - 10} \times \dfrac{3x - 6}{7} \equiv \dfrac{9}{35}$ **d** $\dfrac{2x}{6x - 3} \times \dfrac{4x - 2}{7} \equiv \dfrac{4x}{21}$

5 Show that:

    a $\dfrac{6x}{5} \div \dfrac{3}{4x} \equiv \dfrac{8x^2}{5}$ **b** $\dfrac{9}{2x} \div \dfrac{3x}{4} \equiv \dfrac{6}{x^2}$

    c $\dfrac{6x + 9}{4} \div \dfrac{2x + 3}{12} \equiv 9$ **d** $\dfrac{15x - 5}{10x} \div \dfrac{6x - 2}{3x} \equiv \dfrac{3}{4}$

Stretch – can you deepen your learning?

1 Work out the values of a, b and c in the identity $2x^2 + 8x - 5 \equiv a(x + b)^2 + c$

2 Work out the values of a, b and c in the identity $(x + 1)(x - 2)(x + a) \equiv x^3 + 4x^2 + bx + c$

3 Show that $\dfrac{1}{6x^2 + 7x - 5} \div \dfrac{1}{4x^2 - 1} \equiv \dfrac{ax + b}{cx + d}$ where a, b, c, and d are integers to be found.

4 Show that $\dfrac{8x + 16}{x^2 - 2x - 8} \div \dfrac{2x + 8}{x^3 - 16x} \equiv px$, where p is an integer to be found.

Are you ready? (A)

1 Expand and simplify:

a $3(v + 2)$ **b** $4t(t + 3) - 5t$

c $r(s + 5) - 3rs$ **d** $3(e - 5) - 5(e + 2)$

2 Factorise fully:

a $24t - 16$ **b** $8h + 4h^2$

c $16xy + 8x^2$ **d** $32a^2b - 8ab$

3 Solve the equations.

a $5x - 9 = 41$ **b** $\frac{x}{4} + 6 = 11$

c $\frac{x + 4}{3} = 19$ **d** $5x + 6 = 2(2x + 1)$

A **formula** is a rule connecting variables, for example $s = \frac{d}{t}$

The **subject** is the variable in a formula that is expressed in terms of the other variables.

For example, in the formula $s = \frac{d}{t}$, s is the subject of the formula as s is written **in terms of** d and t.

You can change the subject of the formula by using inverse operations in the same way that you solve an equation by balancing.

Example

Rearrange to make x the subject of each formula.

a $3x + 2y = 17$ **b** $y - 4x^2 = z$ **c** $\frac{x}{4} + y = 2z$

Method

Solution	Commentary
a $\quad 3x + 2y = 17$ $-2y \qquad\qquad -2y$ $\quad 3x = 17 - 2y$ $\div 3 \qquad\qquad \div 3$ $\quad x = \dfrac{17 - 2y}{3}$	Subtract $2y$ from both sides. Divide both sides by 3

b

$y - 4x^2 = z$

$+ 4x^2$ $+ 4x^2$

$y = z + 4x^2$

$- z$ $- z$

$y - z = 4x^2$

$\div 4$ $\div 4$

$\dfrac{y - z}{4} = x^2$

$\sqrt{}$ $\sqrt{}$

$\sqrt{\dfrac{y - z}{4}} = x$

First add $4x^2$ to both sides. This makes the coefficient of x^2 positive.

Now subtract z from both sides.

Then divide both sides by 4

Finally, square root both sides.

You can write this as $x = \sqrt{\dfrac{y - z}{4}}$ (with x on the LHS) to make it more obvious that x is now the subject.

c

$\dfrac{x}{4} + y = 2z$

$- y$ $- y$

$\dfrac{x}{4} = 2z - y$

$\times 4$ $\times 4$

$x = 4(2z - y)$

First subtract y from both sides.

Then multiply both sides by 4

You can choose to expand the bracket to give $x = 8z - 4y$ if you like, but this is not needed.

Practice (A)

1 Rearrange to make x the subject of each formula.

    a $c = ax + b$ **b** $d = \dfrac{x}{3} + b$ **c** $4 = \dfrac{x}{a} - b$ **d** $y = \dfrac{x + b}{3}$

2 Rewrite these to express x in terms of a and b.

    a $\dfrac{x}{a} + 2 = b$ **b** $\dfrac{1}{2}ax = b$ **c** $b^2 = a^2 + 2x$ **d** $\dfrac{5xa}{b} = 8$

3 Rearrange to make y the subject of each formula.

    a $x = y^2$ **b** $x = y^2 + 4$ **c** $3x = 2y^2 + z$

    d $x = \sqrt{y + a}$ **e** $x = \sqrt{y} + 2a$ **f** $x = \sqrt{\dfrac{y}{a}}$

4 Which is the correct rearrangement of $y = 3x + 4$?

    A $x = \dfrac{y + 4}{3}$ **B** $x = \dfrac{y}{3} + 4$ **C** $x = \dfrac{y}{3} - 4$ **D** $x = \dfrac{y - 4}{3}$

5 Ed has made a mistake in his homework.

    Identify his error and write the correct answer.

    Rearrange to make a the subject of the formula $y = a^2 + 2b$

        $\sqrt{y} = a + 2b$

        $\sqrt{y} - 2b = a$

(6) There is a set of formulae that connect initial speed (u), final speed (v), displacement (s), time (t) and acceleration (a).

Rearrange each formula to make u the subject.

a　$v = u + at$　　　　　　**b**　$v^2 = u^2 + 2as$　　　　　**c**　$s = \frac{1}{2}(u + v)t$

(7) The formula $C = \frac{5}{9}(F - 32)$ can be used to convert temperature in Fahrenheit to Celsius.

a　Convert 102°F to Celsius.

b　Rearrange the formula to make F the subject.

c　Show that when C is equal to −40°, F is also equal to −40°

(8) The volume, V, of a cone with base radius r and height h is given by the formula $V = \frac{1}{3}\pi r^2 h$.

a　Rearrange the formula to make r the subject.

b　Rearrange the formula to make h the subject.

c　A cone has height 16 cm and volume 325 cm³.

Work out the diameter of the base of the cone.

What do you think? 💡

(1) Given $v^2 = u^2 + 2as$ and $v = u + at$

a　Rearrange to make u the subject of $v = u + at$

b　Substitute this into $v^2 = u^2 + 2as$

Simplify and rearrange to work out a formula for s in terms of a, v and t.

Are you ready? (B)

1　Factorise fully:

a　$18t + 60$　　　　　**b**　$6k^2 + 2k$　　　　　**c**　$10rt - 100t^2$　　　　　**d**　$15gh^2 - 25h^2g^3$

2　Solve the equations.

a　$5x - 10 = 4x + 13$　　　　　　　**b**　$2x + 9 = 4x - 7$

c　$2(x + 3) = 5(x - 9)$　　　　　　　**d**　$22 + x = 12(4 - x)$

To rearrange a formula where the new subject appears more than once, you need to gather all like terms containing the subject. This often involves expanding brackets and factorising.

Example

Rearrange to make x the subject of each formula.

a $4x + c = ax + b$ **b** $t = \dfrac{x + 1}{y - x}$

Method

Solution	Commentary
a $$4x + c = ax + b$$ $-ax$ $\quad\quad\quad$ $-ax$ $$4x + c - ax = b$$ $-c$ $\quad\quad\quad$ $-c$ $$4x - ax = b - c$$ $$x(4 - a) = b - c$$ $\div (4 - a)$ $\quad\quad\quad$ $\div (4 - a)$ $$x = \dfrac{b - c}{4 - a}$$	Because the x appears twice in this formula, you need to get all the terms with x on one side of the formula and all the terms that do not involve x on the other side. You can do this by subtracting ax and c from both sides.
	Now you can isolate the x by factorising the left-hand side.
	Finally, divide both sides by $4 - a$ to make x the subject.
b $$t = \dfrac{x + 1}{y - x}$$	First multiply both sides by $y - x$ to clear the fraction.
$$t(y - x) = x + 1$$	Then expand the bracket.
$$ty - tx = x + 1$$ -1 $\quad\quad\quad$ -1 $$ty - tx - 1 = x$$ $+tx$ $\quad\quad\quad$ $+tx$ $$ty - 1 = x + tx$$	Now get all the terms with x on one side – you can do this by adding tx and subtracting 1 from both sides.
$$ty - 1 = x(1 + t)$$ $\div (1 + t)$ $\quad\quad\quad$ $\div (1 + t)$ $$\dfrac{ty - 1}{1 + t} = x$$ Hence, $x = \dfrac{ty - 1}{1 + t}$	Now you can isolate the x by factorising the right-hand side. Finally, divide both sides by $1 + t$ to make x the subject.

Practice (B)

1. Rearrange to make x the subject of each formula.

   a $ax + 4 = x$ **b** $ax = bx + c$ **c** $ax + 3 = bx - c$ **d** $ax + b = cx + d$

2. Express t in terms of p and q.

   a $2(p + t) = q + pt$ **b** $p(t - 3) = 2q + t$

   c $p(2 + t) = q(t - 4)$ **d** $pt + 4t = q(t - 1)$

3. Rearrange to make w the subject of each formula.

   a $a = \dfrac{2w}{w - 4}$ **b** $5 = \dfrac{3w}{w + a}$ **c** $2b = \dfrac{w}{w + 1}$ **d** $c = \dfrac{w + a}{w - b}$

4. Rearrange to make x the subject of these formulae.

   a $z = \sqrt{\dfrac{x + y}{x}}$ **b** $yx = z + \dfrac{x}{3}$

5. $q^2 = \dfrac{3p + m^2}{m^2}$

   Express m in terms of p and q.

6. Given $\dfrac{a^2 + b^2}{16} = 1$, which of the following is an **incorrect** rearrangement of the formula?

   A $b = \sqrt{16 - a^2}$ **B** $a = \sqrt{16 - b^2}$ **C** $a^2 6 = 10 \ b^2$

   D $a + b = 4$ **E** $a^2 = (4 - b)(4 + b)$

7. The circumference of an ellipse is given by the formula $C = 2\pi\sqrt{\dfrac{a^2 + b^2}{2}}$

   Rearrange to make a the subject.

Consolidate – do you need more?

1. Given $x = y - z$, determine whether the following are **true** or **false**.

   a $x + y = z$ **b** $x + z = y$ **c** $y - x = z$

   d $x + y + z = 0$ **e** $x - y + z = 0$

2. Rearrange each formula to make t the subject.

   a $2t + w = 3$ **b** $3(t - s) = v$ **c** $2t + 2v + 3s = 0$

   d $\dfrac{t}{4} + u = v$ **e** $\dfrac{t + u}{6} = v$ **f** $\dfrac{t - 5}{w} = u$

3. Rearrange each formula to make c the subject.

   a $c^2 + d = 3$ **b** $3c^2 - 2 = e$ **c** $(c + e)^2 = a$

   d $\sqrt{c} = d$ **e** $\sqrt{c} = 2d + 6$ **f** $\sqrt{\dfrac{c + 3}{d}} = e$

4 Faith has made a mistake in her homework.

Identify the error and write the correct solution.

Rearrange to make a the subject of $\frac{a}{4} + b = c$

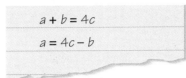

$a + b = 4c$

$a = 4c - b$

5 The volume, V, of a sphere of radius r is given by $V = \frac{4}{3}\pi r^3$

Rearrange to make r the subject of the formula.

6 Rearrange to make n the subject of each formula.

a $4(n + p) = r(n - q)$ **b** $rn + 4p = 2n - 2p + 4$

c $r = \dfrac{n + p}{n + q}$ **d** $q = \dfrac{n + 4}{n - p}$

e $p = \dfrac{nq}{n - q}$ **f** $q = 3 + \dfrac{p}{n}$

Stretch – can you deepen your learning?

1 Rearrange to make x the subject of $\dfrac{x}{y} - \dfrac{a}{b} = \dfrac{x + a}{c}$

2 Rearrange to make v the subject of $\dfrac{1}{u} + \dfrac{1}{v} = \dfrac{1}{f}$, giving your answer in its simplest form.

3 The formula for the period, T, of a pendulum is given by $T = 2\pi\sqrt{\dfrac{l}{g}}$ where l is the pendulum length and g is the acceleration due to gravity.

Rearrange the formula so it can be used to estimate the value of g from given values of T and l.

Are you ready?

1 Simplify $(n^2 + 3n - 6) - (n^2 - 3n + 6)$

2 Expand and simplify:

    a $3n(n + 4) + 2(n - 5)$ **b** $(n - 5)(n - 3)$ **c** $6n(n + 4) - 2n(n - 7)$

3 Factorise fully:

    a $6n^2 + 8n$ **b** $40n - 12n^2$ **c** $n^3 - n$

To **prove** something means to show that it is always true. It is not enough to show this by using a few (or even lots) of numerical examples. You need to use algebra.

Here are some useful reminders of terminology and notation.

Consecutive numbers are whole numbers that follow on from each other.

For example, 6, 7, 8 are consecutive.

Algebraically, consecutive numbers could be written as n, $n + 1$, $n + 2$, etc.

Often, the letter n is used to represent an integer (whole number).

This means $2n$ must be an even number since it is a multiple of 2. Similarly, $2n + 1$ must be an odd number.

You may be asked to prove that an expression is a multiple of some integer. To do this, you need to show that the expression can be written as a product of that integer and another expression.

For example, to prove that $6n^2 + 9n + 3$ is a multiple of 3:

- you can factorise it as $3(2n^2 + 3n + 1)$,
- which is 3 multiplied by an integer,
- and so must be a multiple of 3

Example 1

Prove algebraically that $(2n + 1)^2 - (2n + 1)$ is an even number for all positive integer values of n.

Method

Solution	Commentary
$(2n + 1)^2 - (2n + 1)$ $\equiv (2n + 1)(2n + 1) - (2n + 1)$ $\equiv 4n^2 + 4n + 1 - (2n + 1)$	Start by expanding the first part of the expression.

$\equiv 4n^2 + 2n$	Then simplify by collecting like terms. Be careful with the negative sign before the bracket and note that $1 - +1 = 0$
$\equiv 2(2n^2 + n)$	To show that the expression is always even, you need to show that it is a multiple of 2. So factorise with a 2 outside the bracket.
$2(2n^2 + n)$ is a multiple of 2 and so always even.	State your conclusion clearly.

Example 2

Show that $\dfrac{2x^2 - 2x - 24}{x^2 - 10x + 24}$ can be written in the form $\dfrac{ax + b}{cx + d}$ where a, b, c and d are integers to be found.

Method

Solution	Commentary
$\dfrac{2x^2 - 2x - 24}{x^2 - 10x + 24}$	To simplify algebraic fractions, you need to factorise the numerator and denominator.
$\equiv \dfrac{2(x + 3)(x - 4)}{(x - 4)(x - 6)}$	You can now cancel by the common factor $(x - 4)$.
$\equiv \dfrac{2(x + 3)}{x - 6}$	
$\equiv \dfrac{2x + 6}{x - 6}$	Then you can rewrite in the required form. Hence, $a = 2$, $b = 6$, $c = 1$ and $d = -6$

Example 3

Prove algebraically that the difference between the squares of any two consecutive integers is always an odd number.

Method

Solution	Commentary
Let n be an integer. Then two consecutive integers are n and $n + 1$	Start by defining your variables.
$(n + 1)^2 - n^2$ $\equiv n^2 + 2n + 1 - n^2$	Square each expression and subtract one from the other. This involves expanding $(n + 1)^2$
$\equiv 2n + 1$	Now simplify.
$2n + 1$ is always odd since $2n$ is even and adding 1 gives an odd number.	State your conclusion clearly.

Practice

1 If n is an integer, determine whether each of the following statements is **always true**, **sometimes true** or **never true**.

   a $2n$ is even   **b** $n + 1$ is odd   **c** $2n - 1$ is odd

   d $5n$ is even   **e** $10n$ is odd   **f** $2n + 3$ is odd

2 Prove that $(n - 1)^2 + n^2 + (n + 1)^2 \equiv 3n^2 + 2$

3 Prove that $(n + 4)^2 - (n + 2)^2$ is always a multiple of 4 for all positive integer values of n.

4 Prove that $(2n + 1)^2 - (2n - 1)^2$ is always a multiple of 8 for all positive integer values of n.

5 Prove algebraically that the sum of any three consecutive integers is divisible by 3

6 Prove algebraically that the difference between any two different odd numbers is an even number.

> **Hint:** Use different letters for your odd numbers, such as $2n + 1$ and $2m + 1$

7 Prove algebraically that the difference between the squares of any two consecutive integers is equal to the sum of these two integers.

8 Prove that when any odd integer is squared, the result is always 1 more than a multiple of 4

9 Prove that $\dfrac{8x^3 - 24x}{4x^2 - 12}$ is always an even number.

10 In each case, show that the given fraction can be simplified to a fraction of the form $\dfrac{ax + b}{cx + d}$ where a, b, c and d are integers.

   a $\dfrac{x^2 + 4x + 3}{3x^2 - 3x - 36}$ **b** $\dfrac{2x^2 + 3x - 2}{x^2 - 2x - 8}$ **c** $\dfrac{4x^2 - 1}{2x^2 - 13x - 7}$

What do you think? 💡

1 Explain why $n^3 - n$ must always be a multiple of 6

2 Seb draws a right-angled triangle with integer lengths in which the lengths of the two shorter sides are both even.

He says, "The hypotenuse must also be even."

Prove that Seb is correct.

Consolidate – do you need more?

1 p is an even integer and q is an odd integer.

Decide whether each statement is **true** or **false**.

a $p + q$ is even   **b** $p - q$ is even   **c** $2p + 2q$ is even   **d** $2p + 1$ is odd

2 Prove that $(5n + 2)^2 - (5n - 1)^2$ is always a multiple of 3 for all positive integer values of n.

3 Prove algebraically that the sum of any three consecutive even numbers is always a multiple of 6

4 Prove algebraically that the sum of four consecutive integers is **not** divisible by 4

5 Prove algebraically that when the product of two consecutive positive integers is added to the larger of the two integers, the result is always a square number.

6 Prove algebraically that the sum of the squares of any three consecutive odd numbers is always 11 more than a multiple of 12

7 In each case, show that the given fraction can be simplified to a fraction of the form

$\dfrac{ax + b}{cx + d}$ where a, b, c and d are integers.

a $\dfrac{x^2 + x - 30}{x^2 - 9x + 20}$ **b** $\dfrac{2x^2 + 7x + 3}{x^2 - 9}$

Stretch – can you deepen your learning?

1 k is a multiple of 5

$a = k + 1$

$b = k - 1$

Show that $a^2 - b^2$ is always a multiple of 20

2 The first six terms of a linear sequence are 4, 9, 14, 19, 24, 29

    a Work out an expression for the nth term of the sequence.

    b Beca creates a new sequence by squaring each term in the given sequence and then adding 4

       Prove that all the terms in Beca's sequence are divisible by 5

3 Ed substitutes some values into the expression $x^2 - 10x + 25$

He believes the answer will never be a negative number.

By factorising the expression, explain why Ed is correct.

4 Does the quadratic expression $x^2 + x + 41$ **always**, **sometimes** or **never** produce prime values for positive values of x? Explain your answer.

Identities, formulae and proof: exam practice

1 Show that $3(2a + 5) + 2(4a - 3) \equiv 14a + 9$ [2 marks]

2 Show that $\dfrac{5}{6x} \div \dfrac{2}{3x^2} \equiv \dfrac{5x}{4}$ [2 marks]

3 Rearrange $p = 4q - 1$ to make q the subject. [2 marks]

4 Rearrange to make t the subject of the formula $k = t^2 - 5$ [2 marks]

5 Rearrange $m = \sqrt{\dfrac{p + 2}{3}}$ to make p the subject. [3 marks]

6 p is even.

Explain why $(p + 1)^2$ will always be odd. [2 marks]

7 Work out the values of p and q in the identity

$2(px - 4) + 3(4x + q) \equiv 2(9x - 1)$ [3 marks]

8 Show that $2t(6t + 1) - 3(2t + 1)(2t - 1) \equiv 2t + 3$ [2 marks]

9 Rearrange $a = \dfrac{b + 1}{2b - 3}$ to make b the subject. [3 marks]

10 Prove algebraically that the sum of any two consecutive even integers is always double an odd number. [3 marks]

11 Prove that $(2n + 1)^2 - (2n - 1)^2$

is always a multiple of 8 for all positive integer values of n. [3 marks]

12 Prove that the square of any odd number is always 1 more than a multiple of 8 [3 marks]

4–6

7–9

9 Linear graphs

In this block, we will cover...

9.1 Equations of straight lines

Example 1

Write down the gradient and y-intercept of the l

a $y = 3x - 4$ **b** $2y + 4x = 8$

Method

Solution	Comme
a $y = 3x - 4$	This equ
Gradient $m = 3$	you can
y-intercept $(0, c)$ is at $(0, -4)$	equatio
b $2y + 4x = 8$	Rearran
$-4x$ $-4x$	• subtr

9.2 Parallel and perpendicular lines

Practice

1. Write down the equation of any line paral

   a $y = 4x + 7$ **b** $y = 5 - 3x$

2. Work out the equation of each of the follo

   a   A line parallel to $y = 6x - 3$ that passes

   b   A line parallel to $2y + 3x = 10$ that pass

   c   A line parallel to $y = -3x + 4$ that passe

   d   A line parallel to $3y - x = 12$ that passe

9.3 Interpreting linear graphs

Consolidate – do you need more

1. The graph shows the relationship betwee

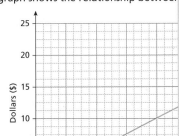

Are you ready? (A)

1 Write down the coordinates of each of the marked points.

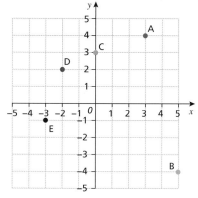

2 Write down the coordinates of three points with:

a an x value of 4   **b** a y value of -3

3 Given that $x = -3$, work out the value of:

a $2x$ **b** $2x - 7$ **c** $-3x$ **d** $4x + 5$ **e** $-9x - 6$

The graph of an equation of the form $x = a$ or $y = b$ is a straight line parallel to the axes.

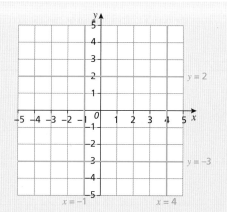

Equations of the form $y = mx + c$ also have graphs that are straight lines.

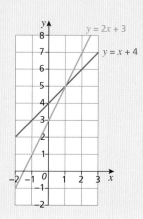

Using your calculator

You can use the table function on some calculators when you need to use a table of values.

Press the 'Menu/Set up' button.

Choose the 'Table' option.

The screen shows 'f(x) =' and here you put your equation, for example $3x - 5$

$$f(x) = 3x - 5$$

Press = when done.

On the next screen, choose your start value, your end value and what step you would like to go up in.

Table Range
Start : 0
End = 5
Step = 1

Press = and you will see the table of values generated.

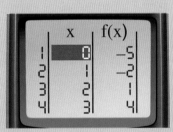

x	f(x)
0	−5
1	−2
2	1
3	4

Example 1

Draw the graph of $y = 3$

Method

Solution	Commentary
(graph showing horizontal line $y = 3$)	You don't need to draw a table of values for equations of this form.
	The equation of the line is read as 'y is equal to 3'.
	So the y value is equal to 3 for all points on the line.
	Some examples of coordinates are (2, 3), (0, 3) and (−1, 3).

Example 2

a Draw the graph of $y = 2x - 4$

b Does the point (10, 16) lie on the line?

Method

Solution	Commentary
a <table><tr><td>x</td><td>-2</td><td>-1</td><td>0</td><td>1</td><td>2</td></tr><tr><td>y</td><td>-8</td><td>-6</td><td>-4</td><td>-2</td><td>0</td></tr></table>	Start by drawing a table of values. If values are not given to you, then you need about five values centred around zero. Each y value is equal to 2 times the x value minus 4 You can also use your calculator – choose the table function (as shown on the previous page), type in your equation ($2x - 4$), choose –2 for the start, 2 for the end and go up in steps of 1
	Then draw suitable axes and plot your points, using the values from your table. These values give you the coordinates to plot. When you join them up, make sure you extend the line as far as your grid allows. If your points do not lie on a straight line, you know you have made a mistake.
b $y = 2x - 4$ Here $x = 10$ and $y = 16$ $2x - 4 = 2 \times 10 - 4 = 16$ This is the correct y value, so the point does lie on the line.	You can answer this part without having to extend the graph as far as $x = 10$ You need to check if the equation holds true for (10, 16).

Practice (A)

1 Draw a coordinate grid with x-axis and y-axis from –5 to 5

On your axes, draw the following graphs.

$y = 2$ $\qquad\qquad$ $x = 5$ $\qquad\qquad$ $y = -3$ $\qquad\qquad$ $x = -5$

2 Amina thinks the equation of the x-axis is $x = 0$

Is Amina correct? Explain your answer.

3 Write down the equations of the lines labelled **a**, **b**, **c** and **d** below.

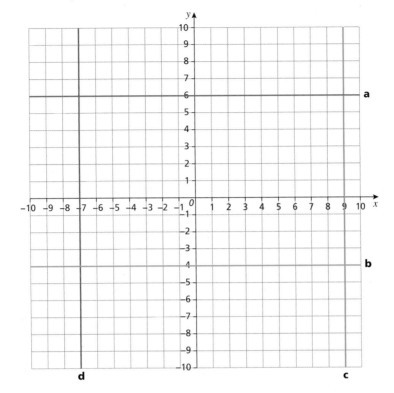

4 **a** Write down four sets of coordinates in which the y value is equal to the x value.

   b Plot these points on a graph.

   c What is the equation of the graph you have drawn?

   d Draw the graph of $y = -x$

5 **a** On the same set of axes, draw the graphs of:

   i $y = x$ **ii** $x = 3$ **iii** $y = 1$

   b Work out the area of the triangle bounded by the lines above.

6 Draw a coordinate grid with x- and y-axes from −10 to 10

   By using a table of values, plot the graphs of:

   $y = 2x + 6$ $y = -x + 4$ $y = 3x - 5$ $y = -2x + 7$

7 Work out whether each point lies on the given graph.

   a (2, 5) and $y = 2x + 1$ **b** (6, 3) and $y = -2x - 2$

   c (−4, 1) and $y = 3x + 13$ **d** (−1, 3) and $y = 4x - 1$

What do you think? 💡

1 Which of these are equations of straight lines? How do you know?
How would you plot the graphs of these equations?

$x + y = 10$ $x = 2y - 4$ $3y + 2x = 6$ $y = x^2 + 3$ $y = \dfrac{5}{x} + 3$

2 Ed says, "You only need the coordinates of two points in order to plot a straight line."
Is Ed correct?

Are you ready? (B)

1 Given that $x = 6$, work out the value of:

   a $3x$ **b** $4x - 6$ **c** $-2x$

   d $6x + 7$ **e** $-8x - 1$

2 Rearrange each formula to make y the subject.

   a $x + y = 8$ **b** $2x + y = 10$ **c** $5x - y = 17$

   d $8x + 2y - 14 = 0$ **e** $3y - 12 = 2x$ **f** $3x - 4y = 15$

3 On the same set of axes, draw the graphs of:

   a $y = 3x + 1$ **b** $y = 3x - 3$ **c** $y = 3x$ **d** $y = 3x + 2$

What do you notice?

The **gradient** of a line measures how steep it is.

The point at which a line meets the y-axis is often referred to as the **y-intercept**.

Always check the
scale. A square
may represent a
different value
on each axis.

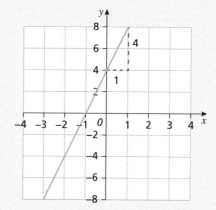

To work out the gradient of a line, you can use the formula, gradient $= \dfrac{\text{change in } y}{\text{change in } x}$

This is sometimes written as $\frac{y_2 - y_1}{x_2 - x_1}$ where (x_1, y_1) and (x_2, y_2)

are the coordinates of the two points.

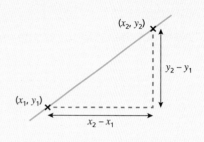

In question 3 of 'Are you ready? (B)' on the previous page, you have seen that the graphs are all parallel, i.e.

- they have the same gradient
- the equations all have 3 as the coefficient of x.

The general equation of a straight line can be expressed in the form $y = mx + c$, where m is the gradient and $(0, c)$ is the point at which the graph intercepts the y-axis (the **y-intercept**). If the value of m is negative, the graph slopes down from left to right.

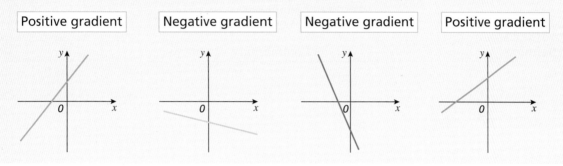

| Positive gradient | Negative gradient | Negative gradient | Positive gradient |

Example 1

Write down the gradient and y-intercept of the lines.

a $y = 3x - 4$ **b** $2y + 4x = 8$

Method

Solution	Commentary
a $y = 3x - 4$ Gradient $m = 3$ y-intercept $(0, c)$ is at $(0, -4)$	This equation is already in the form $y = mx + c$, so you can just read the values of m and c from the equation.
b $2y + 4x = 8$ $-4x$ ($-4x$ $2y = 8 - 4x$ $\div 2$ ($\div 2$ $y = 4 - 2x$ $y = -2x + 4$	Rearrange to make y the subject. You can do this by: • subtracting $4x$ from both sides • then dividing both sides by 2 You can rewrite to make the equation match $y = mx + c$ if it helps, but take care with the signs.
Gradient $m = -2$	Remember that the coefficient of x is the gradient.
y-intercept $(0, c)$ is at $(0, 4)$	Now you can read the values of m and c from the equation.

Example 2

Work out the equation of the line.

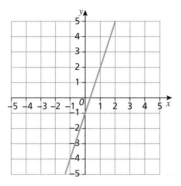

Method

Solution	Commentary
Gradient = $\dfrac{\text{change in } y}{\text{change in } x}$ $= \dfrac{6}{2} = 3$ So $m = 3$ And $c = -1$ $y = 3x - 1$	First work out the gradient and the y-intercept. To work out the gradient, you can use the formula Gradient $= \dfrac{\text{change in } y}{\text{change in } x}$ Pick two points on the graph to help with this. You should choose points with integer coordinates. You can see from the diagram that the graph cuts the y-axis at $(0, -1)$. Now you can write the equation in the form $y = mx + c$

Practice (B)

1. Work out the gradient of each line.

   a 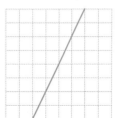   b 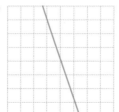   c 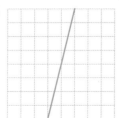   d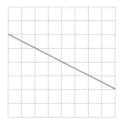

2. Write down the gradient and the y-intercept of each straight line.

   a $y = 5x + 3$
   b $y = x - 3$
   c $y = 4x$
   d $y = 3 - 4x$
   e $y = -3x - 7$
   f $y = \dfrac{1}{2}x + 6$

3 Work out the gradient and the *y*-intercept of each straight line.

    a $2x + y = 7$
    b $x + y = 8$
    c $x - y = 10$

    d $2y = 6x$
    e $6x - 2y = 4$
    f $12y - 6x - 3 = 0$

4 Work out the equation of each line.

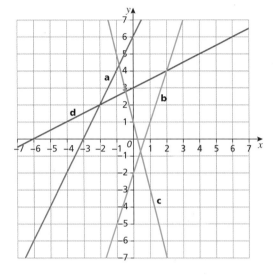

5 Work out the equation of each line.

a   **b**   **c**   **d**

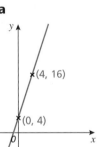
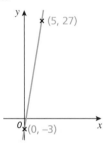
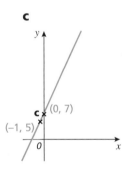
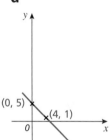

Are you ready? (C)

1 Work out the gradient of each line.

a   **b**   **c**   **d**

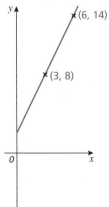
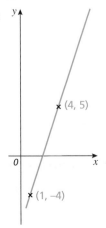
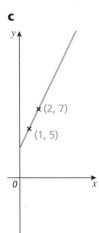
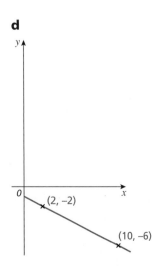

2 Given $x = 5$ and $y = 3$, work out the value of c.

    a $2y - x = c$ **b** $x - c = y$ **c** $y = 2x + c$ **d** $y = c - 3x$

Example 1

Work out the equation of a line with gradient 3 and that passes through the point (3, 4).

Method

Solution	Commentary
$y = 3x + c$	You know that the gradient is 3, so the equation must be of this form.
$4 = 3 \times 3 + c$ $4 = 9 + c$ $-5 = c$	To find c, you can substitute the values of x (3) and y (4) into the equation. You can solve the equation by balancing.
$y = 3x - 5$	Now you can give the equation in the form $y = mx + c$

Example 2

Work out the equation of a line passing through the points (6, 4) and (3, 7).

Method

Solution	Commentary
$m = \dfrac{7 - 4}{3 - 6}$ $\quad = \dfrac{3}{-3} = -1$ So $m = -1$	Work out the gradient first. You can sketch if it helps, or just use the formula gradient $= \dfrac{\text{change in } y}{\text{change in } x}$
$y = -x + c$ $7 = -1 \times 3 + c$ $7 = -3 + c$ $10 = c$	Now you know that the gradient is -1, the equation must be of the form $y = -1x + c$, or just $y = -x + c$ To find c, substitute in the values of x and y that you have been given. You can choose either (6, 4) or (3, 7).
$y = -x + 10$	Now you know both m and c, you can write the equation in the form $y = mx + c$

Practice (C)

1. Work out the gradient of the line passing through each pair of points.

   a (1, 4) and (3, 10)   **b** (−4, 6) and (4, 4)

   c (−3, −1) and (5, 3)   **d** (−5, 6) and (−2, 9)

2. Work out the equation of the straight line with:

   a gradient 4 that goes through (4, 9)

   b gradient 3 that goes through (5, −1)

   c gradient −2 that goes through (3, 8)

   d gradient −4 that goes through (−7, 2)

   e gradient $\frac{1}{2}$ that goes through (9, 1).

3. Work out the equation of the straight line passing through each pair of points.

   a (2, 5) and (7, 10)   **b** (5, 3) and (8, −6)

   c (−2, 9) and (−1, 7)   **d** (6, 0) and (−3, 18)

4. Match each graph to its equation.

   a   **b**   **c**   **d**

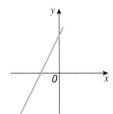   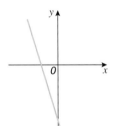   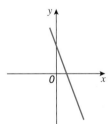   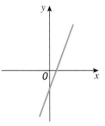

| $y = 3 - 4x$ | $y = 4x - 2$ | $y = 3x + 4$ | $y = -5x - 6$ |

Consolidate – do you need more?

1. On a coordinate grid with x- and y-axes from −10 to 10, draw the graphs of:

   $y = 3x + 5$ $y = -2x - 7$

2. Determine whether each point lies on the given line.

   a (3, 19) and $y = 4x + 7$ **b** (4, −4) and $y = -2x - 4$

3. Work out the gradient and the y-intercept of the lines with these equations.

   a $y = 3x - 8$ **b** $y = 9 - x$ **c** $2y + 6x = 9$ **d** $2y - 4x - 10 = 0$

4 Work out the equation of each line.

a
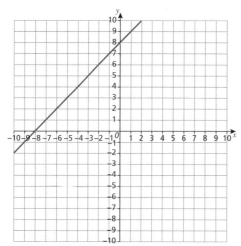

b
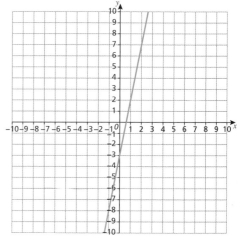

c
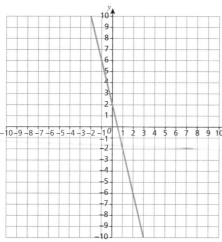

d
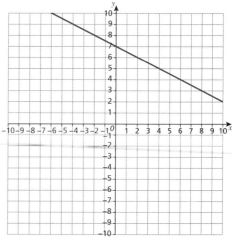

5 Work out the equation of a line with:

   a   gradient 6 that goes through (5, –4)

   b   gradient $-\frac{1}{2}$ that goes through (–6, 10).

6 Work out the equation of a line that goes through each pair of points.

   a   (1, 3) and (–1, 5)   b   (7, –4) and (10, 5)

Stretch – can you deepen your learning?

1 The point A(p, 2p) lies on the line with equation $3y = x + 15$

   Work out the value of p.

2 Work out the coordinates of the points at which each line crosses the x- and y-axes.

   a $y = 2x + 7$ b $y = 5x - 9$ c $x - 3y + 9 = 0$

3 A line has equation $3y + 4x = 15$

   a   Work out the coordinates of the points A and B where the line crosses the x- and y-axes.

   b   Work out the area of the triangle OAB, where O is the origin.

9.2 Parallel and perpendicular lines

Are you ready?

1 Write down the reciprocal of each number.

    a 2   **b** −5   **c** $\frac{1}{3}$

    d $\frac{2}{3}$ **e** $-\frac{5}{7}$

2 Work out the midpoint of the line segment joining each pair of points.

    a (0, 0) and (8, −2)   **b** (3, 4) and (7, 12)   **c** (−2, 5) and (4, −9)

3 Work out the gradient and the y-intercept of the straight lines with these equations.

    a $y = 3x + 6$ **b** $y = -2x + 7$ **c** $y = 9 - 5x$

    d $y + 6x = 8$ **e** $3y - 4x + 12 = 0$

Parallel lines have the same gradient.

For example, $y = 2x - 5$, $y = 2x + 3$ and $2y - 8 = 4x$ are all parallel because they all have gradient 2

Remember that you need to rearrange the equation to the form $y = mx + c$, where m is the gradient of the line.

Note that $2y - 8 = 4x$ can be rearranged to $y = 2x + 4$

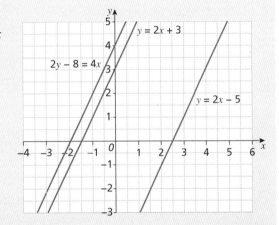

Two straight lines are **perpendicular** if they meet at a right angle.

When straight lines are perpendicular, the product of their gradients is −1

This is sometimes described as the gradients being **negative reciprocals** of one another.

In this example, the line l_1 has gradient 2 and the line l_2 has gradient $-\frac{1}{2}$

Notice that $2 \times -\frac{1}{2} = -1$

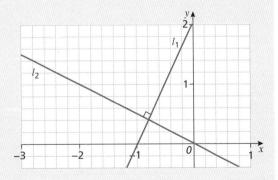

Example 1

The line l_3 has equation $y = 6x + 2$

a Work out the equation of a line parallel to l_3 that goes through the point (0, 4).

b Work out the equation of a line perpendicular to l_3 that goes through the point (0, –8).

Method

Solution	Commentary
a $y = 6x + 2$ has a gradient of 6 $m = 6$	Try to find m and c in the general equation $y = mx + c$ Lines that are parallel have the same gradient, so the line you need will also have a gradient of 6
$c = 4$	(0, 4) is the coordinate of the y-intercept, so c will be 4
$y = 6x + 4$	
b $y = 6x + 2$ has a gradient of 6 Gradient of perpendicular line is $-1 \div 6 = -\dfrac{1}{6}$	Perpendicular lines have gradients with a product of –1 So you want $6 \times m = -1$
$c = -8$	(0, –8) is the coordinate of the y-intercept, so c will be –8
$y = -\dfrac{1}{6}x - 8$	

Example 2

Work out the equation of the line perpendicular to $2y + 4x = 7$ that passes through (4, 6).

Method

Solution	Commentary
$2y + 4x = 7$ $2y = 7 - 4x$ $y = 3.5 - 2x$ $y = -2x + 3.5$	First rearrange to get the equation in the form $y = mx + c$
Gradient of perpendicular line is $-1 \div -2 = \dfrac{1}{2}$ $y = \dfrac{1}{2}x + c$	This line has a gradient of –2, so the perpendicular line will have gradient $\dfrac{1}{2}$ So it will be of the form $y = \dfrac{1}{2}x + c$
$6 = \dfrac{1}{2} \times 4 + c$ $6 = 2 + c$ $4 = c$	You know that the line passes through the point (4, 6) so substitute these values into the equation to find c.
Equation is $y = \dfrac{1}{2}x + 4$	

Example 3

The line l_4 is parallel to $y - 4x = 12$ and passes through the midpoint of the line segment AB, where A is the point $(4, 6)$ and B is the point $(-2, 10)$.

Work out the equation of l_4

Method

Solution	Commentary
$y - 4x = 12$ $y = 4x + 12$ Equation of l_4 will be of the form $y = 4x + c$	To find the gradient of l_4, rearrange $y - 4x = 12$ Since the lines are parallel, they will have the same gradient, so the gradient of l_4 will be 4
$\dfrac{4 + -2}{2} = 1$ $\dfrac{6 + 10}{2} = 8$ Midpoint of AB is $(1, 8)$	To find the midpoint of AB, you need to find the mean values of the x coordinates and the y coordinates.
$8 = 4 \times 1 + c$ $8 = 4 + c$ $4 = c$	Since you know l_4 goes through the point $(1, 8)$, substitute these values into the equation to find c.
So the equation is $y = 4x + 4$	

Practice

1. Write down the equation of any line parallel to the lines with these equations.

    a $y = 4x + 7$ **b** $y = 5 - 3x$ **c** $y = \frac{1}{2}x - 8$ **d** $2x + 5y = 10$

2. Work out the equation of each of the following lines.

    a A line parallel to $y = 6x - 3$ that passes through $(0, 4)$

    b A line parallel to $2y + 3x = 10$ that passes through $(0, -1)$

    c A line parallel to $y = -3x + 4$ that passes through $(4, 10)$

    d A line parallel to $3y - x = 12$ that passes through $(-9, 2)$

3. Write down the equation of any line perpendicular to the lines with these equations.

    a $y = 5x + 2$ **b** $y = \frac{1}{4}x + 3$ **c** $y = -3x + 9$ **d** $2x + y = 13$

4. Work out the equation of each of the following lines.

    a A line perpendicular to $y = 4x - 2$ that passes through $(0, 3)$

    b A line perpendicular to $y = -\frac{1}{2}x + 4$ that passes through $(0, -3)$

    c A line perpendicular to $y + 3x = 8$ that passes through $(6, -4)$

    d A line perpendicular to $5y - x + 10 = 0$ that passes through $(-1, -3)$

5. The line l_5 has equation $y + 3x = 7$ and the line l_6 has equation $9x = 12 - 3y$

    Show that l_5 and l_6 are parallel.

6 Determine whether the lines $2y + 4x - 10 = 0$ and $3y = 4 + 12x$ are perpendicular.

7 The point A has coordinates (4, 5) and the point B has coordinates (2, –7).

    a Work out the gradient of AB.

    b Work out the equation of the line passing through A and B.

    c Work out the midpoint of the line segment AB.

    d Work out the equation of the line perpendicular to AB and passing through the midpoint of AB.

8 The line l_7 passes through the points (3, –2) and (5, 2).

    The line l_8 passes through the points (1, 4) and (–5, 1).

    Determine whether the lines l_7 and l_8 are perpendicular.

9 ABCD is a rhombus.

    The coordinates of A are (4, 10).

    The equation of the line passing through the diagonal DB is $y = \frac{1}{2}x + 5$

    Work out the equation of the line passing through the diagonal AC.

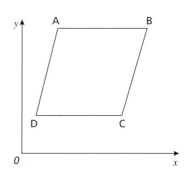

Consolidate – do you need more?

1 Work out the equation of each of the following lines.

    a A line parallel to $y = 6x - 9$ that passes through (0, 1)

    b A line parallel to $y = \frac{1}{2}x + 4$ that passes through (6, 4)

    c A line parallel to $y + 4x = 7$ that passes through (–3, 4)

    d A line parallel to $2x + 4y - 8 = 0$ that passes through (6, –2)

2 Work out the equation of each of the following lines.

    a A line perpendicular to $y = -3x + 2$ that passes through (0, 2)

    b A line perpendicular to $y = \frac{1}{5}x - 5$ that passes through (–1, 4)

    c A line perpendicular to $2y + x = 3$ that passes through (–4, –5)

    d A line perpendicular to $x - 3y + 10 = 0$ that passes through (2, –8)

3 Show that the lines $2y - 12 = 8x$ and $4y + x = 20$ are perpendicular.

4 Line l_1 has equation $y = 3x + 6$

    Line l_2 has equation $2y - 6x + 4 = 0$

    Line l_3 has equation $3y - 3x = 18$

    Line l_4 has equation $y + x = 9$

    Which two lines:

    a are parallel   **b** are perpendicular   **c** have the same y-intercept?

5 The point A has coordinates (5, 9) and the point B has coordinates (−1, 3).

   a Work out the gradient of the line segment AB.

   b Work out the equation of the line passing through A and B.

   c Work out the midpoint of AB.

   d Work out the equation of the line perpendicular to AB and passing through the midpoint of AB.

6 The line l_5 passes through the points (4, 2) and (−1, 12).

   The line l_6 passes through the points (5, −1) and (−3, −5).

   Show that lines l_5 and l_6 are perpendicular.

Stretch – can you deepen your learning?

1 The line l_1 passes through (3, 6) and (2, a).

   The line l_2 passes through (3, −2) and (1, 6).

   Lines l_1 and l_2 are parallel.

   Work out the value of a.

2 The point A has coordinates (1, 5) and the point B has coordinates (p, q).

   The line $3x + 2y = 5$ is perpendicular to the line segment AB.

   Work out an expression for q in terms of p.

3 ABC is a right-angled triangle.

   The angle ABC is 90°.

   A has coordinates (−2, 2).

   B has coordinates (6, 6).

   C has coordinates (4, a).

   Work out the value of a.

4 The line l_3 has equation $4x + 8y = 16$

   The line l_4 is perpendicular to l_3 and passes through the point A(−2, 3).

   a Show that point A also lies on l_3

   b Work out the equation of l_4

   c Work out the coordinates of the points B and C where l_3 and l_4 cross the x-axis respectively.

   d Work out the area of triangle ABC.

Are you ready?

1 Work out the gradient of each line.

a 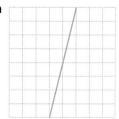 b 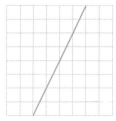 c 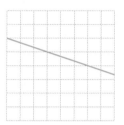 d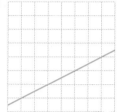

2 Work out the gradient of the line segment joining each pair of points.

   a (1, 2) and (3, 14)   **b** (6, 4) and (3, −5)

   c (2, 7) and (8, 3)   **d** (0, 9) and (7, −5)

3 Use the formula $v = u + at$ to work out:

   a the value of v when $u = 10$, $a = 3$ and $t = 4$

   b the value of u when $v = 30$, $a = 6$ and $t = 3$

   c the value of a when $v = 20$, $u = 5$ and $t = 6$

Example 1

The graph shows the cost, £C, of delivering a parcel a distance of d miles, including an initial administration charge.

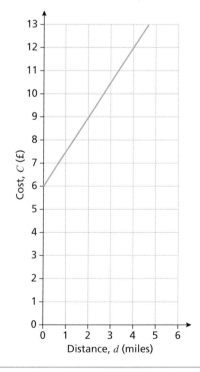

a Work out the cost of delivery for a distance of 4 miles.

b Work out the gradient of the line.

c Interpret the gradient in the context of the question.

d Write a formula for the cost, C, in terms of the distance, d.

Method

Solution	Commentary
a The answer is £12	The horizontal axis represents distance. Draw a vertical line from 4 miles until you reach the graph. Then draw a horizontal line to read the value.
b Gradient $= \dfrac{12-9}{4-2} = \dfrac{3}{2}$ The gradient is $\dfrac{3}{2}$ or 1.5	You know that gradient is $\dfrac{\text{change in } y}{\text{change in } x}$ Choose two coordinates to use, such as (2, 9) and (4, 12).
c The additional cost (£1.50) per extra mile	The gradient represents a rate of change. Here the cost is changing for each extra mile travelled.
d $y = 1.5x + 6$ $C = 1.5d + 6$	This is the same as finding the equation of the line, but with y replaced with C, and x replaced with d. You know the gradient and you can see the y-intercept is 6

Example 2

Water flows into a tank at a constant rate. The graph shows how the volume of water in the tank changes with time.

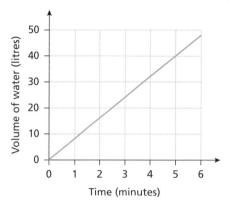

a   Work out the rate of flow in litres per minute.

b   The capacity of the tank is 2000 litres.

How long will it take to completely fill the tank at the rate found in part **a**?

Method

Solution	Commentary
a *(graph showing volume of water (litres) against time (minutes), with dashed lines indicating the point (5, 40))*	The rate of flow is the gradient of the line. Use (0, 0) and a point whose coordinates you can easily read to find the rate. The point (5, 40) is chosen because both of the coordinates are on grid lines.
Rate of flow = $\dfrac{40 \text{ litres}}{5 \text{ minutes}}$ = 8 litres per minute	Divide the number of litres by the time taken.
b 2000 ÷ 8 = 250 minutes = 4 hours 10 minutes	A rate of 8 litres per minute means that 8 litres flow into the tank each minute. Work out how many lots of 8 litres there are in 2000 litres.

Practice

1. Water is poured into an empty pond.

   The graph shows how the volume of water, in litres, in the pond changes over time.

   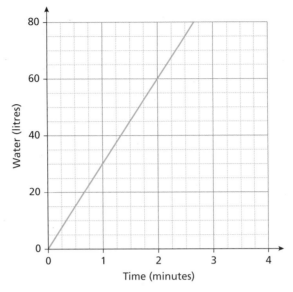

   a Work out the gradient of the graph.

   b Explain what the gradient represents in the context of the question.

2. The graph below shows the price of potatoes at a wholesale market.

   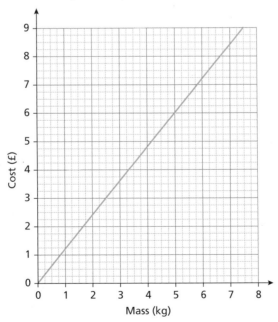

   a Use the graph to work out the cost of 5 kg of potatoes.

   b How could you use the graph to work out the cost of 24 kg of potatoes?

   c Interpret the gradient of the graph in the context of the question.

   d Write a formula for cost, C pounds, in terms of mass, m kilograms.

3 The graph shows the relationship between the cost of a taxi journey and the distance travelled.

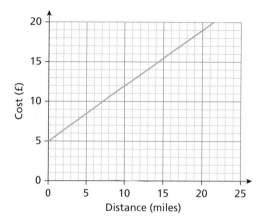

a Work out the cost of a 3-mile journey.

b Work out the cost of a 15-mile journey.

c Work out the gradient of the line and interpret it in the context of the question.

d State the y-intercept and interpret it in the context of the question.

e Write a formula for cost, C pounds, in terms of distance, d miles.

4 Quick Fix plumbing company charge a £40 call-out fee, plus £25 per hour.

Pipes Are Us plumbing company charge a £45 call-out fee, plus £20 per hour.

a Draw graphs on the same set of axes to show this information.

b Which company would be cheaper to use for a job lasting 2 hours?

c Comment on which is the cheaper plumbing company.

5 The graph shows the relationship between miles and kilometres.

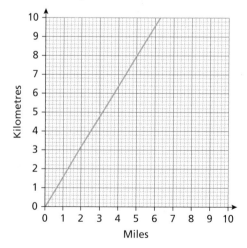

a Use the graph to convert:

   i 6 miles into km   **ii** 7 km into miles.

b Work out the gradient of the line.

c Work out the equation of the line.

6. The graph shows the cost of hiring a bike.

For the first 4 hours there is a flat fee. After this, it costs an extra amount per hour.

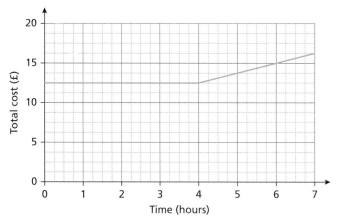

This is an example of a piecewise graph. These have more than one part, with each part represented by a different function.

Use the graph to work out:

a the cost of hiring a bike for 3 hours

b the cost of hiring a bike for 5 hours

c the extra charge per hour after the first 4 hours.

Consolidate – do you need more?

1. The graph shows the relationship between pounds (£) and dollars ($).

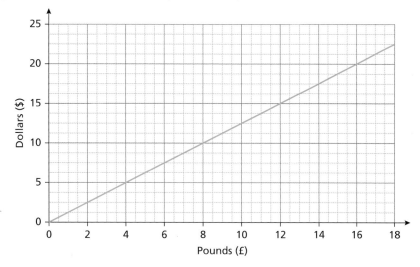

Use the graph to convert:

a £12 into dollars

b $20 into pounds.

2 Ali uses a van to deliver parcels.

For each parcel he delivers, there is a fixed charge plus £1 for each extra mile.

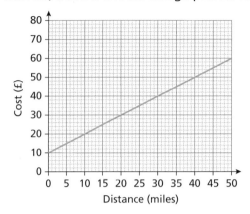

a   How much is the fixed charge?

b   Flo uses her van to deliver parcels.

There is a fixed charge of £5 plus £1.50 for each extra mile.

Draw a graph showing this information and compare the cost of having a parcel delivered by Ali with the cost of having a parcel delivered by Flo.

3 Van rental prices are made up of a fixed rate of £65 plus an extra charge of £3 per mile travelled.

a   Draw a graph to show this information.

b   Interpret the gradient and the y-intercept of the graph in the context of the question.

4 The graph shows how much Seb is paid for working up to 48 hours.

He receives a basic rate of pay for the first 35 hours and a higher rate of pay for any extra hours worked.

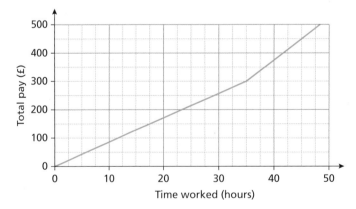

a   Seb works 40 hours in one week.

How much does he get paid?

b   Work out the difference between his basic rate of pay and his higher rate of pay.

Stretch – can you deepen your learning?

1 A bike courier delivers packages within a city centre.

She charges a different amount depending on the mass of the packages.

The graph shows how much she charges.

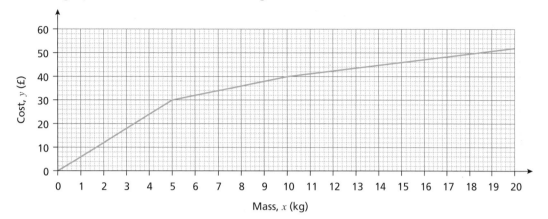

The piecewise graph can be described using equations of three straight lines for each section.

Work out the equation of each straight line.

2 A swimming pool is being drained of water.

The graph shows the volume, V, in the pool after a number of hours.

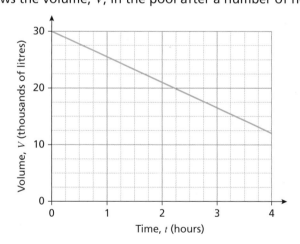

a What is the volume of water in the pool before any water is drained?

b How much water is drained in the first two hours?

c Work out the gradient of the line and explain what it means in the context of the question.

d How long will it take for the swimming pool to empty, assuming it drains at a constant rate?

Linear graphs: exam practice

1 Write down the coordinates of the point of intersection of the straight lines with equations $y = 3$ and $x = 4$ **[1 mark]**

4–6

2 Does the point (4, 9) lie on the straight line with equation $y = 2x + 1$?
Explain your answer. **[2 marks]**

3 The equation of line A is $y = 3x + 4$
Line B is parallel to line A and passes through the point (0, –2).
Work out the equation of line B. **[2 marks]**

4 Show that the lines $2y = x + 5$ and $6y - 3x = 4$ are parallel. **[2 marks]**

5 Line L passes through the points (–2, 1) and (3, 16).
Work out the equation of line L. **[3 marks]**

6 Work out the gradient and the y-intercept of the
line with equation $2y - x - 6 = 0$ **[2 marks]**

7 Work out the equation of the line.

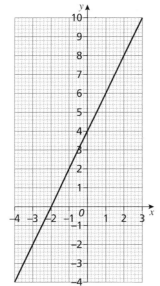

[3 marks]

8 The graph shows the volume, V, of water in a tank after a number of hours.

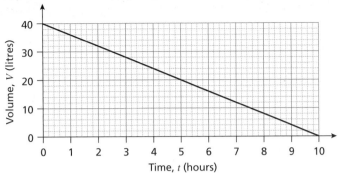

Work out the gradient of the line and explain what it means in the context of this question. **[2 marks]**

9 Work out the equation of the line perpendicular to the graph of $y = -3x + 1$ and which passes through the point $(0, -5)$. **[2 marks]**

10 Here is a sketch of a straight line.

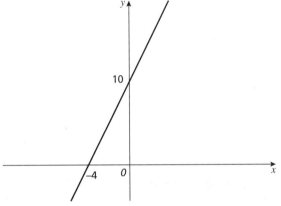

Work out the equation of the line perpendicular to the line shown that passes through the point $(5, 4)$. **[4 marks]**

10 Non-linear functions

In this block, we will cover...

10.1 Quadratic graphs

Example

a Complete the table of values for $y = x^2 - 6x +$

x	−1	0	1	2	3	4
y		4	−1		−5	−4

b Draw the graph of $y = x^2 - 6x + 4$

c Use your graph to work out the value of y wh

d Use your graph to solve the equation $x^2 - 6x +$

e State the coordinates of the turning point or

f Write down the equation of the line of symm

Method

10.2 Cubic and reciprocal graphs

Practice

1 a Copy and complete the table of values

x	−2	−1	0	1	2
y					

b Draw the graph of $y = x^3 + x^2 - 4$

c Work out the value of x for which $y =$ −

d Solve the equation $x^3 + x^2 - 4 = 0$

10.3 Exponential graphs

Consolidate – do you need more

1 a Copy and complete the table of values

x	−2	−1	0	1	2
y					

b Draw the graph of $y = 4^x$

c Use your graph to estimate the value o

d Use your graph to estimate the value o

2 a Copy and complete the table of values

10.4 Graphs of circles

Stretch – can you deepen your le

1 A circle with centre (0, 0) has a tangent wi

   a Work out the value of c.   b

2 A circle with centre at the origin has circu

   Work out the equation of the circle.

3 The diagram shows the circle with equatic

Are you ready?

1 Work out the value of each expression when $x = 2$

   a x^2 **b** $2x$ **c** $3x - 4$

   d $x^2 - 6$ **e** $x^2 + 2x$

2 Work out the value of each expression when $x = -3$

   a $3x$ **b** x^2 **c** $x^2 - 4$

   d $x^2 + x$ **e** $x^2 - 2x$

3 Solve by factorising.

   a $x^2 + 6x + 8 = 0$ **b** $x^2 + 5x - 6 = 0$ **c** $x^2 + 6x + 9 = 0$

   d $x^2 - 16x + 63 = 0$ **e** $2x^2 + 9x + 4 = 0$

Using your calculator 🖩

Remember you can use the table function on a calculator to help you to fill in a table of values. See Chapter 9.1

Quadratic graphs have equations of the form $y = ax^2 + bx + c$, where a, b and c are usually integers. The shape of the quadratic graph is called a **parabola**, which is a symmetrical, U-shaped curve.

Whether a is positive or negative will determine the shape of the graph.

If $a > 0$, the parabola will look like this: If $a < 0$, the parabola will look like this:

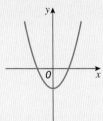   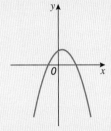

The point at which a quadratic graph changes direction is called its **turning point**. Notice that the turning point of the graph on the left is a **minimum point**. In the same way, the turning point of the graph on the right is a **maximum point**. A turning point of a quadratic graph is also called its **vertex**.

Example

a Complete the table of values for $y = x^2 - 6x + 4$

x	−1	0	1	2	3	4	5	6
y		4	−1		−5	−4		4

b Draw the graph of $y = x^2 - 6x + 4$

c Use your graph to work out the value of y when $x = 2.3$

d Use your graph to solve the equation $x^2 - 6x + 4 = 5$

e State the coordinates of the turning point or vertex of the graph.

f Write down the equation of the line of symmetry of the graph.

Method

Solution	Commentary
a $y = x^2 - 6x + 4$ When $x = -1$ $\qquad y = (-1)^2 - 6 \times -1 + 4 = 11$ When $x = 2$ $\qquad y = 2^2 - 6 \times 2 + 4 = -4$ When $x = 5$ $\qquad y = 5^2 - 6 \times 5 + 4 = -1$ <table><tr><td>x</td><td>−1</td><td>0</td><td>1</td><td>2</td><td>3</td><td>4</td><td>5</td><td>6</td></tr><tr><td>y</td><td>11</td><td>4</td><td>−1</td><td>−4</td><td>−5</td><td>−4</td><td>−1</td><td>4</td></tr></table>	You can find the values of y by substituting or by using the table function on your calculator.
b 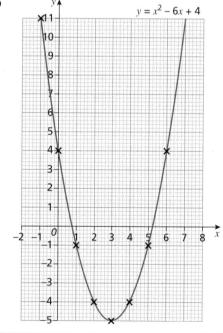	Plot the points and join them with a smooth curve.

c

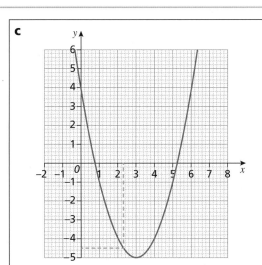

When $x = 2.3$, $y = -4.5$

Draw a vertical line where $x = 2.3$ until it meets the graph, then draw a horizontal line to the y-axis and read off the value, as accurately as you can.

You could also find the value by substituting $x = 2.3$ into the equation of the graph.

d

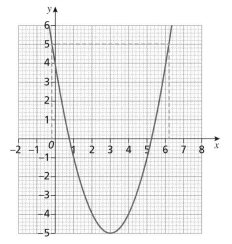

The two solutions are $x = -0.2$ and $x = 6.2$

Here, you are told that $y = 5$ so draw a horizontal line through 5 on the y-axis until it meets the graph. Note that this happens in two places.

In each case, draw a line to the x-axis to read off the values as accurately as you can.

e

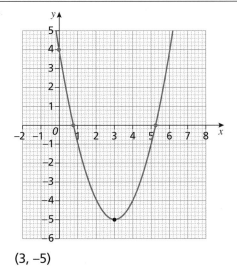

$(3, -5)$

The turning point is where the graph changes direction, in this case the minimum point of the graph.

f

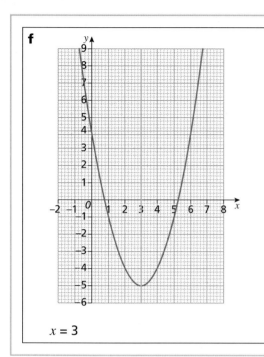

$x = 3$

The line of symmetry is vertical and will have an equation of the form $x = a$

You can see that it goes through 3 on the x-axis, so its equation is $x = 3$

Practice

1 **a** Copy and complete the table of values for $y = x^2 + 6x + 8$

x	−7	−6	−5	−4	−3	−2	−1	0	1
y									

b Draw the graph of $y = x^2 + 6x + 8$

c Using your graph, work out the value of y when $x = -2.5$. Give your answer to 2 decimal places.

d Use your graph to solve the equation $x^2 + 6x + 8 = 2$, giving your answers to 2 decimal places.

e Solve the equation $x^2 + 6x + 8 = 0$ algebraically.

How are your solutions related to the graph you have drawn?

2 Rhys has drawn the graph of $y = x^2 - 3x - 5$

What has Rhys done wrong?

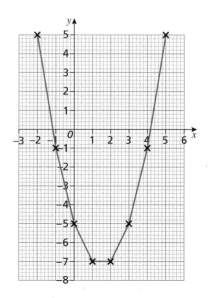

3 **a** Copy and complete the table of values for $y = x^2 - 5x + 2$

x	-2	-1	0	1	2	3	4	5	6	7
y										

b Draw the graph of $y = x^2 - 5x + 2$ Take care with the turning point.

c Using your graph, work out the possible x values when $y = 1$, giving your answers to 1 decimal place.

d Work out the coordinates of the turning point of the graph.

e Write down the equation of the line of symmetry of the graph.

4 **a** Copy and complete the table of values for $y = -x^2 + 3x + 4$

x	-2	-1	0	1	2	3	4	5
y								

b Draw the graph of $y = -x^2 + 3x + 4$

c Using your graph, work out the value of y when $x = 1.5$, correct to 2 decimal places.

d Where does your graph cross the y-axis?

How is this related to the equation of the graph?

e Work out the coordinates of the turning point of the graph.

f Solve the equation $-x^2 + 3x + 4 = 2$, giving your answers to 2 decimal places.

g Ed says that equations of the form $-x^2 + 3x + 4 = k$ (where k is a constant) always have two solutions.

Do you agree? Explain your answer.

5 Match each graph to its equation.

1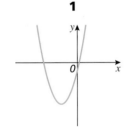

2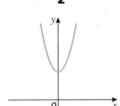

3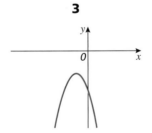

4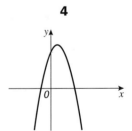

$y = -x^2 - 3x - 5$

$y = x^2 + 4$

$y = -x^2 + 2x + 5$

$y = x^2 + 4x - 1$

6 Draw each pair of graphs on the same set of axes and identify their points of intersection.

a $y = x^2 + 5x$ and $y = 2x + 10$

b $y = x^2 - 4$ and $y = 3x$

199

Consolidate – do you need more?

1 **a** Copy and complete the table of values for $y = x^2 - 3x - 6$

x	-2	-1	0	1	2	3	4	5
y								

b Draw the graph of $y = x^2 - 3x - 6$ Think carefully about the turning point.

c Using your graph, work out the value of y when $x = 2.4$, giving your answer to 1 decimal place.

d Give the coordinates of the turning point of the graph.

e Use your graph to solve the equation $x^2 - 3x - 6 = 0$, giving your answers to 1 decimal place.

2 **a** Copy and complete the table of values for $y = -x^2 + 5x - 1$

x	-1	0	1	2	3	4	5	6
y								

b Draw the graph of $y = -x^2 + 5x - 1$

c Using your graph, solve the equation $-x^2 + 5x - 1 = 3.5$, correct to 1 decimal place.

d For what value of k does $-x^2 + 5x - 1 = k$ have only one solution?

3 **a** Copy and complete the table of values for $y = 2x^2 + 8x - 1$

x	-5	-4	-3	-2	-1	0	1
y							

b Draw the graph and state the coordinates of the turning point.

c State the equation of the line of symmetry of the graph.

d Solve the equation $2x^2 + 8x - 1 = 0$, correct to 1 decimal place.

4 Draw the graphs of $y = 2x^2 + x - 3$ and $y = 3x + 1$ on the same set of axes.

State the coordinates of the points of intersection of the two graphs.

Stretch – can you deepen your learning?

1 Flo draws the graph of $y = -x^2 + 5x + 6$

Explain what Flo has done wrong.

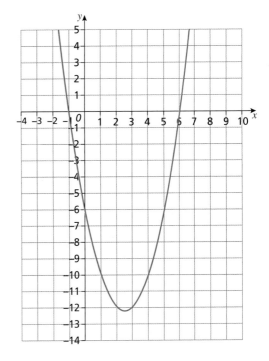

2 Here is the graph of $y = ax^2 + bx + c$

Work out the values of a, b and c.

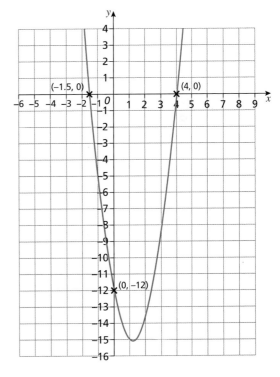

3 How could you use the graph of $y = x^2 - 3x + 2$ to solve the following equations?

You don't need to solve the equations.

a $x^2 - 3x + 1 = 0$ **b** $x^2 - 3x = 0$

c $x^2 - 4x + 2 = 0$ **d** $x^2 - x - 1 = 0$

Are you ready?

1 Work out the value of each expression when $x = 2$

   a x^2 **b** $x^3 + 2$ **c** $x^3 + 2x - 3$ **d** $x^3 + 2x^2 + 7$

2 Work out the value of each expression when $x = -3$

   a $2x^2$ **b** $x^3 - 4$ **c** $x^3 + 2x - 7$ **d** $x^3 - 2x^2 + 5$

3 Work out the value of $\dfrac{1}{x}$ when:

   a $x = 3$ **b** $x = 2$ **c** $x = -7$

   d $x = 0.5$ **e** $x = \dfrac{1}{4}$

You have already met **quadratic** functions, in which the greatest power of x is x^2

Cubic functions are functions where the highest power of x is x^3

The general form of a cubic function is $y = ax^3 + bx^2 + cx + d$

Here are some examples of cubic graphs:

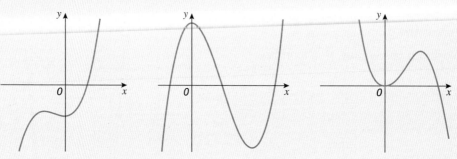

Whether the coefficient of the x^3 term is positive or negative will determine the shape of the graph:

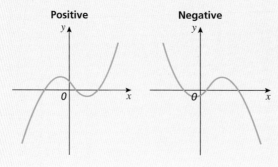

Positive   Negative

Most cubic functions have two turning points, but $y = x^3$, $y = -x^3$ and other functions of the form $y = \pm x^3 + k$ just have one.

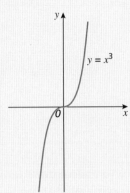

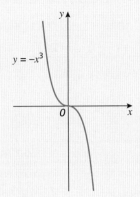

Reciprocal functions are of the form $y = \dfrac{k}{x}$

Here are the graphs of three reciprocal functions:

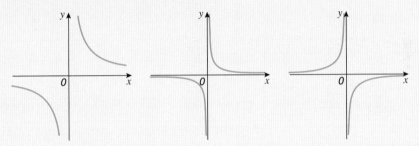

The graph doesn't touch either of the axes. It just gets closer and closer to them.

Whether k is positive or negative will determine which two quadrants contain the sections of the graph.

k positive

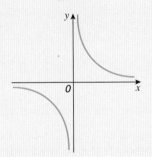

k negative

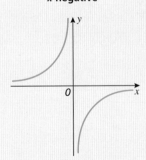

Example 1

a Complete the table of values for $y = x^3 + 4x^2 - 3$

x	-2	-1	0	1	2
y			-1		21

b Draw the graph of $y = x^3 + 4x^2 - 3$

c Use your graph to estimate the values of x for which $y = 3$

Method

Solution	Commentary
a $x = -2$ $y = (-2)^3 + 4 \times (-2)^2 - 3$ $= -8 + 16 - 3 = 5$ $x = -1$ $y = (-1)^3 + 4 \times (-1)^2 - 3$ $= -1 + 4 - 3 = 0$ $x = 1$ $y = (1)^3 + 4 \times 1^2 - 3$ $= 1 + 4 - 3 = 2$	You can use your calculator's table function or just work out the remaining three values by substitution.
b 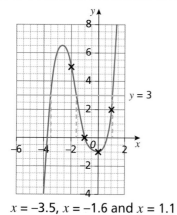	Plot the points and join with a smooth curve.
c $x = -3.5$, $x = -1.6$ and $x = 1.1$ (to 1 d.p.)	Draw the line $y = 3$ and see where it intersects the cubic graph. Use the graph to read off the corresponding x values.

Example 2

a Complete the table of values for $y = \dfrac{1}{x}$

x	−2	−1	−0.5	−0.25	0.25	0.5	1	2	3	4
y	$-\dfrac{1}{2}$	−1		−4		2	1	$\dfrac{1}{2}$		

b Draw the graph of $y = \dfrac{1}{x}$

c What happens to y as x gets greater?

d Work out the value of x for which $y = 2.5$

Method

Solution	Commentary
a $\quad x = -0.5 \qquad y = 1 \div -0.5 = -2$ $\quad x = 0.25 \qquad y = 1 \div 0.25 = 4$ $\quad x = 3 \qquad y = \dfrac{1}{3}$ $\quad x = 4 \qquad y = \dfrac{1}{4}$	You can use your calculator's table function to find the remaining four values or just work them out by substitution.
b	Plot the points and join with a smooth curve.
c As x gets greater, y gets smaller, getting closer and closer to 0	You can see from the graph that as x gets greater, the value of y decreases.
d $x = 0.4$	Draw a horizontal line at $y = 2.5$ until it meets the curve. Then draw a vertical line to the x-axis and read the value as accurately as you can.

Practice

1 **a** Copy and complete the table of values for $y = x^3 + x^2 - 4$

x	−2	−1	0	1	2
y					

b Draw the graph of $y = x^3 + x^2 - 4$

c Work out the value of x for which $y = -3$

d Solve the equation $x^3 + x^2 - 4 = 0$

2 a Copy and complete the table of values for $y = x^3 + 2x^2 - 3x - 4$

x	-3	-2	-1	0	1	2
y						

b Draw the graph of $y = x^3 + 2x^2 - 3x - 4$

c Work out the values of x for which $y = 2$

3 a Copy and complete the table of values for $y = \dfrac{2}{x}$

x	-2	-1	-0.5	0.5	1	2
y						

b Draw the graph of $y = \dfrac{2}{x}$

c Work out the value of x for which $y = 1.5$

d Draw the graph of $y = x + 1$ on the same set of axes.

e Give the coordinates of the points of intersection of the two graphs.

4 a Copy and complete the table of values for $y = -\dfrac{3}{x}$

x	-2	-1	-0.5	0.5	1	2
y						

b Draw the graph of $y = -\dfrac{3}{x}$

c What is the difference between the graph of $y = -\dfrac{3}{x}$ and $y = \dfrac{3}{x}$?

5 Match each equation to the correct graph.

$y = -\dfrac{1}{x}$	$y = x^3$	$y = 3x - 5$	$y = 5 - 3x$	$y = x^3 - 9x^2 + 23x - 15$

1

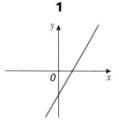

2

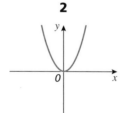

3

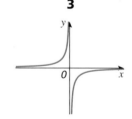

4

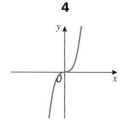

5

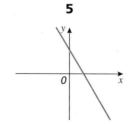

6

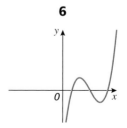

What do you think? 💭

1 Use a graph plotter to investigate the graphs of $y = \dfrac{1}{x + a}$ and $y = \dfrac{1}{x} + a$ for different values of a.

What is the same and what is different?

Consolidate – do you need more?

1 **a** Copy and complete the table of values for $y = 3x^3 - 2x^2 + 3$

x	−1	−0.5	0	0.5	1
y					

b Draw the graph of $y = 3x^3 - 2x^2 + 3$

c Use your graph to work out approximate solutions to the equation $3x^3 - 2x^2 + 3 = 0$

d Use your graph to work out approximate solutions to the equation $3x^3 - 2x^2 + 3 = 2.5$

2 **a** Copy and complete the table of values for $y = x^3 + 5x^2 - 4$

x	−5	−4	−3	−2	−1	0	1
y							

b Draw the graph of $y = x^3 + 5x^2 - 4$

c Estimate the values of x for which $y = 5$

d Benji says that there will always be three values of x for each value of y.

Is he correct? Explain your answer.

3 **a** Copy and complete the table of values for $y = \dfrac{3}{x}$

x	−2	−1	−0.5	−0.25	0.25	0.5	1	2
y								

b Draw the graph of $y = \dfrac{3}{x}$

c State the value of x for which $y = 2$

d On the same axes, draw the line $y = 2x$

e State the coordinates of the points of intersection of the two graphs.

4 Match each graph to the correct equation.

1	**2**	**3**	**4**
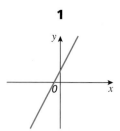	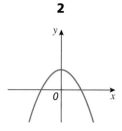	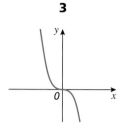	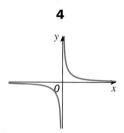

$y = \dfrac{1}{x}$ $y = -x^3$ $y = 2x + 4$ $y = -x^2 + 6$

Stretch – can you deepen your learning?

1 **a** Draw the graph of $y = \dfrac{1}{x^2}$

  b In what way is the graph different to the graph of $y = \dfrac{1}{x}$?

2 **a** Draw the graph of $y = x^3$

  b Given $f(x) = x^3$, find $f^{-1}(x)$

  c On the same set of axes, draw the graph of $y = f^{-1}(x)$

  d What is the relationship between the two graphs?

3 **a** Draw the graph of $y = \dfrac{1}{x}$

  b Given $f(x) = \dfrac{1}{x}$, work out $f^{-1}(x)$

  c What happens? What does the graph of $y = f^{-1}(x)$ look like?

4 The graph of $y = \dfrac{1}{x}$ has two lines of symmetry.
Write down the equations of these lines.

Are you ready?

1 Given that $x = 4$, work out the value of:

    a x^3 **b** 2^x **c** 3^x **d** 0.5^x

2 Given that $x = -2$, work out the value of:

    a 2^x **b** 3^x **c** 0.5^x **d** 0.2^x

3 Evaluate:

    a 3^0 **b** 5^0 **c** 3^{-2}

    d 4^{-1} **e** 5^{-3}

Graphs that have equations of the form $y = k^x$ (where k is a positive number) are called **exponential graphs**.

When $k > 1$, this is **exponential growth**; the value of y increases more rapidly as x increases.

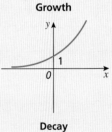
Growth

When $k < 1$, this is **exponential decay**; the value of y decreases more rapidly as x increases.

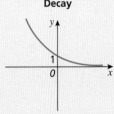
Decay

In both cases, the graph $y = k^x$ passes through the point (0, 1) on the y-axis. This is because $k^0 = 1$ for all values of k.

Note that neither type of graph crosses the x-axis because all values of y are positive.

Example

a Complete the table of values for $y = 3^x$

x	−2	−1	0	1	2	3	4
y		$\frac{1}{3}$			9	27	81

b Draw the graph of $y = 3^x$

c Use your graph to estimate the value of x when $y = 40$

Method

Solution	Commentary
a $x = -2$ $y = 3^{-2} = \dfrac{1}{9}$	To find the missing values, you can substitute the x value into the formula $y = 3^x$ or use the table function on your calculator.
$x = 0$ $y = 3^0 = 1$	
$x = 1$ $y = 3^1 = 3$	
b 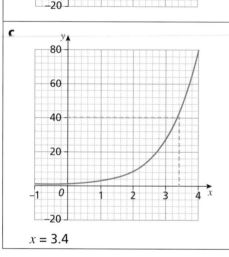	Plot the points in the table and join them up with a smooth curve.
c $x = 3.4$	Draw a horizontal line where $y = 40$ until it reaches the graph. Then draw a vertical line to the x-axis and read off the value as accurately as you can.

Practice

 a   Copy and complete the table of values for $y = 2^x$

x	−2	−1	0	1	2	3	4	5
y								

b   Draw the graph of $y = 2^x$

c   Use your graph to estimate the value of x for which $y = 5$

d   Use your graph to estimate the value of x for which $y = 0.6$

2 **a** Copy and complete the table of values for $y = \left(\frac{1}{2}\right)^x$

x	−4	−3	−2	−1	0	1	2	3
y								

b Draw the graph of $y = \left(\frac{1}{2}\right)^x$

c Use your graph to solve the equation $\left(\frac{1}{2}\right)^x = 7$

d Use your graph to solve the equation $\left(\frac{1}{2}\right)^x = 0.4$

3 The graph shows the growth of a population of rabbits.

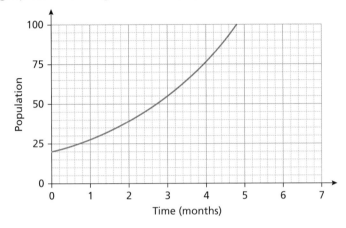

a What size is the initial population?

b How many rabbits are there after 3 months?

c Explain why this model may be unrealistic.

4 The graph shows the concentration of a medicine in a person's body.

a What is the initial amount of medicine in the body?

b What is the amount of medicine in the body after 5 hours?

c How long does it take for the amount of medicine in the body to halve?

d Explain why this model may be unrealistic.

5 The graph $y = ax^b$ passes through the points (1, 4) and (3, 108).

Work out the values of a and b.

6 Ed draws the graph of $y = 4^x$

Explain how you know the graph is incorrect.

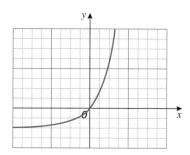

Consolidate – do you need more?

1 a Copy and complete the table of values for $y = 4^x$

x	−2	−1	0	1	2	3
y						

b Draw the graph of $y = 4^x$

c Use your graph to estimate the value of x for which $y = 7$

d Use your graph to estimate the value of x for which $y = 0.9$

2 a Copy and complete the table of values for $y = \left(\dfrac{1}{3}\right)^x$

x	−3	−2	−1	0	1	2	3
y							

b Draw the graph of $y = \left(\dfrac{1}{3}\right)^x$

c Use your graph to solve the equation $\left(\dfrac{1}{3}\right)^x = 7$

d Use your graph to solve the equation $\left(\dfrac{1}{3}\right)^x = 0.4$

3 The graph shows the growth of a type of bacteria over a period of time.

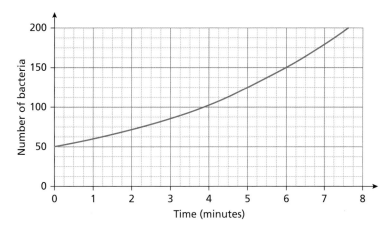

a How many bacteria were there initially?

b How many bacteria were there after 4 minutes?

c How long does it take for the population of bacteria to double?

d Why might this be an unrealistic model?

4 The graph shows the change in the value of a car.

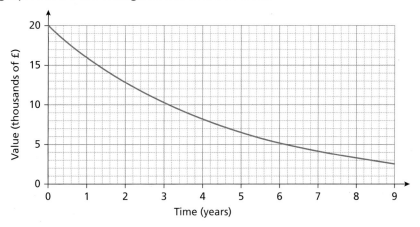

a What is the initial value of the car?

b What is the value of the car after 4 years?

c How long does it take for the value of the car to fall to one-quarter of its original value?

Stretch – can you deepen your learning?

1 The graph shows the growth of an investment over a period of time.

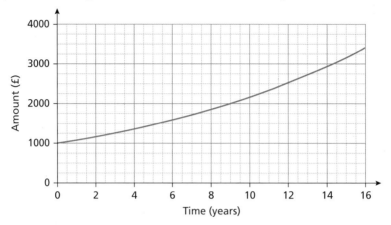

a Approximately how long does it take for the investment to double?

b What is the percentage growth of the investment after 12 years?

c What is the percentage growth of the investment from year 9 to year 12?

d The graph goes through the points (1, 1080) and (0, 1000).

Use this to work out the equation of the graph in the form $y = ab^x$

e Interpret the values of a and b in the context of the question.

2 Match each graph to its equation.

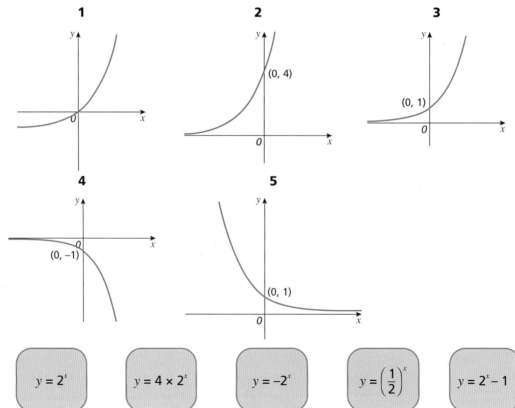

$$y = 2^x \qquad y = 4 \times 2^x \qquad y = -2^x \qquad y = \left(\frac{1}{2}\right)^x \qquad y = 2^x - 1$$

Are you ready?

1 Work out the gradient of the line segments joining the points with these coordinates.

    a (3, 4) and (–2, 14)   **b** (5, –9) and (3, –1)   **c** (4, 4) and (7, –2)

2 Work out the gradient of a line perpendicular to the straight lines with these equations.

    a $y = 3x + 4$ **b** $y = -2x + 6$ **c** $y = \frac{1}{3}x - 4$ **d** $y = -\frac{1}{5}x - 11$

3 Work out the equation of a line with:

    a gradient 4, passing through the point (5, 1)

    b gradient –3, passing through (4, –1).

Here is a circle with centre at (0, 0) and radius 5

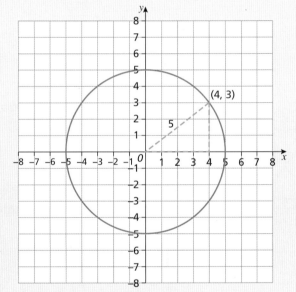

You can see for the point with coordinates (4, 3) that Pythagoras' theorem gives $4^2 + 3^2 = 5^2$

For any point (x, y) on the circle, $x^2 + y^2 = 25$

The general equation of a circle with centre at (0, 0) and radius r is given by $x^2 + y^2 = r^2$

In this chapter, you will use your knowledge of the circle theorem that states that the radius is always perpendicular to a tangent to the circle at a given point. This will allow you to work out the equation of the tangent to a circle. This theorem is studied in detail in *Collins White Rose Maths AQA GCSE 9–1 Higher Student Book 2* (Chapter 3.3).

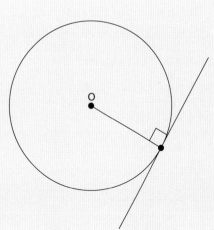

Example 1

The equation of a circle is $x^2 + y^2 = 9$

a State the coordinates of the centre of the circle.

b Work out the length of the radius of the circle.

c Does the point (1, 2) lie on the circle?

Method

Solution	Commentary
a Centre is at (0, 0)	Remember that $x^2 + y^2 = r^2$ is the equation of a circle centred at the origin.
b $r^2 = 9$ so $r = 3$	Use $r^2 = 9$ to find the radius.
c $x^2 + y^2 = 1^2 + 2^2 = 1 + 4 = 5$ $5 \neq 9$, so the point doesn't lie on the circle.	To check if this point is on the circle, substitute $x = 1$ and $y = 2$ into the equation.

Example 2

Work out the equation of the tangent to the circle $x^2 + y^2 = 25$ at the point (4, 3).

Method

Solution	Commentary
   The gradient of the radius is $\dfrac{3}{4}$ So the gradient of the tangent is $-\dfrac{4}{3}$	To find the equation of a straight line you need the gradient. First find the gradient of the radius from (0, 0) to (4, 3). The tangent will be perpendicular to the radius, so its gradient will be the negative reciprocal of $\dfrac{3}{4}$ The relationship between the gradient of perpendicular lines is covered in Chapter 9.2
Equation of tangent: $y = -\dfrac{4}{3}x + c$	Now you know the gradient, you can use the general form of the equation of a straight line $y = mx + c$.
$3 = -\dfrac{4}{3} \times 4 + c$ $3 = -\dfrac{16}{3} + c$ $\dfrac{25}{3} = c$	You know the line passes through (4, 3) so you can substitute in $x = 4$ and $y = 3$ to find the value of c.
So the equation is $y = -\dfrac{4}{3}x + \dfrac{25}{3}$	

Practice

1 State the equation of each circle.

a

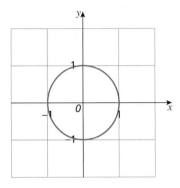

b

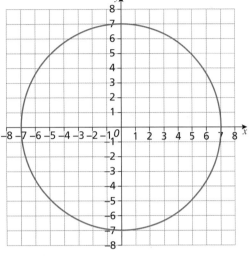

c

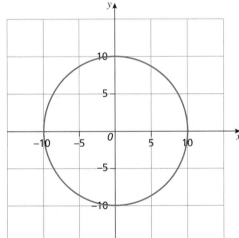

2 Sketch each circle, labelling the points where they meet the x-axis and the y-axis.

a $x^2 + y^2 = 16$ **b** $x^2 + y^2 = 64$

3 **a** State the equation of a circle with radius 10 and centre at the origin.

b For each point, determine whether the point lies **inside** the circle, **on** the circle or **outside** the circle.

i (4, 4)   **ii** (6, 8)   **iii** (−4, 8)   **iv** (−5, 7)   **v** (−9, −6)

4 A circle has equation $x^2 + y^2 = 16$

a State the coordinates of the centre of the circle.

b State the length of the radius of the circle.

c Does the point (4, 2) lie on the circle?

Justify your answer.

5 A circle has equation $x^2 + y^2 = 40$

    a State the length of the radius, giving your answer in surd form.

    b The point P with coordinates $(2, k)$ lies on the circle.

       Work out two possible values for k.

6 The point $(0, -9)$ lies on a circle with centre at $(0, 0)$.

   Give the equation of the circle.

7 The point $(5, 3)$ lies on a circle with centre at $(0, 0)$.

   Give the equation of the circle.

8 **a** Show that the point A$(4, -2)$ lies on the circle with equation $x^2 + y^2 = 20$

    b Work out the gradient of the line segment between $(0, 0)$ and A.

    c Work out the gradient of the tangent to the circle at A.

    d Work out the equation of the tangent to the circle at A.

9 Work out the equation of the tangent to each circle at the given point.

    a $x^2 + y^2 = 25$ at $(4, -3)$ **b** $x^2 + y^2 = 40$ at $(6, 2)$ **c** $x^2 + y^2 = 68$ at $(-2, 8)$

What do you think? 💭

1 Use Pythagoras' theorem to work out the equation of a circle with centre (h, k).

   Check your answer by drawing circles of different radii and with different centres using dynamic geometry software.

Consolidate – do you need more?

1 State the equation of each circle.

a

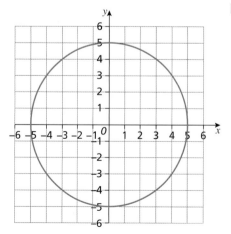

b

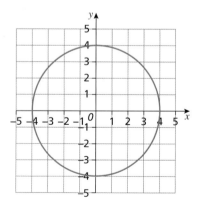

2 Sketch the circle with equation $x^2 + y^2 = 144$, labelling the points where it meets the x-axis and the y-axis.

3 A circle has equation $x^2 + y^2 = 20$

    a State the coordinates of the centre of the circle.

    b Work out the length of the radius of the circle, giving your answer in surd form.

    c Does the point $(4, -2)$ lie on the circle?

4 A circle has equation $x^2 + y^2 = 100$

    a Work out the length of the radius of the circle.

    b The point P with coordinates $(m, 6)$ lies on the circle.

    Work out two possible values for m.

5 Work out the equation of the tangent to the circle $x^2 + y^2 = 72$ at the point $(6, 6)$.

6 Work out the equation of the tangent to the circle $x^2 + y^2 = 10$ at the point $(1, -3)$.

Stretch – can you deepen your learning?

1 A circle with centre $(0, 0)$ has a tangent with equation $y = -1.5x + c$ at the point $(12, 8)$.

    a Work out the value of c.   **b** Work out the equation of the circle.

2 A circle with centre at the origin has circumference 12π.

    Work out the equation of the circle.

3 The diagram shows the circle with equation $x^2 + y^2 = 17$

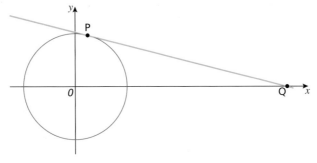

    P lies on the circle and has coordinates $(1, k)$.

    The tangent to the circle at P intersects the x-axis at Q.

    Work out the coordinates of Q.

4 A circle has equation $x^2 + y^2 = 25$

    The line L is a tangent to the circle at the point $A(-3, 4)$.

    The line L crosses the x-axis at the point P and crosses the y-axis at the point Q.

    a Work out the area of triangle OAP.

    b Work out the length of the line segment PQ.

Non-linear functions: exam practice

1 (a) Complete a copy of the table of values for $y = 2x^2 - 4x - 1$ **[1 mark]**

x	-2	-1	0	1	2	3	4
$y = 2x^2 - 4x - 1$	15		-1				15

(b) Draw the graph of $y = 2x^2 - 4x - 1$ for values of x from -2 to 4   **[2 marks]**

2 Sketch graphs of the following using copies of the axes shown.

(a) $y = -x^2$ **(b)** $y = x^3$

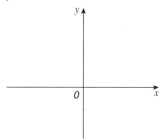

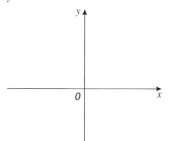

[2 marks]

3 Sketch graphs of the following using copies of the axes shown.

(a) $y = 2^x$ **(b)** $y = x^{-1}$

[2 marks]

4 Here is the graph of $y = -2x^2 + 5x + 17$

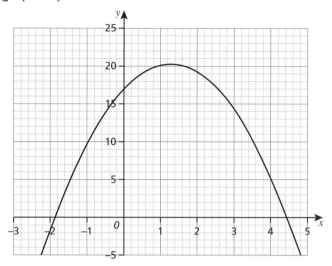

Use the graph to estimate the solutions to the equation $-2x^2 + 5x + 14 = 0$ **[2 marks]**

5 Write down the radius of a circle with equation $x^2 + y^2 = 16$ **[1 mark]**

6 P is the point (3, 1) on the circle with equation $x^2 + y^2 = 10$

Work out the equation of the tangent to the circle at P. **[4 marks]**

7–9

In this block, we will cover...

11.1 Solving by factorisation

Example 1

Here is a rectangle.

The area of the rectangle is 16 cm².

Work out the dimensions of the rectangle.

$(x - 2)$

Method

Solution	Comme
Area of rectangle = 16 cm²	Form a
Area of rectangle = $(x + 4)(x - 2)$	using a
Therefore $(x + 4)(x - 2) = 16$	Equate

11.2 Completing the square

Practice

1 Which of these expressions are perfect squ
   A $x^2 + 6x + 9$ B $x^2 + 6x - 9$

2 Write each expression in the form $(x + a)^2$
   a $x^2 + 8x - 8$ b $x^2 + 10x - 3$
   e $x^2 + 5x - 1$ f $x^2 - 7x + 2$

3 Solve the equations, giving your answers i
   a $x^2 + 4x + 1 = 0$ b $x^2 + 2x - 7 = 0$
   d $x^2 + 12x + 25 = 0$ e $x^2 - 2x - 5 = 0$

11.3 Using quadratic graphs

Consolidate – do you need more

1 Work out the coordinates of the points wh
   a $y = x^2 - 9$ b $y = x^2 - 13x + 40$

2 Work out the coordinates of the turning p
   a $y = x^2 + 8x - 13$ b $y = x^2 - 2x + 3$

3 Sketch the graphs of these equations, clea
   coordinates of the turning point.
   a $y = x^2 + 12x + 11$ b $y = x^2 - 6x - 27$

11.4 The quadratic formula

Stretch – can you deepen your le

1 Ali solves a quadratic equation of the form
   $$x = \frac{-6 \pm \sqrt{3}}{2}$$
   Work out the values of a, b and c for the e

2 A rectangular sheet of paper is 4 cm longe
   Beca cuts out the largest circle she can fro
   The area of the leftover paper is 35 cm².

   What are the dimensions of the paper?

Are you ready? (A)

1 Solve:

    a $x + 3 = 0$ **b** $x - 5 = 0$ **c** $x - 2 = 0$

    d $2x + 1 = 0$ **e** $3x - 2 = 0$

2 Factorise:

    a $x^2 + 8x + 15$ **b** $x^2 - 5x - 36$ **c** $x^2 + 5x - 14$ **d** $x^2 - 10x + 21$

3 Factorise:

    a $x^2 - 16$ **b** $x^2 - 25$ **c** $4x^2 - 1$ **d** $4x^2 - 9$

4 Factorise:

    a $2x^2 - 7x - 4$ **b** $2x^2 + 5x - 3$ **c** $3x^2 + 19x - 14$ **d** $2x^2 - x - 10$

Quadratic equations can be written in the form $ax^2 + bx + c = 0$, where a, b and c are integers.

An example is $2x^2 + 5x - 3 = 0$

They can have two, one or zero **solutions**.

For example, $x = -3$, $x = \frac{1}{2}$ are the solutions to $2x^2 + 5x - 3 = 0$

These solutions are the values of x that make the equation true. They are sometimes referred to as the **roots** of the equation.

One method of solving these equations is by factorising the quadratic expression. You learnt how to do this in Chapter 6.3

Example

a Solve $x^2 - 8x + 15 = 0$ **b** Solve $2x^2 - 3x = 2$

Method

Solution	Commentary
a $x^2 - 8x + 15 = 0$ $(x - 5)(x - 3) = 0$	First factorise the quadratic expression.
So either $x - 5 = 0$ or $x - 3 = 0$	If the product of two brackets is equal to zero, then one of the brackets must equal zero.
When $x - 5 = 0$, $x = 5$ When $x - 3 = 0$, $x = 3$ so $x = 5$ or $x = 3$	Solve the two equations to find the values of x.

b $2x^2 - 3x = 2$ $2x^2 - 3x - 2 = 0$	To solve a quadratic equation, first rearrange it so that the quadratic expression is equal to zero. Here you need to subtract 2 from both sides.
$(2x + 1)(x - 2) = 0$	Factorise the quadratic expression.
$2x + 1 = 0$ or $x - 2 = 0$	If the product of two brackets is equal to zero, then one of the brackets must equal zero.
$2x + 1 = 0$ $2x = -1$ $x = -\dfrac{1}{2}$	Solve the two equations to find the values of x. These are the solutions of the equation.
$x - 2 = 0$ $x = 2$	
So $x = -\dfrac{1}{2}$ or $x = 2$	

Practice (A)

1 Solve by factorising.

a $x^2 + 9x + 20 = 0$ b $x^2 + 11x + 18 = 0$ c $x^2 - 4x - 21 = 0$

d $x^2 - 5x - 66 = 0$ e $x^2 - 5x + 6 = 0$ f $x^2 - 12x = -32$

g $x^2 = 6x + 7$ h $x^2 - 30 = -7x$

2 Solve by factorising.

a $x^2 - 49 = 0$ b $x^2 - 81 = 0$

c $x^2 = 169$ d $4x^2 - 1 = 0$

e $9x^2 - 16 = 0$ f $16x^2 = 25$

> You need to remember how to factorise an expression that is the difference of two squares. See Chapter 6.3

3 Solve by factorising.

a $2x^2 + 5x + 3 = 0$ b $2x^2 + 9x + 4 = 0$ c $2x^2 - 5x - 42 = 0$

d $2x^2 - 13x = -18$ e $3x^2 = 2x + 1$ f $3x^2 + 13x + 14 = 0$

g $3x^2 = 19x + 40$ h $4x^2 + 16x + 15 = 0$

4 Solve by factorising.

a $x^2 + 8x + 7 = 0$ b $x^2 + 12x + 27 = 0$ c $x^2 = 100$

d $2x^2 + 9x = 5$ e $x^2 + 16 = 10x$ f $2x^2 - 9x = 18$

5 Show algebraically that each equation has only one solution.

a $x^2 - 14x + 49 = 0$ b $x^2 + 6x + 9 = 0$ c $4x^2 - 20x + 25 = 0$

Are you ready? (B)

1 Factorise:

    a $x^2 + 6x + 5$ **b** $x^2 - 4$ **c** $x^2 + x - 12$ **d** $x^2 - 13x + 42$

2 Factorise:

    a $2x^2 + 7x + 3$ **b** $2x^2 - 9x + 10$ **c** $2x^2 - 7x - 49$ **d** $3x^2 - 17x - 56$

3 Write as a single fraction.

    a $\dfrac{1}{x+3} + \dfrac{2}{x-4}$ **b** $\dfrac{3}{x+2} + \dfrac{4}{x-7}$

    c $\dfrac{5}{x+1} - \dfrac{6}{x+2}$ **d** $\dfrac{7}{x-9} - \dfrac{3}{x-4}$

Example 1

Here is a rectangle.

The area of the rectangle is 16 cm².

Work out the dimensions of the rectangle.

$(x-2)$ cm

$(x+4)$ cm

Method

Solution	Commentary
Area of rectangle = 16 cm² Area of rectangle = $(x+4)(x-2)$ Therefore $(x+4)(x-2) = 16$	Form an equation for the area of the rectangle using area = length × width Equate your expression to the given area, 16
$x^2 + 2x - 8 = 16$	Expand the brackets.
$x^2 + 2x - 24 = 0$	Subtract 16 from both sides to make the equation equal to zero.
$(x-4)(x+6) = 0$	Factorise the quadratic expression.
When $x - 4 = 0$, $x = 4$	Set each bracket equal to zero and solve.
When $x + 6 = 0$, $x = -6$ So $x = 4$	You can discard −6 in this case because you are finding a length, and negative lengths are not possible.
Length: $x + 4 = 4 + 4 = 8$ cm Width: $x - 2 = 4 - 2 = 2$ cm	Substitute $x = 4$ to find the width and length.

Example 2

Solve the equation $\dfrac{4}{x+1} + \dfrac{5}{x+2} = 2$

Method

Solution	Commentary
$\dfrac{4}{x+1} + \dfrac{5}{x+2} = 2$ $\dfrac{4(x+2)}{(x+1)(x+2)} + \dfrac{5(x+1)}{(x+1)(x+2)} = 2$	First simplify the left-hand side by adding the fractions using a common denominator. Here the common denominator is $(x+1)(x+2)$.
$\dfrac{4x+8}{(x+1)(x+2)} + \dfrac{5x+5}{(x+1)(x+2)} = 2$	Expand and simplify the numerator.
$\dfrac{4x+8+5x+5}{(x+1)(x+2)} = 2$ $\dfrac{9x+13}{(x+1)(x+2)} = 2$	Write as a single fraction and simplify the numerator by collecting like terms.
$9x+13 = 2(x+1)(x+2)$	Multiply both sides by $(x+1)(x+2)$ to remove the fraction.
$9x+13 = 2(x^2+3x+2)$ $9x+13 = 2x^2+6x+4$	Expand and simplify the right-hand side.
$0 = 2x^2-3x-9$	Rearrange to make the equation equal to zero. Here, subtract $9x$ and 13 from both sides.
$(2x+3)(x-3) = 0$	Now factorise the quadratic expression.
$2x+3 = 0$ so $x = -\dfrac{3}{2}$ Or $x-3 = 0$ so $x = 3$ So the solutions are $x = -\dfrac{3}{2}$, $x = 3$	Set each bracket equal to zero and solve.

Practice (B)

1 A rectangle is shown.

The rectangle has area $21\,\text{cm}^2$.

Form and solve an equation to work out the dimensions of the rectangle.

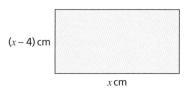

$(x-4)\,\text{cm}$

$x\,\text{cm}$

2 The trapezium has area $36\,\text{cm}^2$.

a Show that $x^2 - x - 56 = 0$

b Work out the value of x.

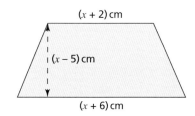

$(x+2)\,\text{cm}$

$(x-5)\,\text{cm}$

$(x+6)\,\text{cm}$

3 ABC is a right-angled triangle.

Work out the value of n.

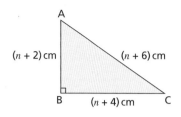

4 Seb is x years old. Faith is 8 years older than Seb.

The product of their ages is 84

a Set up an equation to show this information.

b Work out Faith's age.

5 A rectangular lawn measures 8 m by 4 m.

It is surrounded on all sides by a path that is x m wide.

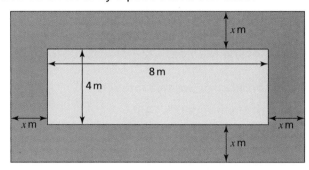

The area of the path is 64 m².

Work out the value of x.

6 Solve the equations.

a $\dfrac{3}{x+2} + \dfrac{4}{x-3} = 2$

b $\dfrac{3}{2x-1} - \dfrac{4}{3x-1} = 1$

c $\dfrac{7}{x+1} - \dfrac{4}{3x-2} = 1$

d $\dfrac{8}{x+3} + \dfrac{3}{x+8} = 1$

e $\dfrac{7}{x+1} + \dfrac{4}{x+4} = 3$

f $\dfrac{7}{x+2} + \dfrac{10}{2x-5} = 3$

Consolidate – do you need more?

1 Solve by factorising.

a $x^2 - 8x + 7 = 0$

b $x^2 - 9x - 22 = 0$

c $x^2 + x - 12 = 0$

d $x^2 - 25 = 0$

e $x^2 - x = 72$

f $x^2 = -16x - 60$

2 Solve by factorising.

a $2x^2 - 21x + 10 = 0$

b $2x^2 + 13x + 18 = 0$

c $3x^2 + 20 = 19x$

d $4x^2 = 25$

e $2x^2 - 7x = 49$

f $9x^2 - 100 = 0$

3 The area of the triangle is 30 cm².

    a Form an equation to show this information.

    b Work out the value of x.

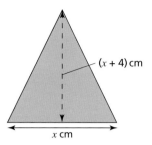

(x + 4) cm

x cm

4 ABC is a right-angled triangle.

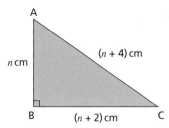

A

n cm

$(n + 4)$ cm

B $(n + 2)$ cm   C

Work out the perimeter of the triangle.

5 Let n and $n + 1$ be two consecutive positive integers.

The sum of the squares of the integers is 145

    a Form an equation to show this information.

    b Work out the values of the two integers.

6 Solve the equations.

    a $\dfrac{2}{x} + \dfrac{2}{x + 3} = 1$ **b** $\dfrac{5}{3x + 2} - \dfrac{3}{2x - 3} = 4$ **c** $\dfrac{8}{3x - 2} + \dfrac{6}{x + 1} = 2$

    d $\dfrac{3}{x + 1} + \dfrac{2}{2x - 3} = 1$ **e** $\dfrac{6}{x - 2} - \dfrac{6}{x + 1} = 1$

Stretch – can you deepen your learning?

1 Given that $2x + 1 : x + 2 = x + 8 : 3x - 4$, work out the possible values of x.

2   **a** Solve $x^2 - 10x + 9 = 0$

    b Use your answers to part **a** to solve the following.

        i $y^4 - 10y^2 + 9 = 0$

        ii $z - 10z^{\frac{1}{2}} + 9 = 0$

        iii $(w + 2)^2 - 10(w + 2) + 9 = 0$

3 Ed drives 300 miles to visit his sister. He travels at an average speed of x mph.

On the way back, his average speed increases by 20 mph, reducing his journey time by 1 hour and 15 minutes.

    a Show that $\dfrac{300}{x + 20} = \dfrac{300}{x} - 1.25$

    b Solve the equation to work out his average speed for each journey.

Are you ready?

1 Multiply out and simplify:

a $(x + 4)(x + 4)$ **b** $(x - 2)(x - 2)$ **c** $(x + 12)^2$ **d** $(x - 8)^2$

2 Solve:

a $x^2 + 4x - 32 = 0$ **b** $x^2 - 6x - 27 = 0$ **c** $x^2 + 8x + 7 = 0$ **d** $x^2 - 14x + 40 = 0$

3 Simplify:

a $\sqrt{20}$ **b** $\sqrt{200}$ **c** $\sqrt{27}$ **d** $\sqrt{80}$

Some quadratic expressions can be written as perfect squares of binomials.

$x^2 + 6x + 9 \equiv (x + 3)^2$

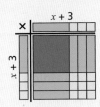

You can see this using algebra tiles.

One x^2 tile, $6x$ tiles and 9 ones tiles can be arranged to form a square.

Other quadratic expressions are not perfect squares, but can be compared to the closest perfect square expression.
This process is called 'completing the square'.

$x^2 + 6x + 11 \equiv (x + 3)^2 + 2$

$x^2 + 6x + 11$
$\equiv x^2 + 6x + 9 + 2$
$\equiv (x + 3)^2 + 2$

Here are the two 'extra' tiles.

$x^2 + 6x + 5 \equiv (x + 3)^2 - 4$

$x^2 + 6x + 5$
$\equiv x^2 + 6x + 9 - 4$
$\equiv (x + 3)^2 - 4$

You would need four more tiles to 'complete the square'.

When you rewrite a quadratic expression like $x^2 + 6x + 5$ as $(x + 3)^2 - 4$, you are writing it in the form $(x + a)^2 + b$. In this case $a = 3$ and $b = -4$

For the examples above, each expression is of the form $x^2 + 6x \pm \dots$

This means that in $(x + a)^2 + b$, a must be 3, since when you expand $(x + 3)^2$ you get two terms of $+3x$, which total to $6x$.

In the same way, for expressions of the form $x^2 - 6x \pm \dots$, a must be -3, since when you expand $(x - 3)^2$ you get two terms of $-3x$, which total to $-6x$.

In general, halving the coefficient of x in an expression of the form $x^2 + px \pm \dots$ will tell you the value of a.

Example 1

Write $x^2 + 8x - 13$ in the form $(x + a)^2 + b$.

Method

Solution	Commentary
$x^2 + 8x - 13 \equiv (x + 4)^2 + b$	Start by working out a.
	You know $a = 4$ as when you expand $(x + 4)^2$ you get two terms of $+4x$, which total $8x$.
$x^2 + 8x - 13 \equiv x^2 + 8x + 16 + b$	Expand $(x + 4)^2$ to give $x^2 + 8x + 16$
$-13 = 16 + b$	Compare the constant terms in the identity to set up an equation in b.
$b = -29$	Solve the equation to find b.
$(x + 4)^2 - 29$	You can now write the equation in the required form.
	Check your answer by expanding.
	$(x + 4)^2 - 29 \equiv x^2 + 8x + 16 - 29 \equiv x^2 + 8x - 13$ as required.

Example 2

Solve the equation $x^2 - 4x - 10 = 0$, giving your answer in the form $p \pm \sqrt{q}$.

> This could also be expressed as 'in simplified surd form'.

Method

Solution	Commentary
$x^2 - 4x - 10 \equiv (x - 2)^2 + b$	Start by writing $x^2 - 4x - 10$ in completed square form.
$x^2 - 4x - 10 \equiv x^2 - 4x + 4 + b$	As in Example 1, work out a first by dividing the coefficient of x, -4, by 2. Then set up and solve an equation in b.
$-10 = 4 + b$	
$b = -14$	
$(x - 2)^2 - 14 = 0$	Rewrite the equation in 'completed square form'.
$(x - 2)^2 = 14$	Add 14 to both sides to find the value of $(x - 2)^2$
$x - 2 = \pm\sqrt{14}$	Square root both sides of the equation.
	You need to write $\pm\sqrt{14}$ so that you get both solutions.
$x = 2 \pm \sqrt{14}$	Now add 2 to both sides to get the solutions.
	The solutions are read aloud as '2 plus or minus the square root of 14'. These are the exact solutions of the equation in the form required.

If the question had asked for solutions to 2 decimal places, you would use your calculator to work them out separately.

$x = 2 + \sqrt{14} = 5.74$ (to 2 d.p.)

$x = 2 - \sqrt{14} = -1.74$ (to 2 d.p.)

Practice

1. Which of these expressions are perfect squares?

   A $x^2 + 6x + 9$ B $x^2 + 6x - 9$ C $x^2 - 6x + 9$ D $x^2 - 6x - 9$

2. Write each expression in the form $(x + a)^2 + b$.

   a $x^2 + 8x - 8$ b $x^2 + 10x - 3$ c $x^2 - 4x + 11$ d $x^2 - 2x + 3$

   e $x^2 + 5x - 1$ f $x^2 - 7x + 2$ *a does not have to be an integer.*

3. Solve the equations, giving your answers in the form $p \pm \sqrt{q}$.

   a $x^2 + 4x + 1 = 0$ b $x^2 + 2x - 7 = 0$ c $x^2 - 4x - 8 = 0$

   d $x^2 + 12x + 25 = 0$ e $x^2 - 2x - 5 = 0$ f $x^2 - x - 4 = 0$

4. Write $2x^2 + 8x + 10$ in the form $a(x + p)^2 + q$.

   Hint: Start by writing the expression as $2(x^2 + 4x) + 10$ and then write $x^2 + 4x$ in completed square form.

5. Write each expression in the form $a(x + p)^2 + q$.

   a $2x^2 - 20x - 7$ b $3x^2 + 6x - 5$ c $5x^2 - 30x + 11$

6. The length of a field is 20 m longer than its width. The area of the field is 1000 m².

   Work out the width and length of the field, giving your answers in simplified surd form.

What do you think? 💡

1. The expression $x^2 + 12x + 25$ has a minimum value.

   Work out this minimum value and the value of x for which this occurs.

   You will explore this idea in more detail in the next chapter.

2. By completing the square, prove that $x^2 + 2x + 5$ is positive for all values of x.

Consolidate – do you need more?

1. Write each expression in the form $(x + a)^2 + b$.

   a $x^2 + 6x - 1$ b $x^2 + 4x + 7$ c $x^2 - 8x - 18$

   d $x^2 - 12x + 4$ e $x^2 + x - 3$

2. Solve the equations, giving your answers in the form $p \pm \sqrt{q}$.

   a $x^2 - 2x - 1 = 0$ b $x^2 + 4x + 2 = 0$ c $x^2 + 10x + 2 = 0$ d $x^2 + 12x - 5 = 0$

Stretch – can you deepen your learning?

1. Write each expression in the form $a(x + b)^2 + c$.

   a $5 + 6x - x^2$ b $7 - 12x - 2x^2$ c $x^2 + 6\sqrt{2}x - 1$

2. Flo solves the equation $x^2 + px + q = 0$

   The solutions are $-4 + \sqrt{21}$ and $-4 - \sqrt{21}$

   Work out the values of p and q.

Are you ready?

1 Use factorisation to solve the equations.

   a $x^2 - 4x - 12 = 0$ **b** $x^2 + 5x + 4 = 0$ **c** $x^2 - 3x + 2 = 0$ **d** $x^2 + 2x - 99 = 0$

2 Write each expression in the form $(x + a)^2 + b$

   a $x^2 + 4x + 3$ **b** $x^2 + 8x - 2$ **c** $x^2 - 6x + 13$ **d** $x^2 - 2x + 9$

You met quadratic graphs in Chapter 10.1. Remember the equation of a quadratic graph is of the form $y = ax^2 + bx + c$ and the graph is a **parabola** ('U' shaped).

In this chapter, you will explore the key features of a quadratic graph without having to draw it using a table of values. For example, you can find where a quadratic graph meets the y-axis by substituting $x = 0$ into the equation of the graph.

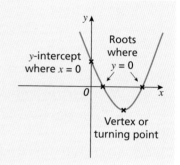

Likewise, you can find the points on the x-axis that the graph passes through by solving the equation $y = 0$. The values of x that are found are called the **roots** of the quadratic equation.

You have seen in Chapter 10.1 that quadratic graphs have a **turning point** where they change direction. This is also known as the **vertex** of the curve, and is the minimum or maximum point of the parabola. You can find the coordinates of the vertex by completing the square, which was explained in Chapter 11.2

Using all this information, you can **sketch** the graph, which means draw a diagram showing the key information, but not to scale and without a table of values.

Example

Sketch the graph of $y = x^2 - 4x - 5$, showing:

a where the graph crosses the y-axis

b where the graph crosses the x-axis

You might see these written as where the graph **meets** the axes, rather than crosses. The meaning is the same.

c the coordinates of the turning point.

Method

Solution	Commentary
a When $x = 0$, $y = 0 - 0 - 5 = -5$ So the graph crosses the y-axis at the point $(0, -5)$.	Set $x = 0$ to find the coordinates of the y-intercept.

b When $y = 0$, $x^2 - 4x - 5 = 0$ $(x - 5)(x + 1) = 0$ $x = 5$, $x = -1$ The graph crosses the x-axis at $(5, 0)$ and $(-1, 0)$.	Solve the equation $x^2 - 4x - 5 = 0$ by factorising. You could use other methods, but always try factorising first as it is usually the most efficient way.
c $y = x^2 - 4x - 5$ $y = (x - 2)^2 - 4 - 5$ $y = (x - 2)^2 - 9$ Minimum value occurs when $(x - 2)^2 = 0$, which is when $x = 2$ At this point $y = 0 - 9 = -9$ so the coordinates of the turning point are $(2, -9)$.	The value of $(x - 2)^2$ is always at least 0 as squaring cannot result in a negative value. You can also think of the graph as a translation of $y = x^2$ by 2 units to the right and 9 units down. This is explored in Chapter 15.2
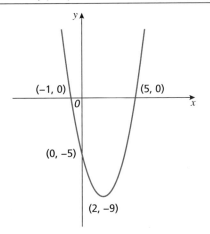	You can now use the points you have worked out to sketch the curve. A quadratic graph is symmetrical about its turning point. In this example, this occurs when $x = 2$, so the straight line with equation $x = 2$ is a line of symmetry of the graph. 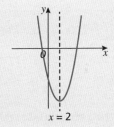

Practice

1 Work out the coordinates of the points where these graphs meet the coordinate axes.

   a $y = x^2 - 11x + 28$ **b** $y = x^2 + 10x + 16$ **c** $y = x^2 + 3x - 54$ **d** $y = 2x^2 + x - 3$

2 Work out the coordinates of the turning point of each graph.

   a $y = x^2 + 6x - 11$ **b** $y = x^2 - 4x + 1$ **c** $y = x^2 + 2x - 7$ **d** $y = x^2 + 3x - 5$

3 Given that $y = x^2 + 8x - 20$

   a Work out the coordinates of the point where the graph crosses the y-axis.

   b Solve $x^2 + 8x - 20 = 0$ to find where the graph crosses the x-axis.

   c Work out the coordinates of the turning point of the graph.

   d Sketch the graph, labelling all key points.

4 Use the same steps as in question 3 to sketch the graphs of:

  a $y = x^2 - 10x + 24$

  b $y = -x^2 + 6x - 8$

  Hint: In part **b**, the graph will have a maximum point instead of a minimum point. Think why.

5 Given $y = x^2 - 6x + 9$

  a Work out the coordinates of the point where the graph crosses the y-axis.

  b Solve the equation $x^2 - 6x + 9 = 0$

  c What does this tell you about the graph?

  d Sketch the graph, labelling all the key features.

6 Given $y = x^2 + 5x + 10$

  a Try to solve the equation $x^2 + 5x + 10 = 0$ by completing the square. What happens?

  b What does this tell you about the graph?

  c Work out the coordinates of the turning point of the graph.

  d Hence sketch the graph of $y = x^2 + 5x + 10$

7 Seb draws a quadratic graph with a minimum point at (2, –3).

  The equation of the graph is of the form $y = x^2 + bx + c$.

  Work out the values of b and c.

Consolidate – do you need more?

1 Work out the coordinates of the points where these graphs meet the coordinate axes.

  a $y = x^2 - 9$ b $y = x^2 - 13x + 40$ c $y = x^2 - 3x - 10$ d $y = 2x^2 + 7x + 5$

2 Work out the coordinates of the turning points of the following graphs.

  a $y = x^2 + 8x - 13$ b $y = x^2 - 2x + 3$ c $y = x^2 + 4x - 9$ d $y = x^2 + 5x - 6$

3 Sketch the graphs of these equations, clearly identifying the roots, y-intercept and coordinates of the turning point.

  a $y = x^2 + 12x + 11$ b $y = x^2 - 6x - 27$

4 Sketch the graph of $y = x^2 + 10x + 25$, labelling all key features.

Stretch – can you deepen your learning?

1 a Work out the coordinates of the turning point of the graph $y = x^2 + 6x - 13$

  b Without completing the square, use your answer to part **a** to work out the coordinates of the turning point of:

    i $y = x^2 + 6x - 14$

    ii $y = x^2 + 6x - 3$

    iii $y = x^2 + 6x + 1$

2 **a** Work out the coordinates of the turning point of the graph $y = x^2 + 2bx + c$.

    b Hence write down the turning points of:

        i $y = x^2 + 2bx + 3c$

        ii $y = x^2 + 4bx + c$

        iii $y = x^2 + 4bx + 2c$

3 Here is a sketch of a quadratic graph:

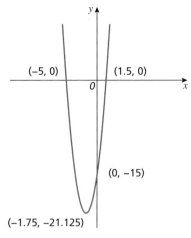

$(-5, 0)$ $(1.5, 0)$

$(0, -15)$

$(-1.75, -21.125)$

Work out the equation of the graph, giving your answer in the form $y = ax^2 + bx + c$.

4 Ali throws a ball up in the air. The ball's height above the ground, h metres, can be modelled by the equation $h = 1.2 + 5t - 5t^2$, where t is the time after throwing the ball, measured in seconds.

    a Write down the height of the ball above the ground when Ali first throws the ball.

    b By completing the square, work out the maximum height that the ball reaches above the ground.

    c Sketch a graph of the height of the ball, labelling all key features on your graph.

Are you ready?

1 Solve:

a $x^2 + 4x - 21 = 0$ **b** $x^2 - 7x - 18 = 0$ **c** $2x^2 + 15x + 18 = 0$ **d** $3x^2 - 13x + 4 = 0$

2 Given $a = 2$, $b = -3$ and $c = 5$, work out the value of:

a $4ac$ **b** $b^2 + 2c$ **c** $\sqrt{b + 5c}$ **d** $\frac{1}{2}(b^2 + ac)$

Many quadratic equations cannot be solved by factorisation. In these cases, you can complete the square as seen in Chapter 11.2, or you can use the **quadratic formula** shown below. This is particularly useful if you are using a calculator and don't know the exact roots of the equation in surd form.

For a general quadratic equation of the form $ax^2 + bx + c = 0$, the solutions can be found using the formula:

$$x = \frac{-b \pm \sqrt{b^2 - 4ac}}{2a}$$

Note that you need to apply the formula twice: once with a + and once with a − in order to get both solutions to the equation.

The symbol ± means 'plus or minus'.

The quadratic formula can be derived by completing the square for the general equation $ax^2 + bx + c = 0$. This is explored in the Stretch section at the end of this chapter.

Example 1

Solve $5x^2 + x - 2 = 0$, giving your answers to 2 decimal places.

Method

Solution	Commentary
$5x^2 + x - 2 = 0$ $a = 5$, $b = 1$ and $c = -2$	First identify a, b and c from your equation, paying attention to the signs.
$x = \dfrac{-b \pm \sqrt{b^2 - 4ac}}{2a}$	Next write out the quadratic formula.
$x = \dfrac{-1 \pm \sqrt{(1^2 - 4 \times 5 \times -2)}}{2 \times 5}$	Substitute the values of a, b and c.
$x = \dfrac{-1 \oplus \sqrt{(1^2 - 4 \times 5 \times -2)}}{2 \times 5} = 0.54$	Then carefully input the calculation with a + in your calculator.
$x = \dfrac{-1 \ominus \sqrt{(1^2 - 4 \times 5 \times -2)}}{2 \times 5} = -0.74$	Then repeat with the −
The solutions are $x = 0.54$ and $x = -0.74$ (to 2 d.p.)	Round your answers to 2 decimal places. Completing the square gives $x = \dfrac{-1 \pm \sqrt{41}}{10}$ Check that these answers are the same.

Using your calculator 🖩

You can use the fraction function on your calculator to enter each calculation all at once.

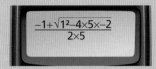

$$\frac{-1 + \sqrt{1^2 - 4 \times 5 \times -2}}{2 \times 5}$$

If your calculator gives the answer in surd form, you may need to use the surd/decimal button to change it to decimal form, which you can then round to the required degree of accuracy.

When you have found the first solution, you can then use the arrow keys to go back and change the sign before the square root part to find the second solution, rather than typing the whole expression.

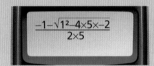

$$\frac{-1 - \sqrt{1^2 - 4 \times 5 \times -2}}{2 \times 5}$$

Example 2

A rectangle measures x m by $(x + 4)$ m.

It has area $20\,\text{m}^2$.

a Show that $x^2 + 4x - 20 = 0$

b Work out the value of x.

Method

Solution	Commentary
a Area $= x \times (x + 4)$	The area of a rectangle is its length multiplied by its width.
$x(x + 4) = 20$	You know the area is 20, so you can form an equation.
$x^2 + 4x = 20$ $x^2 + 4x - 20 = 0$	Now expand, simplify and rearrange to get into the form $ax^2 + bx + c = 0$
b $a = 1$, $b = 4$, $c = -20$	To use the quadratic formula, identify the values of a, b and c.
$x = \dfrac{-4 \pm \sqrt{(4^2 - 4 \times 1 \times -20)}}{2 \times 1}$	Use the formula as in Example 1.
$x = 2.90$ and $x = -6.90$	Remember to apply it twice to get both solutions.
$x = 2.90$ (to 2 d.p.)	In the context of this question, x represents a length and cannot be negative. This means that you can discount the negative solution.

Practice

1. Solve each equation, giving your answers correct to 2 decimal places.

   a $2x^2 + x - 8 = 0$ **b** $x^2 + 7x + 5 = 0$ **c** $6x^2 + 22x + 19 = 0$

   d $5x^2 - 2x - 3 = 0$ **e** $7x^2 + 3x - 2 = 0$ **f** $3x^2 + 5x + 1 = 0$

2. Solve each equation, giving your answers in surd form.

   a $x^2 + 6x - 11 = 0$ **b** $x^2 + 2x - 7 = 0$ **c** $x^2 + 5x + 1 = 0$

   d $5x^2 - 3x - 11 = 0$ **e** $3x^2 + x - 1 = 0$ **f** $5x^2 - 6x - 3 = 0$

3. Solve each equation, giving your answers correct to 3 significant figures.

   a $5x^2 - 10x = -1$ **b** $x^2 = 6 - 3x$ **c** $11x^2 = 7x + 10$ **d** $x(1 + 7x) = 5$

4. A rectangular field measures x m by $(x + 2)$ m.

   The area of the field is $21\,\text{m}^2$.

   a Show that $x^2 + 2x - 21 = 0$

   b Work out the value of x to 1 decimal place.

5. The trapezium has area $64\,\text{cm}^2$.

   a Show that $2x^2 + 5x - 62 = 0$

   b Work out the value of x to 3 significant figures

   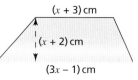
   $(x + 3)$ cm
   $(x + 2)$ cm
   $(3x - 1)$ cm

6. A right-angled triangle has sides $4\,\text{cm}$, $x\,\text{cm}$ and $(2x + 3)\,\text{cm}$, where $(2x + 3)\,\text{cm}$ is the hypotenuse.

   a Show that $3x^2 + 12x - 7 = 0$

   b Work out the value of x, giving your answer in exact form.

7. The difference between a positive number and its reciprocal is 4

   Work out the value of the number correct to 3 significant figures.

Consolidate – do you need more?

1. Solve each equation, giving your answers to 2 decimal places.

   a $x^2 + 2x - 4 = 0$ **b** $x^2 - 6x - 20 = 0$ **c** $3x^2 - 10x + 4 = 0$ **d** $3x^2 = 10 - 2x$

2. Solve each equation, leaving your answers in surd form.

   a $2x^2 + 7x + 2 = 0$ **b** $2x^2 + 5x - 10 = 0$ **c** $7x^2 - 6x + 1 = 0$ **d** $2x^2 - 40 = 9x$

3. A rectangular farm measures x m by $(x + 5)$ m.

   The farm has area $400\,\text{m}^2$.

   a Show that $x^2 + 5x - 400 = 0$

   b The farmer wishes to put new fencing around the whole farm.

   Work out the length of the perimeter of the farm to 1 decimal place.

4 A rectangle measures $(x + 1)$ cm by $(x + 2)$ cm.

The diagonal of the rectangle measures 44.5 cm.

Work out the value of x to 3 significant figures.

Stretch – can you deepen your learning?

1 Ali solves a quadratic equation of the form $ax^2 + bx + c = 0$ to get the solutions

$$x = \frac{-6 \pm \sqrt{3}}{2}$$

Work out the values of a, b and c for the equation Ali solved.

2 A rectangular sheet of paper is 4 cm longer than it is wide.

Beca cuts out the largest circle she can from the paper.

The area of the leftover paper is 35 cm².

What are the dimensions of the paper?

3 Starting with the general quadratic equation $ax^2 + bx + c = 0$, derive the quadratic formula.

Hint: Start by dividing everything by a. Then use the completing the square method.

More quadratics: exam practice

1 Factorise $x^2 + 5x + 6$ **[2 marks]**

2 Solve $x^2 - 3x - 18 = 0$ **[3 marks]**

3 Solve $x^2 + 10x = 24$ **[3 marks]**

4 Work out the coordinates of the points where the graph of $y = x^2 - x - 20$ meets the coordinate axes. **[3 marks]**

5 Here is the graph of $y = x^2 - 6x - 7$

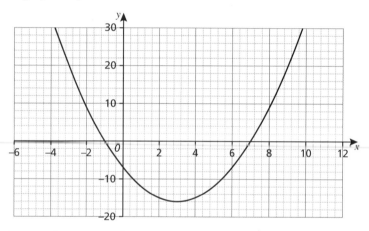

    (a) Use the graph to write down the solutions to $x^2 - 6x - 7 = 0$ **[2 marks]**

    (b) Write down the coordinates of the turning point of $y = x^2 - 6x - 7$ **[1 mark]**

6 Solve $2x^2 + 6x - 11 = 0$, giving your answers correct to 2 decimal places. **[3 marks]**

7 Write $x^2 - 4x + 12$ in the form $(x - a)^2 + b$ **[2 marks]**

8 Solve $x^2 - 6x - 10 = 0$, giving your answers in the form $p \pm \sqrt{q}$ **[3 marks]**

12 Inequalities

In this block, we will cover...

12.1 Linear inequalities

Example 1

Solve the inequalities and show each solution on

a $3x - 5 > 10$

b $-2 \leqslant 2x + 3 \leqslant 5$

c $5 - 3x > 14$

Method

Solution
a $\quad 3x - 5 > 10$
$+5 \left(\right) +5$

12.2 Inequalities on graphs

Practice

1. On a set of coordinate axes, shade the reg

   a $x \geqslant 3$ **b** $y < -2$

2. On a set of coordinate axes, identify the r
   In each case, shade the unwanted region.

   a $x < 3$ and $y > -2$ **b** $x \leqslant -2, x \geqslant$

   c $-3 < x < 1$ and $y > 2$ **d** $y \geqslant 4$ and

3. On a set of coordinate axes, shade the reg

   a $y \leqslant x$ **b** $y > 3x + 1$

12.3 Quadratic inequalities

Consolidate – do you need more

1. Solve each inequality, giving your answers

   a $x^2 + 7x + 12 > 0$ **b** $x^2 - 10x + 2$

   d $x^2 - 3x + 2 < 0$ **e** $x^2 - 12 < 4x$

2. Solve each inequality, giving your answers

   a $x^2 + 8x + 11 \leqslant 0$ **b** $x^2 + 11x +$

   d $x^2 + x \geqslant 3$ **e** $x^2 < 7 - 2x$

3. Solve each inequality.

Are you ready?

1 Solve:

   a $2x + 3 = 15$ **b** $6 - 2x = 10$ **c** $2x + 7 = 4x - 1$ **d** $3(x + 1) = 4x - 19$

2 Give three numbers which satisfy each inequality.

   a $x < 4$ **b** $x \geqslant 7$ **c** $4 < x < 10$ **d** $x \leqslant -3$

3 Copy and complete the statements using < or >

   a $-2.6 \boxed{} -2.7$ **b** $\frac{1}{3} \boxed{} \frac{1}{4}$ **c** $\frac{1}{4} \boxed{} \frac{1}{3}$ **d** $-\frac{1}{5} \boxed{} -\frac{1}{6}$

An **inequality** compares two values, showing if one is less than, greater than, or not equal to the other value.

Solutions to inequalities are called **solution sets**, and can be represented on number lines.

A filled circle shows that a value is included in the solution set.

An open circle shows that a value is not included in the solution set.

This shows 'x is greater than or equal to 2'

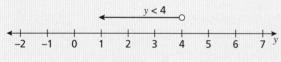

This shows 'y is less than 4'

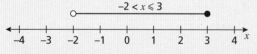

This shows 'x is greater than −2 and less than or equal to 3'

You can solve inequalities using a similar approach to solving equations. However, if you multiply or divide both sides of an inequality by a negative number, the direction of the inequality needs to be reversed to keep the statement true.

For example, 3 > 2. If you multiply by −1, then −3 > −2 is no longer true; you need to reverse the sign to give −3 < −2

Example 1

Solve the inequalities and show each solution on a number line.

a $3x - 5 > 10$

b $-2 \leqslant 2x + 3 \leqslant 5$

c $5 - 3x > 14$

Method

Solution	Commentary
a $3x - 5 > 10$ $+5 \quad\quad +5$ $\quad 3x > 15$ $\div 3 \quad\quad \div 3$ $\quad x > 5$	Add 5 to both sides of the inequality. Divide both sides of the inequality by 3
	The solution can be shown on a number with an open circle since 5 is not included in the solution set.
b $-2 \leqslant 2x + 3$ $\quad$ $2x + 3 \leqslant 5$ $-3 \quad\quad -3 \quad -3 \quad\quad -3$ $\quad -5 \leqslant 2x \quad\quad\quad 2x \leqslant 2$ $\div 2 \quad\quad \div 2 \quad \div 2 \quad\quad \div 2$ $\quad -2.5 \leqslant x \quad\quad\quad x \leqslant 1$	Solve by writing the given inequality as two separate inequalities, $-2 \leqslant 2x + 3$ and $2x + 3 \leqslant 5$ Solve each one separately.
$-2.5 \leqslant x \leqslant 1$	Combine your solutions to write the answer as a single inequality.
	Show your answer on a number line.
c $5 - 3x > 14$ $-5 \quad\quad -5$ $\quad -3x > 9$ $\div -3 \quad\quad \div -3$ $\quad x < -3$	Subtract 5 from both sides of the inequality. Divide both sides of the inequality by -3, remembering to reverse the direction of the inequality.
	The solution can be shown on a number line with an open circle since -3 is not included in the solution set.

Example 2

a Write down the inequality shown on the number line.

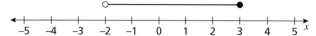

b List all integer solutions to the inequality.

Method

Solution	Commentary
a $-2 < x \leqslant 3$	Note that 3 is included in the solution set, but -2 is not.
b The integer solutions are $-1, 0, 1, 2, 3$	The integer solutions are the whole numbers included in the range of the inequality.

Practice

1 Show each inequality on a number line.

    a $x < 3$
    b $x \geqslant 5$
    c $-2 \leqslant x < 3$
    d $-4 < x \leqslant 0$

2 Write down the inequality shown on each number line.

    a

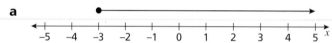

    b

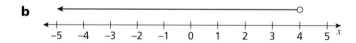

    c

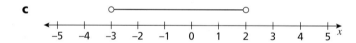

    d

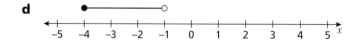

3 List all the integers which satisfy each inequality.

    a $4 < n \leqslant 7$
    b $-1 \leqslant n < 3$
    c $-3 \leqslant n \leqslant 1$
    d $5 < n < 9$

4 Solve:

    a $6x + 2 \geqslant 8$
    b $5x + 9 \leqslant -1$
    c $2(x - 4) \geqslant -3$
    d $3 + 2x < 7$

    e $\dfrac{2x - 5}{3} > 5$
    f $2x + 2 > 5x - 7$
    g $2x + 3 \geqslant x + 1$

5 Solve:

    a $3 - 2x \leqslant 1$
    b $3 \leqslant 2(3 - x)$
    c $4 - 6x > 22$
    d $-1 \leqslant 5 - 3x$

    e $11 - 3x \geqslant 20$
    f $8x - 10 \geqslant 3(4 - x)$
    g $11 - 2x \leqslant 3 - 4x$

6 Solve:

   a $9 < 2x + 1 < 15$ **b** $3 < 4x - 5 < 11$ **c** $-1 \leqslant 3x + 2 \leqslant 17$ **d** $-7 < 2x + 5 \leqslant 1$

7 Solve each inequality and show the solution set on a number line.

   a $4x + 2 \leqslant 28$ **b** $\dfrac{x}{5} - 1 \leqslant -2$ **c** $32 - 9x > -4$ **d** $-5 \leqslant 2x - 1 < 0$

8 The perimeter of the rectangle is less than 40 cm.

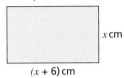

x cm

$(x + 6)$ cm

   a Form an inequality to show this information.

   b Work out the range of possible values of x.

9 The area of a circle is greater than 15 cm² and less than 25 cm².

   Work out the range of possible values for r, the radius of the circle, in centimetres.

Consolidate – do you need more?

1 Show each inequality on a number line.

   a $x \geqslant 7$ **b** $x < 1$

   c $-3 \leqslant x < 0$ **d** $-8 \leqslant x \leqslant -2$

2 Write down the inequality shown on each number line.

   a

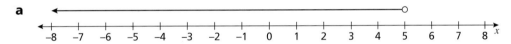

   b

   c

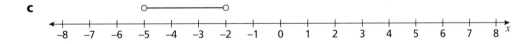

   d

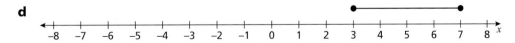

3 Solve:

   a $3x - 1 \leqslant 17$ **b** $4x + 12 > 28$

   c $3(x - 2) < 12$ **d** $14 - 3x \geqslant 8$

4 Solve:

    a $8x - 2 > 3x + 38$ **b** $3x + 4 < 19 - 2x$

    c $-14 < x + 3 \leqslant -4$ **d** $3 \leqslant \dfrac{1 + 2x}{2} < 5$

5 The triangle has perimeter less than 34 but greater than or equal to 19

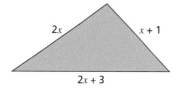

Work out the range of possible values of x.

6 Work out the integer solutions of the inequality $-3 < 2n + 1 < 7$

Stretch – can you deepen your learning?

1 Solve:

    a $x + 5 < 3x + 6 < x + 14$

    b $5 - x < 2(x - 2) \leqslant x$

2 Work out the set of values of x which satisfy both $-3 < x \leqslant 3$ and $5 \leqslant x + 5 < 10$

3 Given that $-4 \leqslant x \leqslant 2$ and $-1 \leqslant y \leqslant 4$, work out:

    a   the greatest possible value of y^2

    b   the greatest possible value of x^2

    c   the least possible value of $x + y$

    d   the least possible value of $x - y$.

Are you ready?

1 On one set of axes from –5 to 5 in each direction, draw the graphs of:

   a $x = 3$ **b** $y = -2$ **c** $x = -1$ **d** $y = 4$

2 On separate sets of axes from –5 to 5 in each direction, draw the graphs of:

   a $y = x$ **b** $y = 2x + 1$ **c** $y = 3x - 1$ **d** $y = 4 - 2x$

An inequality can be represented as a region on a coordinate grid.

The line with equation $x = 3$ is shown.

On the line, all the points have an x coordinate of 3

To the left of the line, all points have an x coordinate less than 3, so you can describe that region as $x < 3$

To the right of the line, all points have an x coordinate greater than 3, so you can describe that region as $x > 3$

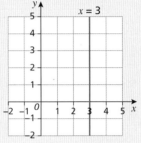

When using number lines, you use empty or filled circles to distinguish between inequalities involving <, > or ⩽, ⩾

When working with regions, you can use solid or dashed lines to distinguish the two cases. Use a dashed line for < or > (the line is not included in the region) and use a solid line for ⩽ or ⩾ (the line is included in the region).

Example 1

Show the region defined by the inequalities $x > -2$ and $y ⩽ 3$

Method

Solution	Commentary
a	To start with, sketch your axes.
	Then draw the line $x = -2$ (you draw this dashed as it is not included).
	It may be easier to shade the unwanted region. You want $x > -2$, which is to the right of the line, so you can shade the left-hand side.

<table>
<tr><td>

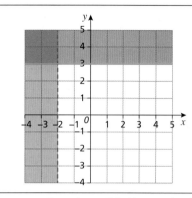

</td><td>

Next draw the line $y = 3$ (you draw this as a solid line because it is included).

Since you want the region where $y \leqslant 3$, you want the part below the line. Therefore, shade the region above.

The region you want is unshaded.

</td></tr>
<tr><td>

$x > -2$ and $y \leqslant 3$ is the unshaded region.

</td><td>

Identify your choice of region clearly.

</td></tr>
</table>

Example 2

Show the region where $3x + 4y > 24$

Method

Solution	Commentary
$3x + 4y = 24$ When $x = 0$, $4y = 24$, $y = 6$ When $y = 0$, $3x = 24$, $x = 8$ 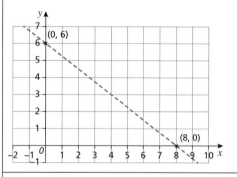	First draw the line $3x + 4y = 24$ using a dashed line. Setting x and y equal to 0 gives you two key points through which to draw the straight line. Alternatively, you can use a table of values or the table function on your calculator.
Choosing $(2, 3)$, which is below the line, as a test point gives: $3x + 4y = 3 \times 2 + 4 \times 3 = 18 < 24$ and so does not lie in the region. 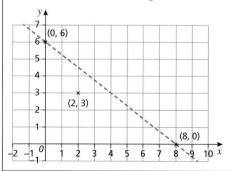	To determine which side to shade in, choose a test point on one side of the line. Here $(2, 3)$ is an easy point to work with. $(0, 0)$ is often a great point to use as a test point.

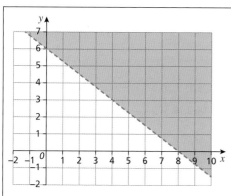

Since (2, 3) does not satisfy the inequality, you can shade the region above the line.

The question asks you to show the region, rather than shading the unwanted region. Always make it clear what your shading means.

The required region is the shaded region.

Practice

1 On a set of coordinate axes, shade the region defined by each inequality.

    a $x \geqslant 3$ **b** $y < -2$ **c** $-3 \leqslant x < 5$ **d** $-2 < y \leqslant 4$

2 On a set of coordinate axes, identify the region(s) defined by each set of inequalities. In each case, shade the unwanted region.

    a $x < 3$ and $y > -2$ **b** $x \leqslant -2$, $x \geqslant 5$ and $y < 4$

    c $-3 < x < 1$ and $y > 2$ **d** $y \geqslant 4$ and $-2 \leqslant x < 3$

3 On a set of coordinate axes, shade the region defined by each inequality.

    a $y \leqslant x$ **b** $y > 3x + 1$ **c** $x - 3y \leqslant 9$ **d** $2x + 3y \leqslant 12$

4 On a set of coordinate axes, label with the letter R the region which satisfies the given set of inequalities.

    a $y \geqslant x + 1$, $y \leqslant -2x$ and $x \geqslant -1$

    b $y \leqslant 3$, $y > 2x - 1$ and $y \geqslant 3 - x$

    c $y + x \leqslant 3$, $y > -x - 1$ and $x \leqslant 1$

    d $3y + 2x < 12$, $y - 2x \leqslant 4$ and $x - 3y < 6$

5 State the inequalities which define each **shaded** region.

a
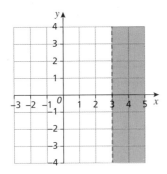

b
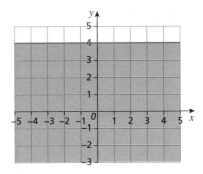

c
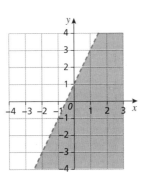

d
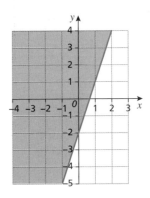

6 State the sets of inequalities which define each **unshaded** region.

a
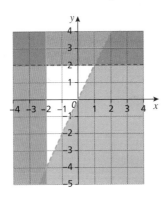

b
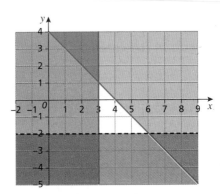

7 State the sets of inequalities which define the **unshaded** region.

a
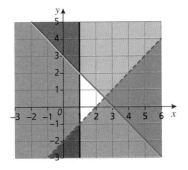

b
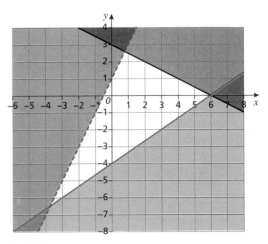

8 x and y satisfy the inequalities $y < 2x + 1$, $3y + 2x < 12$ and $y > 1$

   a On a set of coordinate axes, show the region satisfied by the inequalities and shade the unwanted region.

   b Given that x and y are both integers, state the possible values of x and y.

Consolidate – do you need more?

1 On a set of coordinate axes, shade the region defined by each inequality.

   a $x > 2$ **b** $y < 1$ **c** $-5 \leqslant y < -2$ **d** $-3 \leqslant x < 2$

2 On a set of coordinate axes, shade the region defined by each inequality.

   a $y > 2x + 1$ **b** $y \geqslant 2x + 2$ **c** $x + y < 4$ **d** $y > 1 - 2x$

3 On a set of coordinate axes, label with the letter R the region which satisfies the given set of inequalities.

   a $y > x - 1$, $x \geqslant -2$ and $y \leqslant 2$

   b $y \leqslant 2 - 2x$, $x > 0$ and $y > x - 3$

   c $y \leqslant 5x - 4$, $y - x > -4$ and $2y + x \leqslant 4$

4 State the inequalities which define each **shaded** region.

a

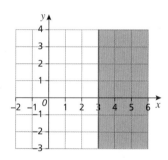

b

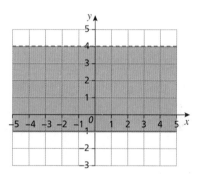

c

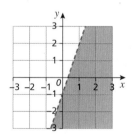

d

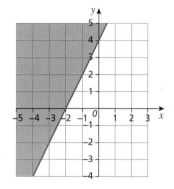

5 State the sets of inequalities which define each **unshaded** region.

a

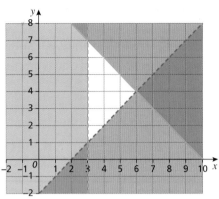

b

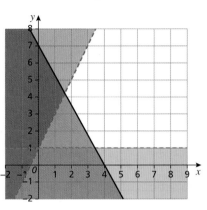

Stretch – can you deepen your learning?

1 Benji wants to buy some chocolate bars.

He decides to buy x choco-wafer bars and y cocoa-caramel bars.

- He wants at least two cocoa-caramel bars.

- He wants at least one choco-wafer bar.

- His mum says he can have no more than 5 bars altogether.

Illustrate the problem above using inequalities on a graph and list the different ways he can buy the chocolate bars.

2 At a theme park, one ride on the water slide costs £4 and one ride on the Ferris wheel costs £3. Flo has £20 to spend. She has W rides on the water slide and F rides on the Ferris wheel.

a Explain why the following inequalities hold.

   i $W \geqslant 0$ and $W \leqslant 5$

   ii $F \geqslant 0$ and $F \leqslant 6$

   iii $4W + 3F \leqslant 20$

b Show the region satisfied by the inequalities on a graph, using the x-axis for W and the y-axis for F. Shade the unwanted region.

c Mark with a cross (X) all possible combinations of rides that satisfy the inequalities.

d Which combination gives a mixture of both rides and means Flo spends the whole £20?

Are you ready?

1 Solve each equation by factorising.

   a $x^2 + 11x + 30 = 0$ **b** $x^2 + x - 6 = 0$ **c** $x^2 - 3x - 28 = 0$

 2 Solve each equation by completing the square, giving your answers in surd form.

   a $x^2 + 4x + 1 = 0$ **b** $x^2 - 6x - 3 = 0$ **c** $x^2 + 2x - 11 = 0$

3 Sketch each of the following graphs, labelling the coordinates of the points where the graph meets the axes, and its turning point.

   a $y = x^2 + 4x - 5$ **b** $y = x^2 + 8x + 7$ ▦ **c** $y = x^2 + 3x - 2$

A **quadratic inequality** is an inequality containing a quadratic expression. For example, $x^2 - 3x + 1 \leqslant 0$

The solution set to a quadratic inequality may have one or two parts.

For example, $x^2 < 16$ is true for all values of x between -4 and 4. The solution set can be written as $\{x: -4 < x < 4\}$. This is read aloud as "the set of values of x such that x is between -4 and 4".

The inequality $x^2 \geqslant 16$ is true both when $x \geqslant 4$ and when $x \leqslant -4$. The solution would be written as $\{x: x \leqslant -4\} \cup \{x: x \geqslant 4\}$, using the $\cup$ symbol to represent the union of the two sets. This is read aloud as "the set of values of x such that x is less than or equal to -4 and the set of values of x such that x is greater than or equal to 4".

To decide which values are in the solution set of a quadratic inequality, you should always sketch the graph to help.

Example 1

Solve the inequality $x^2 + 6x + 8 < 0$, giving your answer in set notation.

Method

Solution	Commentary
When $x^2 + 6x + 8 = 0$ $(x + 2)(x + 4) = 0$ The critical values are $x = -2$ and $x = -4$	First, set the quadratic equal to 0 and solve the equation to find the **critical values**. These are the x values where the quadratic graph crosses the x-axis.

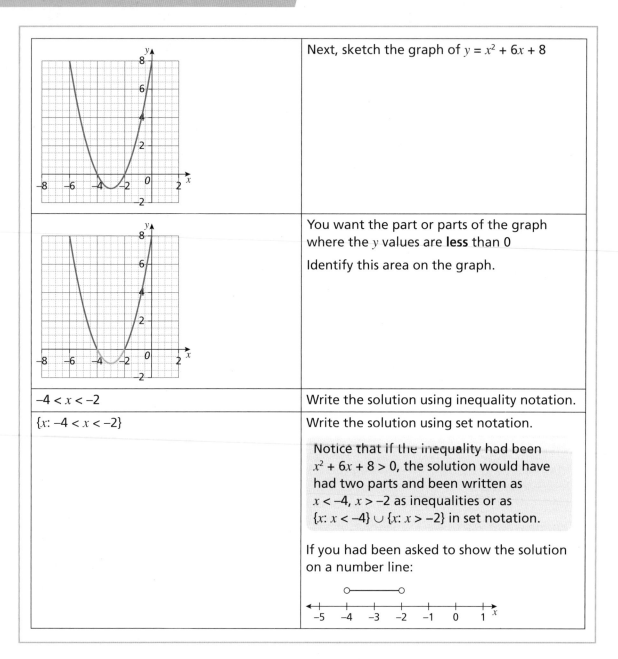

	Next, sketch the graph of $y = x^2 + 6x + 8$
	You want the part or parts of the graph where the y values are **less** than 0. Identify this area on the graph.
$-4 < x < -2$	Write the solution using inequality notation.
$\{x: -4 < x < -2\}$	Write the solution using set notation.

> Notice that if the inequality had been $x^2 + 6x + 8 > 0$, the solution would have had two parts and been written as $x < -4$, $x > -2$ as inequalities or as $\{x: x < -4\} \cup \{x: x > -2\}$ in set notation.

If you had been asked to show the solution on a number line:

Example 2

Work out the set of values of x for which $x^2 - 2x - 24 \leqslant 0$ **and** $6x - 3 \leqslant 0$

Method

Solution	Commentary
When $x^2 - 2x - 24 = 0$ $(x - 6)(x + 4) = 0$ $x = 6$ and $x = -4$ are the critical values.	You need to solve both inequalities. Start by solving the quadratic to find the critical values.

	Next sketch the graph and identify the required regions – the parts where the y values are less than or equal to 0
$-4 \leqslant x \leqslant 6$	Give your answers using inequalities.
$6x - 3 \leqslant 0$ $\qquad 6x \leqslant 3$ $\qquad x \leqslant 0.5$	Now solve the linear inequality.
	Draw a number line to show the solutions of both inequalities.
Answer is $-4 \leqslant x \leqslant 0.5$	The parts that satisfy both inequalities are where $-4 \leqslant x \leqslant 0.5$ Showing this single inequality on a number line:

Practice

1. Solve each inequality.

   a $x^2 + 2x - 35 > 0$ b $x^2 - x - 30 \geqslant 0$ c $x^2 + 2x - 48 < 0$

   d $x^2 - 16 > 6x$ e $x^2 > 32 - 4x$

   Hint: Rearrange to an inequality with > 0 or < 0 first.

2. Solve each inequality, giving your answers using set notation.

   a $x^2 - 9x + 14 \leqslant 0$ b $x^2 + 7x + 12 \geqslant 0$ c $x^2 - 5x - 24 < 0$

   d $x^2 - 9x + 20 > 0$ e $x^2 + 12 < 7x$

3. Solve each inequality, giving your answers as inequalities using surd form.

   a $x^2 - 2x - 1 < 0$ b $x^2 + 4x - 2 \geqslant 0$ c $x^2 + 12x - 5 > 0$

   d $x^2 + 10x + 2 \leqslant 0$ e $x^2 - 6x - 3 < 0$

4. Work out the sets of values of x that satisfy both inequalities.

   a $x^2 - 6x - 40 \leqslant 0$ and $3x - 5 > 1 - x$ b $x^2 - 5x - 24 > 0$ and $2x + 10 < 34$

   c $x^2 + 30 < 13x$ and $5x - 8 \leqslant 3x + 5$

5 Solve each inequality, giving your answers using set notation.

a $2x^2 + 7x + 5 > 0$ **b** $3x^2 + 8x - 3 \leqslant 0$ **c** $2x^2 - 9x + 4 < 0$

6 **a** Solve $x^2 + 6x + 2 = 0$, giving your answers correct to 2 decimal places.

     b State the integer values of x which satisfy $x^2 + 6x + 2 < 0$

Consolidate – do you need more?

1 Solve each inequality, giving your answers using set notation.

a $x^2 + 7x + 12 > 0$ **b** $x^2 - 10x + 24 \leqslant 0$ **c** $x^2 - x - 20 \geqslant 0$

d $x^2 - 3x + 2 < 0$ **e** $x^2 - 12 < 4x$

2 Solve each inequality, giving your answers in surd form.

a $x^2 + 8x + 11 \leqslant 0$ **b** $x^2 + 11x + 1 > 0$ **c** $x^2 + 7x - 3 < 0$

d $x^2 + x \geqslant 3$ **e** $x^2 < 7 - 2x$

3 Solve each inequality.

a $3x^2 + 13x + 4 < 0$ **b** $2x^2 - 5x - 3 \leqslant 0$ **c** $x^2 - 36 \geqslant 0$

4 Work out the integer values of x which satisfy each inequality.

a $x^2 + 7x + 5 \leqslant 0$ **b** $x^2 + 2x - 1 < 0$ **c** $2x^2 + x - 4 < 0$

5 Work out the set of values of x which satisfy both $x^2 + 8x \leqslant 48$ and $2x + 12 < -8$

6 A rectangle measures x cm by $(x - 3)$ cm.

Its area is greater than $10 \, \text{cm}^2$.

a Show that $x^2 - 3x - 10 > 0$

b Work out the range of possible values of x.

Stretch – can you deepen your learning?

1 A rectangle measures $(2x + 1)$ cm by $(x - 1)$ cm.

Its area is less than $90 \, \text{cm}^2$.

Its perimeter is greater than $20 \, \text{cm}$.

a Write down two inequalities to show this information.

b Work out the range of possible values of x.

2 Ali solves a quadratic inequality to get the solution $-4 < x < 1$

Write down a possible quadratic inequality he could have solved.

3 Flo solves a quadratic inequality to get the solution $x \leqslant -4$ and $x \geqslant 8$

Write down a possible quadratic inequality that she could have solved.

Inequalities: exam practice

1 Show the inequality $-3 < x \leqslant 4$ on a number line. **[3 marks]**

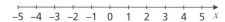

2 Write down all the integers that satisfy the inequality $-8 \leqslant 2x < 4$ **[2 marks]**

3 Solve the inequality $2x - 12 \leqslant 3$ **[2 marks]**

4 Solve $3(x + 2) > 7(x - 2)$ **[3 marks]**

5 On a copy of the grid, identify the region represented by
$x \leqslant 3$, $y \geqslant 2$ and $y \leqslant x + 1$ **[3 marks]**

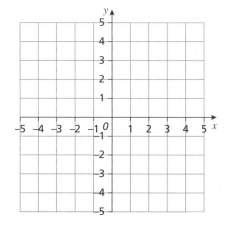

6 Write down the three inequalities that define the shaded region. **[3 marks]**

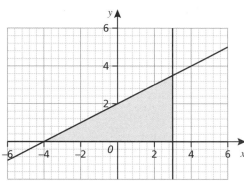

7 The length of a rectangle is no more than 30 cm.

The width of the rectangle is 10 cm less than the length of the rectangle.

(a) Draw a graph to represent this information, using a copy of the axes below. **[2 marks]**

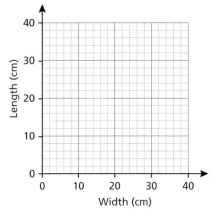

(b) Should your graph touch either of the axes? Explain your answer. **[1 mark]**

8 Solve the inequality $x^2 - 4x - 12 \leqslant 0$ **[3 marks]**

9 Write down the integer values that satisfy the inequality $x^2 + x - 6 < 0$ **[4 marks]**

10 Solve the inequality $2x^2 + 11x + 12 \geqslant 0$ **[3 marks]**

4–6

7–9

In this block, we will cover...

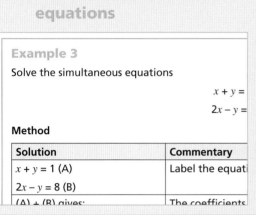

13.1 Linear simultaneous equations

Example 3

Solve the simultaneous equations

$$x + y =$$
$$2x - y =$$

Method

Solution	Commentary
$x + y = 1$ (A)	Label the equati
$2x - y = 8$ (B)	
(A) + (B) gives:	The coefficients

13.2 Linear and non-linear simultaneous equations

Practice

1. Solve each pair of simultaneous equations
   Suggested x values have been given in eac

   a $y = x^2 - 4$
   $y = 3x$ (x values from -2 to $\,$

   b $y = 2x^2 + x - 3$
   $y = 3x + 1$ (x values from -2 to $\,$

   c $y = x^2 + 2$
   $y = -3x$ (x values from -3 to $\,$

White Rose
MΛTHS

Are you ready? (A)

1 Solve:

a $2x + 11 = 17$ **b** $3x - 5 = 10$ **c** $5(2x + 4) = 35$ **d** $3x + 7 = 5x - 2$

2 Given $x = 4$, work out the value of y.

a $y = 2x + 3$ **b** $2y = x - 4$ **c** $2y + 3 = 5x$ **d** $3y + 2x = 12$

3 For the graph of each equation, state the gradient and the y-intercept.

a $y = 6x + 3$ **b** $y = 4 - 3x$ **c** $2y = 4x - 3$ **d** $y + x = 17$

Simultaneous equations involve more than one variable.

One way of solving a pair of linear simultaneous equations is to draw the straight lines given by their equations and find the coordinates of the point of intersection.

You can also solve pairs of simultaneous equations algebraically. The first step is to substitute, add or subtract to get an equation in only one variable.

These bar models show the equations $3h + 2j = 14$

$h + 2j = 10$

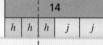

By 'cutting off' the ends of the bars, you can see that $2h = 4$

This result can also be found by subtracting the equations term by term.

You can solve $2h = 4$ to give $h = 2$ and then substitute this into either of the given equations to find the value of j ($= 4$).

Example 1

By drawing graphs, solve the simultaneous equations

$$y = 3x - 5$$
$$2x + y = 5$$

Method

Solution	Commentary
For $y = 3x - 5$	Start by working out some coordinates for each line. You can use a table of values or the table function on your calculator.
$x = 0, y = 0 - 5 = -5$	
$x = 1, y = 3 - 5 = -2$	
$x = 2, y = 6 - 5 = 1$	
For $2x + y = 5$	
$x = 0, y = 5$	
$y = 0, 2x = 5$ so $x = 2.5$	

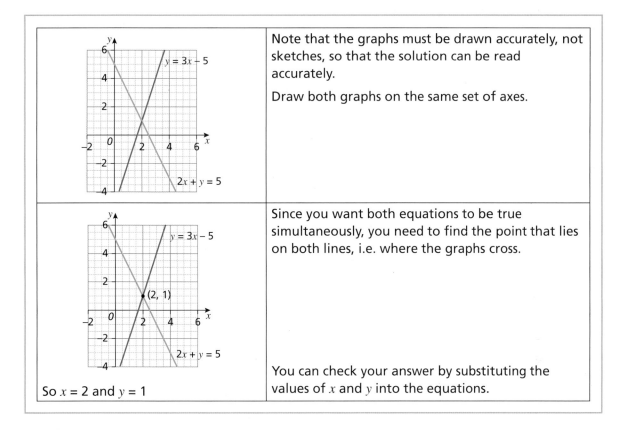

	Note that the graphs must be drawn accurately, not sketches, so that the solution can be read accurately.
	Draw both graphs on the same set of axes.
	Since you want both equations to be true simultaneously, you need to find the point that lies on both lines, i.e. where the graphs cross.
So $x = 2$ and $y = 1$	You can check your answer by substituting the values of x and y into the equations.

One algebraic approach to solving a pair of linear simultaneous equations is to eliminate one of the variables, as shown in Example 2.

Example 2

Solve the simultaneous equations

$$4x + 3y = 27$$
$$5x + 2y = 25$$

Method

Solution	Commentary
$4x + 3y = 27$ (A) $5x + 2y = 25$ (B)	Label the equations to make your working clear.
$8x + 6y = 54$ (C) $15x + 6y = 75$ (D)	Multiply equation A by 2 and equation B by 3 so that the coefficients of one of the variables (in this case y) are the same. You may not need to do this (see Example 3).
(D) – (C) gives $7x = 21$ $x = 3$	Subtract equation C from equation D to eliminate y. Solve to find x.

Substitute in (A): $4(3) + 3y = 27$ $12 + 3y = 27$ $3y = 15$ $y = 5$	Substitute $x = 3$ into any of the equations A–D to find y. Here equation A is probably easiest as the numbers are smaller. Solve this as you would any linear equation.
Check in (B): $5x + 2y = 5 \times 3 + 2 \times 5 = 15 + 10$ $= 25$, which is correct. The solution is $x = 3$, $y = 5$	Check your answer using the other original equation.

Example 3

Solve the simultaneous equations

$$x + y = 1$$
$$2x - y = 8$$

Method

Solution	Commentary
$x + y = 1$ (A) $2x - y = 8$ (B)	Label the equations to make your working clear.
(A) + (B) gives: $3x = 9$ $x = 3$	The coefficients of y are 1 and -1, so you don't need to multiply the equations to eliminate a variable. This time add the equations as $y + -y = 0$, leaving an equation in x only. Solve in order to find x.
Substitute in (A): $3 + y = 1$ $y = -2$	You can use either equation to find y, but the one with positive signs is likely to be easier.
Check in (B): $2x - y = 2 \times 3 - -2 = 6 + 2 = 8$, which is correct. The solution is $x = 3$, $y = -2$	Check your answer using the other equation.

Practice (A)

1 Solve the simultaneous equations graphically.

    a $y = 2x + 2$ **b** $y = x$ **c** $y = 3x + 1$

          $y + x = -4$ $2y = x + 4$ $x + y = 7$

2 Solve the simultaneous equations by elimination.

a $3x + y = 17$
   $x + y = 7$

b $-2x + y = 2$
   $2x + 4y = 18$

c $4x - 3y = 18$
   $x - 3y = 9$

d $x + 3y = 13$
   $x + y = 7$

e $5x + y = 22$
   $2x - y = 6$

3 Solve the simultaneous equations by elimination.

a $2x + y = 9$
   $x + 2y = 6$

b $-5x + y = 9$
   $2x + 8y = 30$

c $2x - 5y = 13$
   $4x - 2y = 10$

d $x + 2y = 5$
   $9x + 3y = 30$

e $-5x + y = 14$
   $-2x - 4y = -12$

4 Solve the simultaneous equations by elimination.

a $3x + 2y = 16$
   $2x - 3y = 2$

b $3x + 5y = 1$
   $2x - 3y = 7$

c $2x - 5y = 9$
   $4x + 3y = 5$

d $2x + 9y = 43$
   $6x + 4y = 14$

e $5x - 3y = 24$
   $4x - 8y = 8$

5 Jakub pays £10 for 2 cups of tea and 4 cups of coffee.

Samira pays £8.50 for 2 cups of tea and 3 cups of coffee.

Work out the cost of 1 cup of tea and the cost of 1 cup of coffee.

> **Hint:** Start by letting the cost of a cup of tea be £t and the cost of a cup of coffee be £c. You can then use the information in the question to form a pair of simultaneous equations in c and t.

6 At a local cinema, 2 adult tickets and 4 child tickets cost £44

1 adult ticket and 3 child tickets cost £28

Work out the cost of 1 adult ticket and the cost of 1 child ticket.

7 Rob is older than Seb.

The sum of their ages is 33

The difference between their ages is 5

Work out Rob's age and Seb's age.

8 The diagram shows a rectangle.

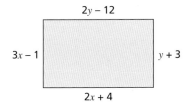

Work out the perimeter of the rectangle.

Are you ready? (B)

1 Given that $x = -3$, work out the value of:

a $3x$ **b** $4x - 3$ **c** $5 - 2x$

2 Expand and simplify:

a $2(x + 4) + 3x$ **b** $3(x - 2) - 5$ **c** $6(2x + 3) - 7x$

Sometimes it is easier to solve a pair of simultaneous equations using substitution rather than elimination. This is particularly useful if one of the equations shows one variable expressed in terms of the other variable, such as $y = 3x + 2$ or $a = 5 - 2b$.

Example

Solve the simultaneous equations

$$y = x - 11$$
$$3x + 7y = 13$$

Method

Solution	Commentary
$y = x - 11$ (A) $3x + 7y = 13$ (B)	Start by labelling your equations to make your working clear.
Substitute (A) in (B): $3x + 7(x - 11) = 13$	Substitute $x - 11$ in place of y in equation B.
$3x + 7x - 77 = 13$ $10x - 77 = 13$	Expand and simplify.
$10x = 90$ $x = 9$	Solve the equation to find the value of x.
$y = 9 - 11 = -2$	Substitute $x = 9$ into equation A to find the value of y.
So $x = 9$ and $y = -2$	

Practice (B)

1 Solve the simultaneous equations by substitution.

a $y = 2x + 3$
     $2x + 3y = 17$
b $y = x - 7$
     $2x - 3y = 13$
c $x = y + 2$
     $3y - 7x = 2$

d $x + y = 4$
     $3x + y + 4 = 10$
e $x - y = 1$
     $2x - 3y = -3$

2 Ali and Benji have £15 between them.

Ali has £4.50 more than Benji.

How much money do they each have?

3 The sum of Chloe's and Samira's ages is 27

Chloe is twice as old as Samira.

How old are Chloe and Samira?

Consolidate – do you need more?

1 Solve the simultaneous equations using a graphical method.

a $y = x - 1$

$y + 4x = 15$

b $y = x + 2$

$y + 3x = 14$

2 Solve the simultaneous equations by elimination.

a $5x + 2y = 12$

$3x - 2y = 4$

b $2x + y = 3$

$6x - y = 17$

c $3x - y = 23$

$2x + 3y = 8$

d $2y - x = 10$

$y - 2x = -7$

3 Solve the simultaneous equations by elimination.

a $5x + 7y = 6$

$3x + 2y = -3$

b $4x + 3y = -3$

$6x + 5y = -6$

4 Solve the simultaneous equations by substitution.

a $x = 2 + y$

$2x + 3y = 9$

b $y = 5 - x$

$3x - y = 11$

5 In a bakery, 4 pastries and 2 slices of cake cost £16

3 pastries and 1 slice of cake cost £10.50

Work out the cost of 1 slice of cake and the cost of 1 pastry.

6 A parallelogram is shown.

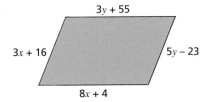

3y + 55

3x + 16

5y – 23

8x + 4

Work out the perimeter of the parallelogram.

Stretch – can you deepen your learning?

1 A sequence is generated by multiplying the previous term by a and then adding b.

The first three terms in the sequence are 4, 13, 31

a Show that $4a + b = 13$

b Write another equation involving a and b.

c Solve to find the values of a and b.

d Work out the fourth term of the sequence.

2 Decide whether each pair of equations has a **unique solution**, **no solutions** or **many solutions**. Use graphs to help explain your choices.

a $2x + 3y = 15$ and $5x - 2y = 8$

b $4x - 2y = 10$ and $2x - y = 5$

c $5x + y = 13$ and $10x + 2y = 15$

3 Work out the area of the triangle enclosed by the straight lines
$2y = x - 5$, $y = -2x + 4$ and $x + y = 5$

Give your answer to 1 decimal place.

Are you ready?

1 Solve each pair of simultaneous equations using the substitution method.

    a $2x + 3y = 30$ **b** $x = 3y + 1$ **c** $2x + 7y = 41$

         $y = 3x - 1$ $7x + 2y = 53$ $x = 3y + 1$

2 Solve:

    a $x^2 - 2x - 24 = 0$ **b** $x^2 - 13x + 40 = 0$ **c** $2x^2 + 13x + 15 = 0$

Simultaneous equations are **not** always linear.

Example 1

Solve the simultaneous equations using a graphical method. Use x-axis values from -2 to 7

$$y = x^2 - 3x - 6$$

$$y = 2x$$

Method

Solution	Commentary
$y = 2x$ <table><tr><td>x</td><td>-2</td><td>-1</td><td>0</td><td>1</td><td>2</td><td>3</td><td>4</td><td>5</td><td>6</td><td>7</td></tr><tr><td>y</td><td>-4</td><td>-2</td><td>0</td><td>2</td><td>4</td><td>6</td><td>8</td><td>10</td><td>12</td><td>14</td></tr></table>	To solve graphically means you need to draw both graphs on the same set of axes.
$y = x^2 - 3x - 6$ <table><tr><td>x</td><td>-2</td><td>-1</td><td>0</td><td>1</td><td>2</td><td>3</td><td>4</td><td>5</td><td>6</td><td>7</td></tr><tr><td>y</td><td>4</td><td>-2</td><td>-6</td><td>-8</td><td>-8</td><td>-6</td><td>-2</td><td>4</td><td>12</td><td>22</td></tr></table>	You can use a table of values or the table function on your calculator to work out coordinates.
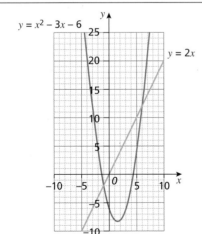	Plot these points and draw both graphs on one set of axes.

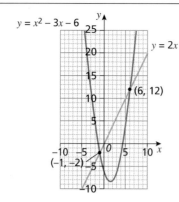

$y = x^2 - 3x - 6$

$y = 2x$

(6, 12)

(−1, −2)

The solutions are where both equations are true. This is at the points of intersection.

One solution is when $x = -1$ and $y = -2$

The other solution is when $x = 6$ and $y = 12$

Notice that there are two possible sets of solutions. Remember to give the y value for each x value.

Example 2

Solve the simultaneous equations.

$$x - y = 1$$

$$x^2 + y^2 = 13$$

Method

Solution	Commentary
$x - y = 1$ (A) $x^2 + y^2 = 13$ (B)	Label the equations to make your working clear. Note that equation A is linear and equation B is non-linear.
From (A): $x = y + 1$ (C)	To use the substitution method, you need to make either x or y the subject of the linear equation. Notice that you cannot use the elimination method as the terms in the two equations are not like terms.
Put this in (B): $(y + 1)^2 + y^2 = 13$	Then substitute this expression for x into the non-linear equation.
$y^2 + 2y + 1 + y^2 = 13$ $2y^2 + 2y - 12 = 0$ $y^2 + y - 6 = 0$	Expand and simplify. Rearrange to get a quadratic that is equal to 0. Here you can divide all terms by 2 to make the quadratic easier to factorise.
$(y + 3)(y - 2) = 0$ So $y = -3$, $y = 2$	Solve this quadratic to get two y values.
Put these values in (C): When $y = -3$, $x = -3 + 1 = -2$ When $y = 2$, $x = 2 + 1 = 3$	Now use either linear equation to find the corresponding x value for each y value.
So the solutions are: $x = -2$, $y = -3$ $x = 3$, $y = 2$	State your solutions in pairs.

Practice

1. Solve each pair of simultaneous equations using a graphical method.

   Suggested x values have been given in each case.

   a $y = x^2 - 4$

   $y = 3x$ (x values from -2 to 5)

   b $y = 2x^2 + x - 3$

   $y = 3x + 1$ (x values from -2 to 3)

   c $y = x^2 + 2$

   $y = -3x$ (x values from -3 to 1)

   d $x^2 + y^2 = 20$

   $y = \dfrac{1}{2}x$ (x values from -4 to 4)

2. Solve each pair of simultaneous equations.

   a $y = x^2 + 5x$ **b** $y = 3x^2 - 4$ **c** $y = x^2 + 2x - 3$

   $y = 2x + 10$ $y = 2x - 3$ $y = 2x + 1$

3. Solve each pair of simultaneous equations.

   a $x^2 + 8y = 13$ **b** $x^2 + y^2 = 20$ **c** $x^2 + y^2 = 25$

   $x + 2y = 2$ $x + y = 6$ $x + y = 7$

4. Solve each pair of simultaneous equations.

   a $xy = 24$ **b** $x + 2y = 3$ **c** $2x - y = 7$

   $x + 2 = y$ $x^2 + 2xy = 10$ $xy = 15$

5. Solve each pair of simultaneous equations, giving your answers to 2 decimal places.

   a $y = x^2 - 3x + 1$ **b** $x^2 + y^2 = 9$

   $y = 2x - 1$ $y = x - 1$

6. **a** Solve the simultaneous equations $y = x^2 + 3x - 4$

   $y + 5 = 5x$

   b What does this tell you about the graphs of $y = x^2 + 3x - 4$ and $y + 5 = 5x$?

   c Sketch the graphs to show the relationship between them.

Consolidate – do you need more?

1. Solve each pair of simultaneous equations using a graphical method.

   a $y = x^2 + x$

   $y = x + 1$ (x values from -2 to 2)

   b $y = 2x^2 - 7x + 4$

   $y = 4x - 1$ (x values from 0 to 6)

   c $x^2 + y^2 = 25$

   $y = 3x + 13$ (x values from -6 to 6)

2 Solve each pair of simultaneous equations.

a $y = x^2 - 3x - 6$
$y = x - 1$

b $y = x^2 - 2x$
$y = 2x - 3$

c $y = 3x^2$
$y = 3x + 6$

3 Solve each pair of simultaneous equations.

a $x^2 + y^2 = 5$
$y = 3x + 5$

b $x^2 + y^2 = 13$
$5x + y = 13$

c $x^2 + y^2 = 25$
$2x - y = 5$

4 Work out the coordinates of the points where the graphs of these equations intersect.

a $x^2 - y + 3 = 0$
$x - y + 5 = 0$

b $x^2 + xy = 4$
$3x + 2y = 6$

Stretch – can you deepen your learning?

1 **a** Work out the points of intersection P and Q of the graphs of $5 - x = y$ and $y = x^2 - 3x + 2$

b Work out the length of the line segment PQ, giving your answer in surd form.

2 **a** Work out the points of intersection A and B of the graphs of $2x - y = 3$ and $2x^2 + y - y^2 = 8$

b Work out the midpoint of the line segment AB.

3 **a** What happens when you try to solve the simultaneous equations $x^2 + y^2 = 20$ and $x - 2y = 12$?

b What does this tell you about the graphs of these equations?

c Sketch the graphs to show the relationship between them.

Simultaneous equations: exam practice

4–6

1 $x + y = 18$

$x - y = 11$

Work out the values of x and y. **[3 marks]**

2 Solve the simultaneous equations $\quad 2x - 3y = 18$

$3x + 2y = 1$ **[2 marks]**

3 In a café, 2 cups of coffee and 3 cups of tea cost £10.60

In the same café, 1 cup of coffee and 5 cups of tea cost £11.95

Work out the price of 1 cup of coffee and the price of 1 cup of tea. **[4 marks]**

4 The graphs of the straight lines with equations $y - x = 3$ and $2x + y = 6$ are shown.

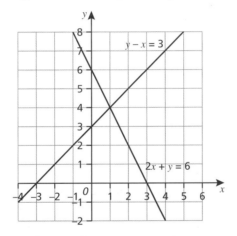

Use the graphs to solve the simultaneous equations $\quad y - x = 3$

$2x + y = 6$ **[2 marks]**

5 Each side of a square is of length $(2a + 3b)$ cm.

The perimeter of the square is 64 cm.

Each side of a pentagon is of length $(a + 2b)$ cm.

The perimeter of the pentagon is 45 cm.

Work out the perimeter of a regular hexagon with sides
of length $(3a - b)$ cm. **[5 marks]**

7–9

6 Solve the simultaneous equations $\quad x^2 + y^2 = 41$

$x - y = 1$ **[4 marks]**

7 Work out the coordinates of the points where the straight line with
equation $y = x - 5$ intersects the circle with equation $x^2 + y^2 = 13$ **[4 marks]**

14 Sequences

White Rose
M▲THS

In this block, we will cover...

14.1 Linear sequences

Example 1

The nth term of a sequence is $4n + 3$

a Work out the first two terms of the sequence

b Is 85 a term in the sequence?

Method

Solution	Commentary
a When $n = 1$, $4n + 3 = 4 \times 1 + 3 = 7$	To find the first tv the rule for the se
When $n = 2$, $4n + 3 = 4 \times 2 + 3 = 11$	

14.2 Quadratic sequences

Practice

1 Write down the next two terms in each qu

  a 0, 3, 8, 15, 24 ...   b 4, 7, 12, 19

2 Work out the first four terms in the seque

  a $n^2 + 4$ b $3n^2$

3 The nth term of the sequence 1, 4, 9, 16, 2

  Use this to work out the nth term rule for

  a 2, 5, 10, 17, 26 ...   b

  c −2, 1, 6, 13, 22 ...   d

14.3 Other sequences

Consolidate – do you need more?

1 Write down the first three terms of the se

  a 4^n b $2 \times 6^{n-1}$

2 Work out a formula for the nth term of ea

  a 4, 16, 64, 256   b $\frac{1}{5}, \frac{1}{25}, \frac{1}{125}$

  d 3, 18, 108, 648   e 5, 10, 20, 4

3 Write down the next two terms in each Fib

  a 6, 7, 13, 20   b

14.4 Iterative sequences

Stretch – can you deepen your le

1 Show that $2x^2 - 5x + 1 = 0$ can be rearrang

  a $x = \frac{1 + 2x^2}{5}$ b $x = \frac{1}{5 - 2x}$

  c $x = \sqrt{\frac{5x - 1}{2}}$ d $x = \frac{1}{2}\left(5 - \right.$

  e Use each iterative formula with $x_1 = 3$
     correct to 2 decimal places, explaining
     Compare the answers to using the qua

2 a Show that the equation $2x^3 - 6x - 11 =$

Are you ready?

1 Write down the next two terms in each of these linear sequences.

    a 2, 5, 8, 11 …
    b 12, 7, 2, −3 …
    c −5, 1, 7, 13 …
    d 2, 1.4, 0.8, 0.2 …

2 Given $n = 3$, work out the value of:

    a $3n + 6$
    b $5n − 9$
    c $4 − 3n$
    d $\frac{1}{2}n + 4$

3 Solve the inequalities.

    a $3n + 4 > 10$
    b $4n − 5 < 15$
    c $6n + 3 > 33$
    d $8n − 4 > 102$

In a **linear** sequence (also known as an **arithmetic** sequence), the terms increase or decrease by the same amount every time. This is known as the **common difference**.

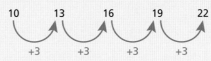

The letter n is often used to represent the position of a term in a sequence, and the rule to find a particular term is often known as the **nth term** rule.

Example 1

The nth term of a sequence is $4n + 3$

a Work out the first two terms of the sequence.

b Is 85 a term in the sequence?

Method

Solution	Commentary
a When $n = 1$, $4n + 3 = 4 × 1 + 3 = 7$ When $n = 2$, $4n + 3 = 4 × 2 + 3 = 11$	To find the first two terms, substitute $n = 1$ and $n = 2$ into the rule for the sequence.
The first two terms are 7 and 11	
b $4n + 3 = 85$ $4n = 82$ $n = 20.5$	If 85 is in the sequence then for some value of n, $4n + 3 = 85$ Solve the equation. If you get an integer value of n, this tells you the position of the number in the sequence.
n is not an integer so 85 is not a term in the sequence.	State your conclusion clearly.

Example 2

a Work out the rule for the nth term of the arithmetic sequence that starts 5, 8, 11, 14 ...

b What is the 100th term of the sequence?

c Work out the first term in the sequence that is greater than 600

Method

Solution	Commentary
a 5 8 11 14 +3 +3 +3	Find the common difference between the terms in the sequence. The common difference is 3, so the nth term rule will be $3n + ...$ or $3n - ...$
<table><tr><td>n</td><td>1</td><td>2</td><td>3</td><td>4</td></tr><tr><td>$3n$</td><td>3</td><td>6</td><td>9</td><td>12</td></tr><tr><td>Term</td><td>5</td><td>8</td><td>11</td><td>14</td></tr></table> The nth term of the sequence is $3n + 2$	Now compare the sequence to multiples of 3 You can see that each term is 2 more than a multiple of 3 This gives you the rule for the nth term of the sequence.
b When $n = 100$, $3n + 2 = 3 \times 100 + 2 = 302$	Substitute $n = 100$ into the rule to find the 100th term.
c $3n + 2 > 600$ $3n > 598$ $n > 199.333$	Form an inequality that uses the nth term rule and 600 Solve the inequality.
The first term greater than 600 is the 200th term. When $n = 200$, $3n + 2 = 3 \times 200 + 2 = 602$	Use your solution to find the first integer value of n that gives a term greater than 600. Then substitute this into the rule to find the value of the term.

Practice

1 Work out the first three terms of the sequence given by each of these rules.

     a $4n + 1$ **b** $3n - 1$ **c** $4 - 5n$ **d** $\frac{1}{2}n + 7$

2   **a**   Is 114 a term in the sequence with nth term rule $2n - 6$?

     b   Is 112 a term in the sequence with nth term rule $3n + 8$?

     c   Is 98 a term in the sequence that begins 15, 22, 29, 36?

     d   Is −61 a term in the sequence that begins −1, −6, −11, −16?

3 Work out the rule for the nth term of each sequence.

     a   9, 12, 15, 18, 21 ...   **b**   11, 20, 29, 38, 47 ...

     c   6, 16, 26, 36, 46 ...   **d**   −10, −3, 4, 11, 18 ...

4   Work out the rule for the nth term of each sequence.

    a   7, 6, 5, 4, 3 …   **b**   11.5, 11, 10.5, 10, 9.5 …

    c   –2, –8, –14, –20, –26 …   **d**   –1, –4, –7, –10, –13 …

5   Work out the 100th term of each sequence.

    a   9, 11, 13, 15, 17 …   **b**   15, 19, 23, 27, 31 …

    c   3, 0, –3, –6, –9 …   **d**   10, 18, 26, 34, 42 …

6   Here are four sequences:

    i   –1, 5, 11, 17, 23 …   **ii**   5, 9, 13, 17, 21 …

    iii   18, 31, 44, 57 …   **iv**   5, 13, 21, 29, 37 …

    a   Work out the rule for the nth term of each sequence.

    b   Work out the first term in each sequence above that is greater than 500

7   Seb makes patterns using sticks.

    Pattern 1   **Pattern 2**   **Pattern 3**

       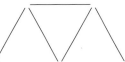

    a   Draw Pattern 4

    b   How many sticks will Seb use in Pattern 8?

    c   Which pattern number would use 51 sticks?

    d   Will Seb be able to create a pattern using exactly 100 sticks?

    e   Work out a rule for the number of sticks in Pattern n.

8   Flo makes patterns using triangles and squares.

    Pattern 1   **Pattern 2**   **Pattern 3**

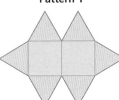   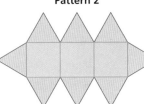   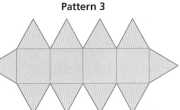

    a   Draw Pattern 4

    b   How many squares will be in Pattern 7?

    c   How many triangles will be in Pattern 9?

    d   Work out a rule for the number of squares in Pattern n.

    e   Work out a rule for the number of triangles in Pattern n.

What do you think? 💭

1 By considering the numerator and denominator separately, work out a rule for the nth term of each sequence.

a $\dfrac{3}{3}, \dfrac{5}{4}, \dfrac{7}{5}, \dfrac{9}{6}, \dfrac{11}{7} \dots$

b $\dfrac{2}{7}, \dfrac{5}{11}, \dfrac{8}{15}, \dfrac{11}{19} \dots$

2 Here is the start of a sequence:

1, 2, 2, 3, 3, 3, 4, 4, 4, 4, 5, 5, 5, 5, 5 …

1 appears once, 2 appears twice, 3 appears three times, and so on.

The sequence continues until 9 appears nine times and then repeats.

a What is the 100th number in the sequence?

b What is the 1000th number in the sequence?

Consolidate – do you need more?

1 Work out a rule for the nth term of each sequence.

a 10, 13, 16, 19, 22 …   **b** 2, 8, 14, 20, 26 …

c 7, 15, 23, 31, 39 …   **d** 5, 10, 15, 20, 25 …

e −1, −6, −11, −16, −21 …   **f** 3.5, 4, 4.5, 5, 5.5 …

g 1.5, 1, 0.5, 0, −0.5 …

2 a Is 43 a term in the sequence with nth term $3n + 7$?

b Is 132 a term in the sequence with nth term $7n + 9$?

c Is 55 a term in the sequence that begins 13, 19, 25, 31?

d Is −116 a term in the sequence that begins −4, −12, −20, −28?

3 Work out the 150th term in each sequence.

a 13, 19, 25, 31, 37 …   **b** −7, −4, −1, 2, 5 …

c −9, −14, −19, −24, −29 …   **d** 7, 4, 1, −2, −5 …

4 Here are four sequences:

i 17, 25, 33, 41, 49 …

ii −2, 8, 18, 28, 38 …

iii 17, 28, 39, 50, 61 …

iv 18, 21, 24, 27, 30 …

a Work out a rule for the nth term of each sequence.

b Work out the first term in each sequence that is greater than 1000

5 Patterns are made using blue squares and red squares.

Pattern 1   Pattern 2   Pattern 3

 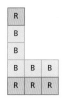 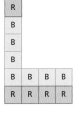

    a   Draw Pattern 4

    b   How many blue squares will be in Pattern 8?

    c   How many red squares will be in Pattern 10?

    d   Ali has 37 blue squares.

         How many red squares will he need to create a complete pattern?

    e   Work out a rule for the number of red squares in Pattern n.

    f   Work out a rule for the number of blue squares in Pattern n.

Stretch – can you deepen your learning?

1 An arithmetic sequence has second term 13 and fifth term 46

Work out a rule for the nth term of the sequence.

2 An arithmetic sequence has first term 17 and fifth term 27

    a   Work out a rule for the nth term of the sequence.

    b   Is 132 in this sequence?

3 An arithmetic sequence has second term $(m + 2)$, third term $(2m + 3)$ and fourth term $(4m - 2)$.

    a   Work out the value of m.

    b   What is the first term of the sequence?

Are you ready?

1 Work out a rule for the nth term of each linear sequence.

   a 4, 7, 10, 13, 16 …   **b** −2, 2, 6, 10, 14 …   **c** 3, 1, −1, −3, −5 …   **d** 7, 6, 5, 4, 3 …

2 Work out the first three terms of the sequences given by each rule.

   a $6n + 5$ **b** $3 - 5n$ **c** $n^2 + 4$ **d** $3n^2 + n$

In Chapter 14.1 you worked with linear sequences in which the difference between each successive term was a constant.

In a **quadratic sequence**, the differences between the differences, called the **second differences**, are constant.

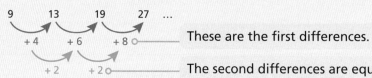

These are the first differences.

The second differences are equal.

The rule for the nth term of a quadratic sequence will always contain a term in n^2.

The general form of a quadratic sequence is $an^2 + bn + c$, like a quadratic expression.

You can work out a, the coefficient of n^2, by halving the value of the second differences.

Example 1

Write down the next two terms in the sequence that starts 5, 16, 33, 56, 85 …

Method

Solution	Commentary

5   16   33   56   85
  + 11 + 17 + 23 + 29
     + 6 + 6 + 6
	First you need to work out the differences between the terms.
	
	Then work out the second differences to help. They are all the same, so the sequence is quadratic.

6th term: 85 + 35 = 120 7th term: 120 + 41 = 161	The next first difference in the sequence will be 29 + 6 = 35
	The following difference will be 35 + 6 = 41

Example 2

Work out, in terms of n, a rule for the nth term of the sequence that starts 4, 13, 26, 43, 64 …

Method

Solution	Commentary
 4 13 26 43 64 +9 +13 +17 +21 +4 +4 +4 $4 \div 2 = 2$, so the rule involves $2n^2$	First, work out the first differences and the second differences. As the second differences are constant, the sequence must be a quadratic sequence. Half of 4 is 2. This tells you the coefficient of n^2 in the nth term rule.

n	1	2	3	4	5
Sequence	4	13	26	43	64
$2n^2$	2	8	18	32	50
Difference	2	5	8	11	14

Write out the sequence and the corresponding terms of $2n^2$, and subtract each term from your sequence to see what's left.

This gives a linear sequence.

2   5   8   11   14
+3   +3   +3   +3

nth term of the linear part is $3n - 1$

Find the nth term rule for the linear sequence as you did in Chapter 14.1. You may by now be able to spot that the common difference is 3 and the rule is $3n - 1$

nth term is $2n^2 + 3n - 1$

The given sequence is the sum of the quadratic part and the linear part.

You can check your answer by substituting in some values of n. For example, when $n = 2$, $2n^2 + 3n - 1 = 2 \times 4 + 3 \times 2 - 1 = 13$, which is correct.

Practice

1 Write down the next two terms in each quadratic sequence.

  a   0, 3, 8, 15, 24 …   **b**   4, 7, 12, 19, 28 …   **c**   10, 17, 26, 37 …

2 Work out the first four terms in the sequences with these rules.

  a $n^2 + 4$ **b** $3n^2$ **c** $n^2 - n$ **d** $n^2 + 2n - 1$

3 The nth term of the sequence 1, 4, 9, 16, 25 … is given by n^2.

  Use this to work out the nth term rule for each of the following sequences.

  a   2, 5, 10, 17, 26 …   **b**   2, 8, 18, 32, 50 …

  c   −2, 1, 6, 13, 22 …   **d**   4, 9, 16, 25, 36 …

4 Work out, in terms of n, a rule for the nth term of each sequence.

  a   8, 11, 16, 23, 32 …   **b**   5, 11, 19, 29, 41 …

  c   2, 9, 18, 29, 42 …   **d**   20, 25, 32, 41, 52 …

  e   −10, −10, −8, −4, 2 …

5 Work out, in terms of n, a rule for the nth term of each sequence.

    a 8, 14, 24, 38, 56 …   **b** 0, 7, 18, 33, 52 …   **c** 7, 20, 39, 64, 95 …

    d 1, 7, 19, 37, 61 …   **e** 5, 4, 1, –4, –11 …

6 **a** A sequence has nth term rule $n^2 + n + 13$

      Which term in the sequence has value 25?

    b A sequence has nth term rule $n^2 - 4n - 45$

      Which term in the sequence has value 15?

    c Show, using algebra, that 35 is **not** a term in the sequence given by the nth term rule $n^2 + 2n + 5$

7 Faith makes patterns using square tiles.

    Pattern 1   **Pattern 2**   **Pattern 3**

    a How many tiles will be needed for Pattern 4?

    b Work out a rule, in terms of n, for the number of tiles in Pattern n.

    c Faith has 85 square tiles.

      Is she able to make a complete pattern using them?

8 The first three triangular numbers are shown.

    a Work out the next two numbers in the sequence.

    b Work out a rule, in terms of n, for the nth triangular number.

9 A sequence has nth term rule $n^2 + bn + c$, where b and c are integers.

The fourth term of the sequence is 6

Work out two possible values of b and c.

What do you think? 💡

1 Linear sequences have rules of the form $an + b$ and have constant first differences.

Quadratic sequences have rules of the form $an^2 + bn + c$ and have constant second differences.

Investigate the rules of cubic sequences. Start with the cube numbers 1, 8, 27, 64 …

Consolidate – do you need more?

1 Work out the first four terms in each sequence.

    a $n^2 - 3$ **b** $n^2 + n$ **c** $2n^2 + 3$

    d $n^2 + 3n - 5$ **e** $(n + 1)^2$

2 A sequence has nth term rule $n^2 + 6n - 3$

    Work out the difference between the fifth and eighth terms in the sequence.

3 Work out, in terms of n, a rule for the nth term of each sequence.

    a   7, 12, 19, 28, 39 …   **b**   –3, –3, –1, 3, 9 …

    c   9, 13, 19, 27, 37 …   **d**   2, 3, 6, 11, 18 …

4 Work out, in terms of n, a rule for the nth term of each sequence.

    a   2, 8, 18, 32, 50 …   **b**   5, 9, 17, 29, 45 …

    c   3, 13, 29, 51, 79 …   **d**   3, 16, 37, 66, 103 …

5 Patterns are made using square and triangular tiles.

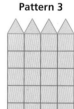

Pattern 1   Pattern 2   Pattern 3

    a   How many triangular tiles will be in Pattern 7?

    b   How many square tiles will be in Pattern 5?

    c   Work out a rule, in terms of n, for the number of triangular tiles in Pattern n.

    d   Work out a rule, in terms of n, for the number of square tiles in Pattern n.

    e   Ali has 9 triangular tiles.

         How many square tiles will he need to create a pattern using all 9 of his triangular tiles?

6 A sequence has nth term rule given by $n^2 + 2n - 12$

    Which term in the sequence has value 23?

Stretch – can you deepen your learning?

1 A sequence has nth term rule $n^2 + bn + c$.

    The fourth term in the sequence is 26

    The fifth term in the sequence is 38

    Work out the values of b and c.

2 A sequence has nth term $n^2 + 6n - 8$

    Which term in the sequence will be the first term greater than 100?

3 A sequence has nth term $n^2 + 5n$.

    Work out which two consecutive terms in the sequence have a difference of 32

14.3 Other sequences

Are you ready?

1 Work out the next two terms in each sequence.

    a 4, 7, 10, 13 …   **b** 5, 10, 20, 40 …   **c** 1, 4, 9, 16 …   **d** 6, 2, –2, –6 …

2 Simplify:

    a $3a + 2b + a$ **b** $2a + 3a + 2b + a$

    c $2a + b + 3a + 2b + a$ **d** $2a + 4a + 6b + 2a + 5b$

In **geometric sequences**, each term is found by multiplying the previous term by a **constant multiplier**.

Here are two examples of geometric sequences:

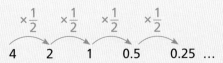

The constant multiplier is also known as the **common ratio**. This can be an integer, a fraction or even a surd.

The nth term of a geometric sequence with first term a and common ratio r is given by the rule ar^{n-1}

In a **Fibonacci sequence**, the next term is found by adding the previous two terms.

Here are two examples of Fibonacci sequences:

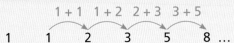 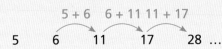

Example 1

A geometric sequence begins 2, 6, 18, 54 …

a Work out the next two terms in the sequence.

b Work out a rule for the nth term of the sequence.

Method

Solution	Commentary
a 2 6 18 54 ×3 ×3 ×3	First find the common ratio.
	You can do this by dividing any successive pair of terms.
	$2 \times r = 6$, so $r = 6 \div 2 = 3$
$54 \times 3 = 162$ $162 \times 3 = 486$	You can get the next two terms by multiplying by 3
b 2 6 18 54	To get the second term, calculate 2×3
The nth term is $2 \times 3^{n-1}$	To get the third term, calculate $2 \times 3 \times 3 = 2 \times 3^2$
	To get the nth term, you need $2 \times 3^{n-1}$

Example 2

Work out an expression for the fifth term of the Fibonacci sequence that begins a, b ...

Method

Solution	Commentary
1st term: a	To generate the terms of a Fibonacci sequence, you add the previous two terms.
2nd term: b	
3rd term: $a + b$	Generate each expression until you have the fifth one.
4th term: $a + b + b = a + 2b$	
5th term: $a + 2b + a + b = 2a + 3b$	

Practice

1 Match each sequence to the correct description.

   a 6, 12, 24, 48　　**b** 2, 8, 18, 32　　**c** 5, 8, 11, 14　　**d** 3, 4, 7, 11

　　　Fibonacci　　　Geometric　　　Linear　　　Quadratic

2 Write down the next two terms in each geometric sequence.

   a 3, 9, 27 ...　　**b** 5, 10, 20 ...　　**c** 8, 16, 32 ...　　**d** 10, 5, 2.5 ...

3 Write down the first three terms in the sequence given by each rule.

   a 3^n　　　　　　**b** 2^n　　　　　　**c** 10^n
   d $2 \times 4^{n-1}$　　　**e** $3 \times 5^{n-1}$

4 Work out a formula for the nth term of each geometric sequence.

   a 5, 25, 125, 625　　**b** 6, 36, 216, 1296　　**c** 3, 9, 27, 81
   d $\dfrac{1}{2}, \dfrac{1}{4}, \dfrac{1}{8}, \dfrac{1}{16}$　　**e** 2, 20, 200, 2000　　**f** 3, 12, 48, 192
   g 5, 10, 20, 40

5 A ball is dropped from a height of 3 m.

After each bounce, it reaches 80% of its previous height.

   a Work out the height of the ball after the first bounce.

   b Work out the height of the ball after the first 4 bounces.

   c Work out a rule to describe the height of the ball after n bounces.

   d What is the height of the ball after 50 bounces?

   e Is this a reasonable model? Explain your answer.

6 Write down the next two terms in each Fibonacci sequence.

a 1, 3, 4, 7

b 2, 5, 7, 12

c 6, 10, 16, 26

d $2a$, $3a$, $5a$

e a, $2b$, $a + 2b$

7 A Fibonacci sequence has first term a and second term $2b$.

a Work out an expression for the third term of the sequence.

b Work out an expression for the sixth term of the sequence.

8 A geometric sequence begins 15, 7.5, 3.75 ...

Will this sequence ever contain a negative term? Explain your answer.

9 Write down the next two terms in each geometric sequence.

a 2, $2\sqrt{2}$, 4, $4\sqrt{2}$

b $\sqrt{3}$, 6, $12\sqrt{3}$, 72

10 At the start of day 1, there are 100 bacteria in a Petri dish.

The population of bacteria grows at a rate of 30% per day.

a Work out the number of bacteria in the dish on the second day.

b Explain why the number of bacteria in the dish at the start of each day forms a geometric sequence.

c The population of bacteria at the start of day 15 is k times as large as the population of the bacteria at the start of day 8

Work out the value of k to 2 decimal places.

Consolidate – do you need more?

1 Write down the first three terms of the sequence with each rule.

a 4^n

b $2 \times 6^{n-1}$

c $\left(\dfrac{1}{3}\right)^n$

d $3 \times 4^{n-1}$

2 Work out a formula for the nth term of each geometric sequence.

a 4, 16, 64, 256

b $\dfrac{1}{5}$, $\dfrac{1}{25}$, $\dfrac{1}{125}$

c 7, 49, 343, 2401

d 3, 18, 108, 648

e 5, 10, 20, 40

3 Write down the next two terms in each Fibonacci sequence.

a 6, 7, 13, 20

b –3, 2, –1, 1

c $3a$, $2b$, $3a + 2b$

d $4a$, $-2b$, $4a - 2b$

4 A sequence, S, begins 3, 6 ...

a Work out the next two terms if S is a linear sequence.

b Work out the next two terms if S is a geometric sequence.

c Work out the next two terms if S is a Fibonacci sequence.

5 Jakub invests £500 in a bank account earning 5% compound interest per year.

He does not withdraw any money from the account.

   a How much money will be in the account after 1 year?

   b How much money will be in the account after 5 years?

   c Explain why the amount of money in the account after n years forms a geometric sequence.

   d In which year will Jakub have more than £1000?

Stretch – can you deepen your learning?

1 The first three terms of a Fibonacci sequence sum to 24

   a What is the third term of the sequence?

   b Work out possible values for the first and second terms of the sequence.

2 A Fibonacci sequence begins $3a$, $4a$ …

The difference between the fifth and third terms is 44

Work out the value of a.

3 Work out the next two terms in the geometric sequence that begins

$$2 + \sqrt{3}, \ 3 + 2\sqrt{3}, \ 6 + 3\sqrt{3} \ …$$

4 Work out the missing terms in the Fibonacci sequence

$$\_\_\_\_, \ 3a - b, \ \_\_\_\_, \ 4a$$

Are you ready? (A)

1 Work out the next two terms in each sequence using the given term-to-term rule.

    a 3, 7, 11   (add 4 to the previous term)

    b 6, 12, 24   (double the previous term)

    c 2, 5, 11, 23   (double the previous term and add 1)

    d 7, 4, −2, −14   (subtract 5 from the previous term and double)

2 Given that $x = 1.3$, work out the value of each expression, giving your answer to an appropriate degree of accuracy.

    a $3x + 7$ **b** $8 - 2x$ **c** $\sqrt{3 + 2x}$ **d** $\dfrac{3}{2 - x}$

The terms of a sequence can be described using subscript notation.

u_1 is the first term, u_2 is the second term and u_n is the nth term.

You can describe term-to-term rules using this notation:

$u_{n+1} = u_n + 7$ 'to get the next term, add 7 to the previous term'

A rule like this, together with one of the terms, will fully define a sequence.

Example

A sequence has a term-to-term rule given by the formula $u_{n+1} = 3u_n + 2$

$u_1 = 4$

Work out the values of u_2, u_3 and u_4

Method

Solution	Commentary
$u_2 = 3u_1 + 2$ $u_2 = 3 \times 4 + 2 = 14$	Replacing n with 1 gives you the formula you need to start with. Use $u_1 = 4$ to find u_2
$u_3 = 3u_2 + 2$ $u_3 = 3 \times 14 + 2 = 44$	Now repeat the process using u_2 to find u_3
$u_4 = 3u_3 + 2$ $u_4 = 3 \times 44 + 2 = 134$	Finally use u_3 to find u_4

Using your calculator

In many models, you can use your calculator to make this easier.

Enter 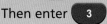 4 as the first answer. Press =

Then enter 3 × ANS + 2

Next press = to get u_2, then press = to get u_3, and so on.

Practice (A)

1 For each sequence, work out the next four terms using the term-to-term rule and the first term.

a $u_{n+1} = u_n + 3$, $u_1 = 7$
b $t_{n+1} = 3t_n - 4$, $t_1 = 4$
c $x_{n+1} = (x_n)^2 - 1$, $x_1 = 2$

2 A sequence begins   5, 14, 23, 32 ...

Write a term-to-term rule for the sequence in the form $u_{n+1} = u_n + a$, where a is a constant.

> Remember to give the value of u_1

3 A sequence begins   3, 6, 12, 24 ...

Write a term-to-term rule for the sequence in the form $t_{n+1} = bt_n$

4 A sequence has term-to-term rule $u_{n+1} = u_n + 10$, with $u_1 = 6$

Will any of the terms in the sequence be a multiple of 10? Explain your answer.

5 The number of rabbits, R, in a colony after month n is modelled by the formula $R_{n+1} = 1.1R_n$

a At the start, there are 35 rabbits.

According to the model, how many rabbits should there be after the first month?

b How many rabbits should there be after 1 year?

c Comment on the suitability of this model.

What do you think?

1 A sequence begins   4, 11, 25, 53 ...

Work out a term-to-term rule for the sequence in the form $u_{n+1} = au_n + b$

Are you ready? (B)

1 Work out $\sqrt[3]{2 + 3x}$ correct to 2 decimal places when:

a $x = 1$
b $x = -3$
c $x = 2.8$
d $x = -3.4$

2 Rearrange each equation to make x the subject.

a $2y = 3x + 4$
b $4x - y = 10$
c $x^2 + a = b$
d $2x^3 - c = 3d$

You don't have a method or formula for solving an equation like $x^3 - 7x - 12 = 0$

You can confirm the location of a solution by looking for a **change of sign**.

If you substitute $x = 3$ into $x^3 - 7x - 12$, you get $27 - 21 - 12 = -6$

If you substitute $x = 4$ into $x^3 - 7x - 12$, you get $64 - 28 - 12 = 24$

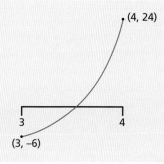

The value of the expression has changed from negative to positive (this is the change of sign). The value you want, 0, is in between -6 and 24. This means the x value you need is between 3 and 4

You then need to rearrange the equation to form an **iterative formula** (this is usually given in the exam question).

Here is one way of rearranging the formula $x^3 - 7x - 12 = 0$:

$$x^3 - 7x - 12 = 0$$
$$x^3 = 7x + 12$$
$$x = \sqrt[3]{7x + 12}$$

The formula is then written using subscript notation $x_{n+1} = \sqrt[3]{7x_n + 12}$

You then choose a sensible starting value (somewhere between 3 and 4 in this case) as x_1, your starting value.

You then work out x_2 using the formula and repeat the process, using x_2 as x_n to get x_3, and so on.

If you have chosen a good starting value, the successive values of x_n get closer and closer together, and **converge** to the root you are looking for.

Using your calculator

The easiest way to get the values of each iteration is by using a key like ANS on your calculator.

For the example above, enter 3 . 5 as your first answer, press = and then enter $\sqrt[3]{\square}$ (7 × ANS + 1 2)

Next press = to get x_2, then press = to get x_3, and so on.

If you keep pressing = , with this example you see the answers get closer and closer together.

Example

a Show that $x^2 - 4x + 2 = 0$ has a root between 3 and 4

b Show that $x_{n+1} = \sqrt{4x_n - 2}$ can be used to solve the equation $x^2 - 4x + 2 = 0$

c Using $x_1 = 4$, work out a solution correct to 2 decimal places.

Method

Solution	Commentary
a $\quad x = 3$ $3^2 - 4 \times 3 + 2 = -1$ $x = 4$ $4^2 - 4 \times 4 + 2 = 2$ Since there is a change of sign, the solution must lie between $x = 3$ and $x = 4$	Substitute $x = 3$ and $x = 4$ into the equation to show there is a change of sign. State your conclusion clearly.

b $x^2 - 4x + 2 = 0$	Rearrange to make x the subject.	
$x^2 = 4x - 2$	Since the iterative formula has a square root in it, this suggests working to find an expression for x^2 first.	
$x = \sqrt{4x - 2}$	Then square root both sides.	
$x_{n+1} = \sqrt{4x_n - 2}$	Finally, write using the subscript notation.	
c $x_2 = \sqrt{4x_1 - 2}$ $x_2 = \sqrt{4 \times 4 - 2} = 3.741\,65\ldots$	Replacing n with 1 gives the formula you need to start with. Use $x_1 = 4$ to find x_2	
$x_3 = 3.600\ldots$ $x_4 = 3.5218\ldots$ $x_5 = 3.4767\ldots$ $x_6 = 3.4506\ldots$ $x_7 = 3.4354\ldots$ $x_8 = 3.426\ldots$ $x_9 = 3.4214\ldots$ $x_{10} = 3.418\ldots$	Repeat this process or use the ANS key on your calculator (as shown earlier) until you have two solutions in a row that give the same value correct to 2 decimal places.	
$x = 3.42$ (to 2 d.p.)		

Practice (B) 🖩

1. Show that the equation $x^3 - 6x + 3 = 0$ has a root between 0 and 1

2. Show that the equation $4x^3 + x - 48 = 0$ has a root between 2 and 3

3. Given the iterative formula $x_{n+1} = \sqrt[3]{x_n + 12}$ with $x_1 = 3.5$, work out the values of x_2, x_3 and x_4 correct to 2 decimal places.

4. Given the iterative formula $x_{n+1} = \sqrt[3]{5 - \dfrac{1}{x_n}}$ with $x_1 = 1.5$, work out the values of x_2, x_3 and x_4 correct to 4 decimal places.

5. **a** Show that the equation $2x^3 - 4x - 9 = 0$ can be rearranged to give the iterative formula $x_{n+1} = \sqrt[3]{2x_n + 4.5}$

   b Using the formula and $x_1 = 2$, work out a solution to the equation correct to 2 decimal places.

6. **a** Show that the equation $x^3 + 2x - 1 = 0$ can be rearranged to give the iterative formula $x_{n+1} = \dfrac{1}{2}(1 - x_n^3)$

   b Use the iterative formula three times, together with $x_1 = 0.5$, to estimate a solution to $x^3 + 2x - 1 = 0$

7. **a** Show that the equation $x^3 - 7x - 11$ has a root between 3 and 4

   b Show that the equation $x^3 - 7x - 11 = 0$ can be rearranged to give $x = \sqrt{7 + \dfrac{11}{x}}$

   c Use the iterative formula $x_{n+1} = \sqrt{7 + \dfrac{11}{x_n}}$, together with $x_1 = 3.2$, to work out the values of x_2, x_3 and x_4 correct to 5 decimal places.

Consolidate – do you need more? 🖩

1 For each sequence, work out the next four terms using the term-to-term rule and the first term.

    a $u_{n+1} = 6u_n - 4$, $u_1 = 2$

    b $u_{n+1} = 2u_n{}^2 - 3$, $u_1 = 0$

2 A substance decays so that after n hours the amount remaining, S, is modelled by $S_{n+1} = 0.9S_n$

    a What is the percentage decrease in the amount of substance each hour?

    b Given that there are initially 100 g of the substance, work out the mass remaining after 3 hours.

    c Work out the mass remaining after 10 hours.

    d Comment on the suitability of this model.

3 Show that the equation $2x^3 - x - 4 = 0$ has a root between $x = 1.35$ and $x = 1.40$

4 **a** Show that the equation $x^3 - 3x - 7 = 0$ can be rearranged to give the iterative formula $x_{n+1} = \sqrt[3]{3x_n + 7}$

    b Using the formula and $x_1 = 2$, work out a solution to the equation correct to 2 decimal places.

5 **a** Show that the equation $x^3 - 2x^2 - 7 = 0$ can be rearranged to give the iterative formula $x_{n+1} = \dfrac{7}{(x_n)^2} + 2$

    b Using this formula and $x_1 = 3$, work out the values of x_2, x_3 and x_4 to 4 decimal places.

Stretch – can you deepen your learning? 🖩

1 Show that $2x^2 - 5x + 1 = 0$ can be rearranged in the following ways.

    a $x = \dfrac{1 + 2x^2}{5}$ **b** $x = \dfrac{1}{5 - 2x}$

    c $x = \sqrt{\dfrac{5x - 1}{2}}$ **d** $x = \dfrac{1}{2}\left(5 - \dfrac{1}{x}\right)$

    e Use each iterative formula with $x_1 = 3$ to work out a solution to the equation correct to 2 decimal places, explaining what happens in each case.

        Compare the answers to using the quadratic formula to solve the equation.

2 **a** Show that the equation $2x^3 - 6x - 11 = 0$ can be rearranged to give the iterative formula $x_{n+1} = \sqrt{a + \dfrac{b}{x_n}}$ where a and b are contants to be found.

    b Use the formula, together with $x_1 = 0.5$, to work out a solution to the equation correct to 2 decimal places.

Sequences: exam practice

4–6

1 Here are the first five terms of an arithmetic sequence.

5   8   11   14   17

Is 1000 a term in this sequence? Explain your answer. **[3 marks]**

2 The nth term of a sequence is $n^2 + 3n$

Work out the difference between the 10th and 12th terms in the sequence. **[2 marks]**

3 Here are the first five terms of a Fibonacci sequence.

2   5   7   12   19

Work out the eighth term in the sequence. **[2 marks]**

4 The nth term of a sequence is $2n^2 + 3n + 11$

Work out the seventh term in the sequence. **[2 marks]**

5 The nth term of a sequence is $31 - 5n$

Which term in the sequence is the first term with a negative value? **[2 marks]**

6 Write down the next **two** terms in the following quadratic sequence.

60   57   52   45   36 **[3 marks]**

7–9

7 The nth term of a sequence is $\left(\sqrt{3}\right)^n$

Work out the fifth term of the sequence, giving your answer in the form $a\sqrt{3}$ **[2 marks]**

8 The nth term of a sequence is $\dfrac{3n}{n+1}$

Work out an expression for the difference between the $(n+1)$th term and the nth term.

Give your answer as a single fraction. **[3 marks]**

9 Work out the formula for the nth term of the quadratic sequence that starts

10   19   32   49   70 **[3 marks]**

10 $P_{n+1} = 1.3(P_n - 20)$

Given that $P_1 = 400$, work out the value of P_4 **[3 marks]**

In this block, we will cover...

15.1 Trigonometric graphs

Example

a Use a copy of the cosine graph to work out th

b Which other angles between 0° and 360° hav

c Use the graph to estimate the solutions to co

15.2 Translations and reflections

Practice

1 Describe the following transformations of
   a f(x + 1)   b f(x) + 1

2 The graph of $y = x^2$ is shown.
   Sketch the graphs of:
   a $y = x^2 + 3$ b $y = x^2 - 2$
   c $y = (x + 4)^2$ d $y = (x - 1)^2$

15.3 Gradient at a point

Consolidate – do you need more?

1 The distance–time graph shows informatic
   bus journey.

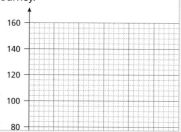

15.4 Area under a curve

Stretch – can you deepen your le

1 The speed–time graph shows information

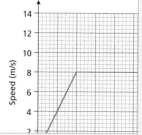

Are you ready? 🖩

1 Work out the following, giving answers to 2 decimal places.

    a   cos 40°   **b**   sin 35°   **c**   tan 143°   **d**   sin 23°

2 Work out the value of x, giving answers to 1 decimal place.

    a   cos x = 0.4   **b**   tan x = 1.2   **c**   sin x = 0.8

There are values for the trigonometric ratios sine, cosine and tangent for angles greater than 90°, and if you study maths at A-level you will find out more about them. For GCSE, you need to recognise and know the key features of the graphs of $y = \sin x$, $y = \cos x$ and $y = \tan x$. These functions are introduced in Chapter 6.2 of *Collins White Rose Maths AQA GCSE 9–1 Higher Student Book 2*.

$y = \sin x$	$y = \cos x$	$y = \tan x$
Has period 360° Maximum value = 1 Minimum value = −1 Crosses the x-axis at 0°, 180°, 360°, etc.	Has period 360° Maximum value = 1 Minimum value = −1 Crosses the x-axis at 90°, 270°, etc.	Has period 180° Has no minimum or maximum value Crosses the x-axis at 0°, 180°, 360°, etc.

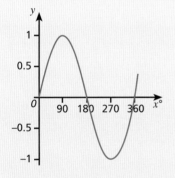 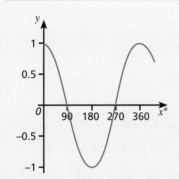 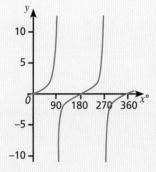

The **period** of the graph tells you how often the shape repeats itself.

Example

a   Use a copy of the cosine graph to work out the value of cos 60°

b   Which other angles between 0° and 360° have the same cosine as 60°?

c   Use the graph to estimate the solutions to cos x = −0.6

Method

Solution	Commentary
a $\cos 60° = 0.5$	Find 60 on the x-axis. Draw a line up to the curve and read off the corresponding y value.
b $\cos 300°$ is also equal to 0.5	Start from 0.5 on the y-axis and draw a horizontal line to meet the curve. The first time it meets will be at 60°. The second time will be the other solution. You can also work this out using the symmetry of the curve.
c The solutions are $x = 127°$ and $x = 233°$	Find −0.6 on the y-axis. Draw a horizontal line. Find all solutions between 0° and 360°

Practice

1. Draw each graph on a new set of axes. For each, use a table of values going up in intervals of 10 degrees from 0° to 360°. You will need to use these to answer the rest of the Practice questions.

   a $y = \sin x$ $\qquad 0° \leqslant x \leqslant 360°$

   b $y = \cos x$ $\qquad 0° \leqslant x \leqslant 360°$

   c $y = \tan x$ $\qquad 0° \leqslant x \leqslant 360°$

2. **a** If $\cos 35° = 0.82$, work out another angle whose cosine is 0.82

   b If $\sin 67° = 0.92$, work out another angle whose sine is 0.92

   c If $\tan 78° = 4.7$, work out another angle whose tangent is 4.7

3. Use your graphs to work out estimates of the solutions to these equations in the range $0° < x < 360°$, giving answers to 1 decimal place.

   a $\sin x = 0.3$ $\qquad$ **b** $\cos x = 0.7$ $\qquad$ **c** $\tan x = 1.2$

   d $\sin x = -0.8$ $\qquad$ **e** $\cos x = -0.3$ $\qquad$ **f** $\tan x = -2.1$

4 Given that the graph of $y = \cos x$ has a period of 360°, work out four positive values of x such that $\cos x = 0.45$

5 Given that the graph of $y = \tan x$ has a period of 180°, work out four positive values of x such that $\tan x = -0.85$

6 Determine whether each statement is **true** or **false**.

a $\sin x = 0.6$ has two solutions in the range $0 < x < 180°$

b $\cos x = 0.6$ has two solutions in the range $0° < x < 180°$

c $\tan x = 0.6$ has two solutions in the range $0° < x < 180°$

d $\cos 60° = \sin 30°$

e $\tan 90° = 1$

f $\sin x = -0.8$ means that x cannot be an acute angle

g $\cos x = -0.8$ means that x cannot be an acute angle

What do you think?

1 Compare the graphs of $y = \sin x$ and $y = \cos x$.
What is the same and what is different?

2 **a** Are these statements **true** or **false**?

$\sin x = \sin(360° - x)$

$\cos x = \cos(360° - x)$

$\tan x = \tan(x + 180°)$

$\sin x = \cos(90° - x)$

b Investigate other relationships between the values of trigonometric ratios.

Consolidate – do you need more?

1 Identify the coordinates of the points labelled A and B on the graph of $y = \sin x$.

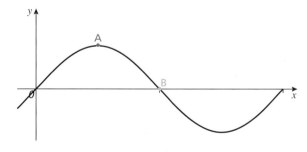

2 Identify the coordinates of the points labelled A and B on the graph of $y = \cos x$.

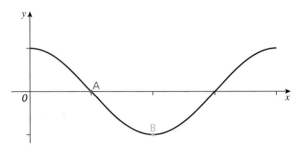

3 Use graphs to work out estimates of the solutions to these equations in the range $0° < x < 360°$, giving answers to 1 decimal place.

a $\sin x = 0.7$ b $\cos x = 0.2$ c $\tan x = 2$

d $\sin x = -0.9$ e $\cos x = -0.1$ f $\tan x = -0.6$

4 Given $\cos x = k$ where $0° \leqslant x \leqslant 360°$ and $-1 \leqslant k \leqslant 1$

a   For what value of k is $x = 270°$?

b   For what value of k is $x = -180°$?

c   Faith says, "If x is an acute angle, then $0 \leqslant k < 1$"

   Is she correct? Explain your answer.

d   Jakub says, "If $0 < k \leqslant 1$, then x is an acute angle."

   Is he correct? Explain your answer.

Stretch – can you deepen your learning?

1 Using graphs to help, express each of the following in terms of the cosine or tangent of an acute angle.

a $\cos 335°$ b $\tan 215°$

2 The depth, d, of water in a tidal river t minutes after midnight is modelled by the following equation.

$$d = 18 + 15\cos t$$

According to the model:

a   what will the depth of the river be at 2 am

b   at what time will the depth be 6 m?

3 Use graphs to estimate solutions in the range $0° < x < 360°$ to the following equations.

a $2\cos x + 3 = 4$ b $3\tan x - 5 = 2$ c $(\sin x)^2 = \dfrac{1}{4}$

   Hint: Rearrange the equations first.

4 Work out the four values of x in the range $0° < x < 360°$ such that $\cos 2x = 0.5$

Are you ready?

1 Sketch the graph of each function, identifying all key features.

   a $y = x^2$ **b** $y = \sin x$ **c** $y = \dfrac{1}{x}$

   d $y = x^3$ **e** $y = \cos x$

Varying the form of a function will affect its graph. Here are four variations, resulting in **translations** and **reflections** of the original graph. Translations and reflections are studied in Chapters 4.1 and 4.2 of *Collins White Rose Maths AQA GCSE 9–1 Higher Student Book 2.*

For a graph of $y = f(x)$,

$y = f(x) + a$ represents a translation of the graph by the vector $\begin{pmatrix} 0 \\ a \end{pmatrix}$

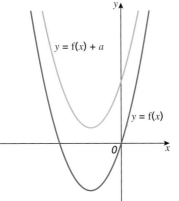

For a graph of $y = f(x)$,

$y = f(x + a)$ represents a translation of the graph by the vector $\begin{pmatrix} -a \\ 0 \end{pmatrix}$

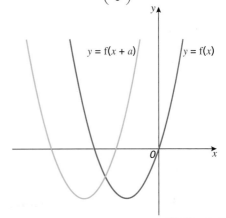

For a graph of $y = f(x)$,

$y = -f(x)$ represents a reflection of the graph in the x-axis.

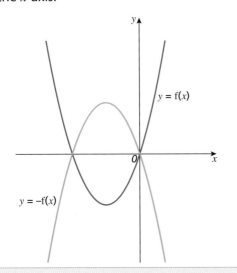

For a graph of $y = f(x)$,

$y = f(-x)$ represents a reflection in the y-axis.

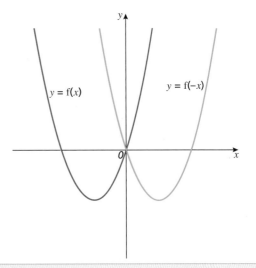

Example 1

The graph of $y = f(x)$ is shown.

Sketch the graphs of:

a $y = f(x - 3)$

b $y = -f(x)$

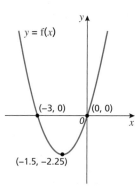

Method

Solution	Commentary
a 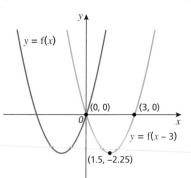	$f(x - 3)$ is a translation, three units to the right. So each point on the graph moves three units horizontally to the right.
b 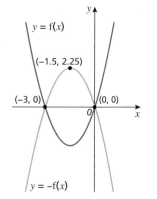	$-f(x)$ is a reflection in the x-axis. The parts that were above the x-axis will now be below and the parts that were below the x-axis will now be above.

Example 2

The graph of $y = f(x)$ is shown.

The point P(2, 5) is shown on the graph.

What are the coordinates of the new position of P when the graph is transformed as follows?

a $y = f(-x)$

b $y = f(x) + 2$

c $y = f(x + 4)$

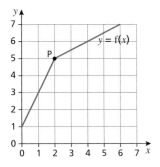

Method

Solution	Commentary
a (−2, 5)	$f(-x)$ is a reflection in the y-axis, so the y value will remain the same whilst the x value will be the negative of its current value.
b (2, 7)	$f(x) + 2$ is a translation up by two units. So all the points on the graph move up two units. This means the x value will remain the same but the y value will increase by 2
c (−2, 5)	$f(x + 4)$ is a translation of four units to the left. The y value won't change, but the x value will reduce by 4

Practice

1. Describe the following transformations of $y = f(x)$ in words.

   a $f(x + 1)$ **b** $f(x) + 1$ **c** $f(x - 1)$ **d** $f(x) - 1$

2. The graph of $y = x^2$ is shown.

   Sketch the graphs of:

   a $y = x^2 + 3$ **b** $y = x^2 - 2$

   c $y = (x + 4)^2$ **d** $y = (x - 1)^2$

   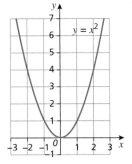

3. The graph of $y = x^3$ is shown.

   Sketch the graphs of:

   a $y = (x + 1)^3$ **b** $y = x^3 - 2$

   c $y = -x^3$ **d** $y = (-x)^3$

   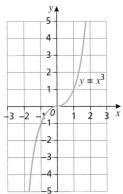

4 The graph of $y = g(x)$ is shown.

The point A(2, 3) is on the graph.

Sketch the graphs of the following functions, showing the coordinates of the point corresponding to A.

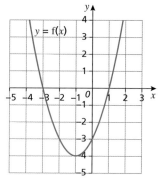

a $y = -g(x)$ **b** $y = g(x + 1)$

c $y = g(x) + 2$ **d** $y = g(x - 3)$

5 The graph of $y = f(x)$ is shown.

The vertex of the curve is at the point (−1, −4).

Write down the coordinates of the vertex of the curve with equation:

a $y = f(x - 4)$ **b** $y = f(x) + 2$

c $y = f(-x)$ **d** $y = -f(x)$

6 The curve with equation $y = f(x)$ is translated so that the point (1, 0) is mapped onto the given points. In each case, give the equation of the translated curve.

a (3, 0)   **b** (1, −4)   **c** (−1, 0)

7 The graph of $y = \sin x$ is shown.

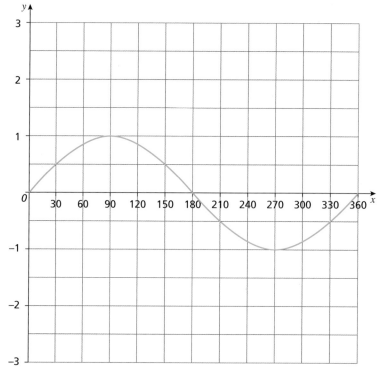

Sketch the graphs of:

a $y = \sin x + 2$ **b** $y = \sin(x + 60°)$

8 The graph of $y = \cos x$ is shown.

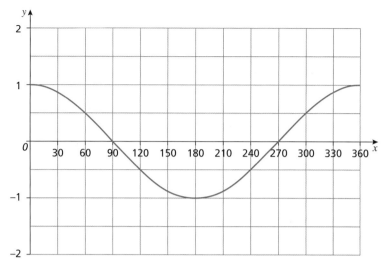

Sketch the graphs of:

a $y = \cos(x - 30°)$ **b** $y = \cos x - 1$

What do you think?

1 The graph of $y = g(x)$ passes through the point A(4, 5).

Work out the coordinates of the point corresponding to A on the graph with equation $y = g(x - 3) + 2$

2 The graph of $y = x^2 + 6x - 2$ is translated three units to the left.

Work out the equation of the translated graph in the form $y = x^2 + bx + c$.

Consolidate – do you need more?

1 The graph of $y = f(x)$ is shown.

Sketch graphs with the given equations.

a $y = f(x) + 3$

b $y = f(x) - 4$

c $y = f(x + 5)$

d $y = -f(x)$

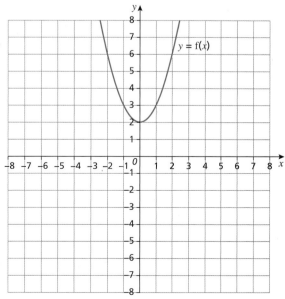

2 The graph of $y = g(x)$ is shown.

Sketch graphs with the given equations, labelling the new positions of the given points.

a $y = g(-x)$

b $y = g(x + 2)$

c $y = -g(x)$

d $y = g(x) - 3$

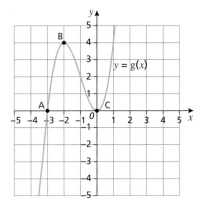

3 The graph of $y = f(x)$ is shown.

The vertex of the graph is at $(4, -1)$.

Write down the coordinates of the vertex of the curve with equation:

a $y = f(x - 2)$

b $y = f(x) + 3$

c $y = f(-x)$

d $y = -f(x)$

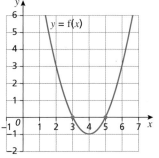

4 The curve with equation $y = f(x)$ is translated so that the point $(1, 0)$ is mapped onto the point $(4, 0)$.

Write down the equation of the translated curve.

5 The curve with equation $y = g(x)$ is translated so that the point $(3, 2)$ is mapped onto the point $(-3, 2)$.

Write down the equation of the translated curve.

Stretch – can you deepen your learning?

1 Which of the following translations of $y = \cos x$ is equivalent to the graph of $y = \sin x$?

A $y = \cos x + 90°$ **B** $y = \cos(x + 90°)$

C $y = \cos x - 90°$ **D** $y = \cos(x - 90°)$

2 The graph of $y = x^2$ is translated to give graphs with the following equations.

By completing the square, describe each set of translations and give the coordinates of the vertex of each graph.

> Completing the square was covered in Chapter 11.2

a $y = x^2 + 4x - 3$ **b** $y = x^2 - 6x + 2$ **c** $y = x^2 + 2x - 2$

3 The graph with equation $y = x^2 + 5x - 3$ is reflected in the x-axis.

Work out the equation of the reflected curve.

4 Sketch the graph of $f(x) = x^2$

Then sketch graphs with each of the following equations.

a $y = f(-x) + 2$ **b** $y = -f(x) + 3$ **c** $y = -f(x + 2)$ **d** $y = -f(x - 3)$

Are you ready?

1 Work out the gradient of each line segment.

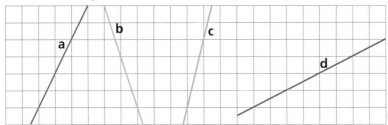

2 Work out the gradient of the line segment joining each pair of points.

a (3, 4) and (2, 7)   **b** (4, 5) and (6, 9)   **c** (3, 9) and (5, −3)   **d** (−2, 4) and (−5, 1)

To estimate the gradient at a point on a curve, you can draw a **tangent** to the curve at the point. A tangent is a straight line that touches a curve at one point only and so it has the same gradient as the curve at this point.

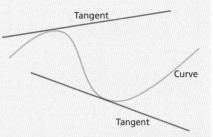

You can then draw a right-angled triangle and work out the gradient using the formula, gradient = $\dfrac{\text{change in } y}{\text{change in } x}$

The gradient of the tangent can be interpreted as a rate of change.
As the gradient of a curve is always changing, this is the rate of change at that instant.

For a distance–time graph, the rate of change is the speed.

For a speed–time graph, the rate of change is the acceleration.

Example

The graph shows the speed of a car for the first 10 seconds of a journey.

a Work out the average rate of change of speed from $t = 0$ seconds to $t = 6$ seconds.

b Interpret the average rate of change in the context of the question.

c Estimate the rate of change of speed at $t = 5$ seconds.

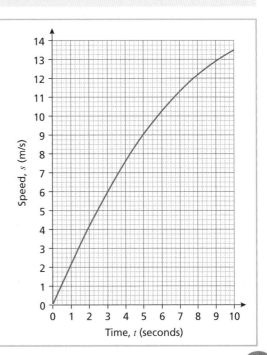

Method

Solution	Commentary
a When $t = 0$, $s = 0$ When $t = 6$, $s = 10.2$ Gradient $= \dfrac{10.2 - 0}{6 - 0} = 1.7$	Use the graph to identify the speeds at $t = 0$ and $t = 6$ Use $\dfrac{\text{change in } y}{\text{change in } x}$ to work out the gradient.
b The average acceleration was 1.7 m/s²	The gradient of a speed–time graph represents the acceleration.
c Gradient $= \dfrac{13 - 5}{8 - 2} = 1.33$ m/s² (to 3 s.f.)	Draw a tangent at $t = 5$ Draw a right-angled triangle onto the tangent to work out the gradient.

Practice

1 A toy rocket is projected from a ledge 5 m above the ground.

The graph shows the height of the rocket as time changes.

a Estimate the speed of the ball after 2 seconds.

b Work out the average rate of change of distance between $t = 5$ and $t = 10$ seconds.

c Describe the motion of the rocket at 4 seconds.

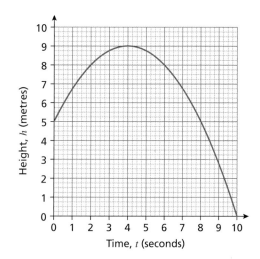

2 The distance–time graph shows information about a journey.

   a Work out the average velocity for the first 7 seconds.

   b Estimate the velocity at 3 seconds.

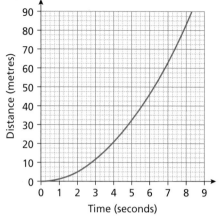

3 The graph shows information about the speed of a car.

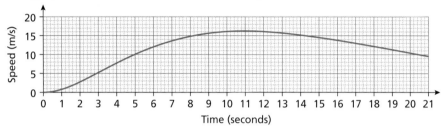

   a Estimate the acceleration at 7 seconds.

   b Estimate the acceleration at 13 seconds and interpret your answer.

   c Describe the motion of the car at 11 seconds.

4 The graph shows the temperature of a bowl of soup as it cools.

   a What is the initial temperature of the soup?

   b Work out the average rate of change of temperature over the 15 minutes.

   c Estimate the gradient of the curve at 3 seconds and explain what this means in the context of the question.

   d Will the gradient of the curve continue to decrease? Explain your answer.

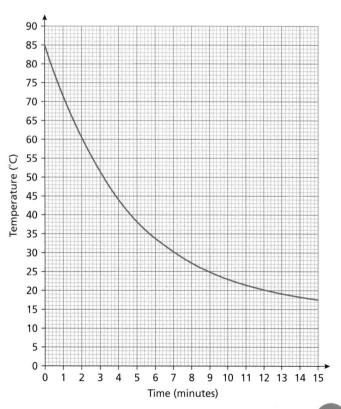

5 The graph of $y = x^2 + 3x - 5$ is shown.

a Estimate the gradient of the graph when $x = -3$

b Estimate the gradient of the graph when $x = -1$

c Why are your answers to parts **a** and **b** only estimates?

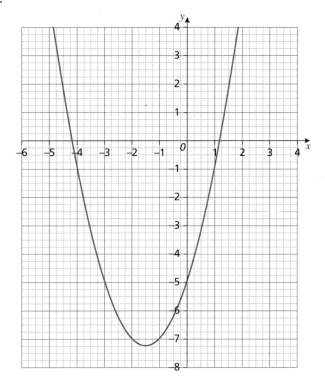

Consolidate – do you need more?

1 The distance–time graph shows information about the first few seconds of a bus journey.

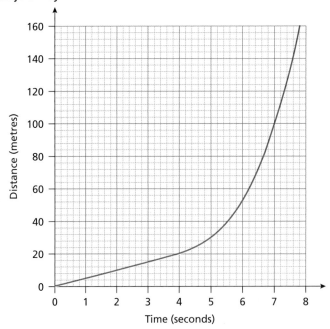

a Estimate the speed of the bus at 6 seconds.

b Work out the average speed of the bus for the first 5 seconds.

2 The graph shows information about a vase being filled with water.

   a Estimate the gradient when $t = 2.5$ seconds.

      Explain what this means in the context of the question.

   b Work out the average rate of change of the depth of water between $t = 2$ and $t = 5$ seconds.

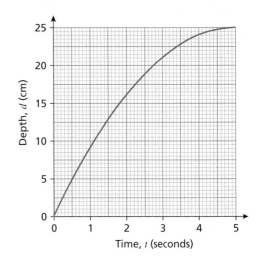

3 The speed–time graph shows information about a bicycle journey.

   a Estimate the gradient at 6 seconds and explain what this means in the context of the question.

   b Describe the motion of the bicycle when $t = 8.5$

   c Estimate the deceleration of the bicycle when $t = 9$

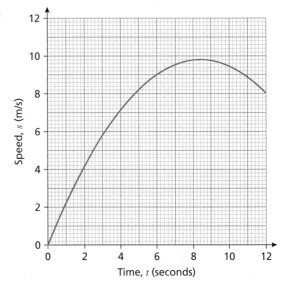

4 The graph of $y = x^3 + 3x^2$ is shown.

   a Estimate the gradient of the curve when $x = 0.5$

   b Estimate the gradient of the curve when $x = -1.5$

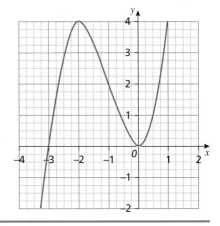

Stretch – can you deepen your learning?

1 Using graphing software or otherwise, investigate:

   a the gradient of the curve $y = x^2 + k$ at the point where $x = 2$ for various values of k

   b the gradient of the curve $y = x^2 + kx$ at the point where $x = 2$ for various values of k.

Are you ready?

1 Work out the area of each shape.

a

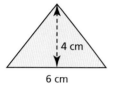

4 cm

6 cm

b

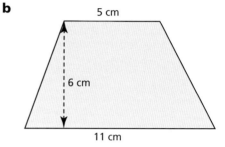

5 cm

6 cm

11 cm

c

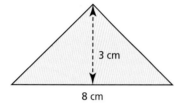

3 cm

8 cm

d

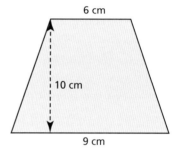

6 cm

10 cm

9 cm

When a graph is made up of straight line sections, you can work out the area under the graph accurately. However, this is not the case if the graph is a curve.

You can estimate the area under a curved graph by splitting it into triangles, rectangles and trapezia. You can then work out the area of each part and add them together.

The area under a speed–time graph represents the distance travelled.

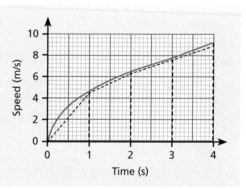

Example

Here is a speed–time graph for a drone.

a Use four strips of equal width to work out an estimate for the distance travelled by the drone over the 8 seconds.

b Is your answer to part **a** an overestimate or an underestimate of the actual distance travelled? Justify your answer.

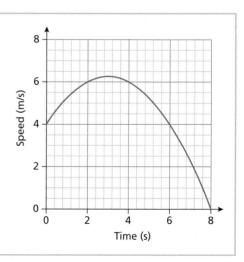

Method

Solution	Commentary
a 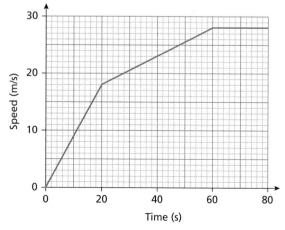	First split the area into strips to make trapezia, a rectangle and a triangle. Label each part and then work out the areas.
Section 1: $\frac{1}{2} \times (4 + 6) \times 2 = 10$ Section 2: $6 \times 2 = 12$ Section 3: $\frac{1}{2} \times (6 + 4) \times 2 = 10$	Sections 1 and 3 are trapezia and section 2 is a rectangle.
Section 4: $\frac{1}{2} \times 2 \times 4 = 4$	Section 4 is a triangle.
$10 + 12 + 10 + 4 = 36$ m	Now add the areas together to get your estimate.
b An underestimate, as the areas of all four strips are below the curve, so the actual area is greater, meaning the actual distance is greater.	Make sure you justify your answer.

Practice

1 The speed–time graph shows information about a car journey.

Work out the total distance travelled by the car, giving your answer in metres.

2 The speed–time graph shows information about a projectile.

a Using three strips of equal width, work out an estimate for the distance travelled in the first 6 seconds.

b How could you improve your estimate in part **a**?

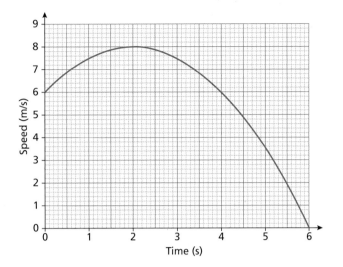

3 Here is a speed–time graph.

a Using four strips of equal width, work out an estimate for the distance travelled in the first 4 seconds.

b Is your answer to part **a** an overestimate or an underestimate? Justify your answer.

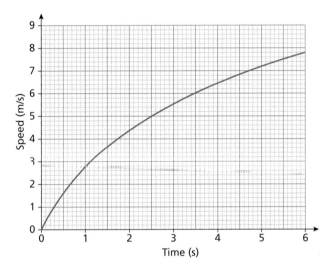

4 Here is a speed–time graph.

a Using three strips of equal width, work out an estimate for the distance travelled in the first 6 seconds.

b Is your answer to part **a** an overestimate or an underestimate? Justify your answer.

c How could you improve your estimate in part **a**?

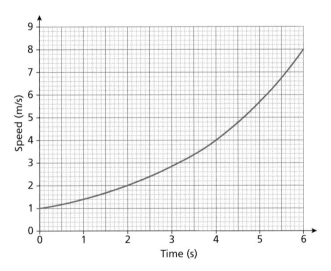

5 The graph of $y = 9 - x^2$ is shown.

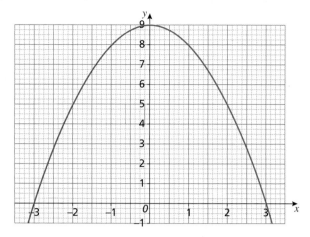

Work out an estimate for the area under the curve between $x = 0$ and $x = 3$

6 The graph of $y = x^3 + 4x^2$ is shown.

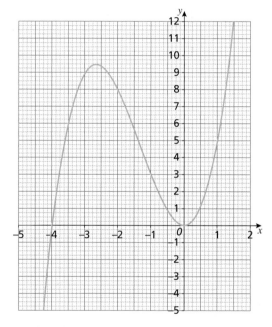

Work out an estimate for the area enclosed between the curve and the x-axis between $x = -4$ and $x = -1$

Consolidate – do you need more?

1 A speed–time graph is shown.

a Using five strips of equal width, work out an estimate for the distance travelled in the first 5 seconds.

b How could you improve your estimate in part **a**?

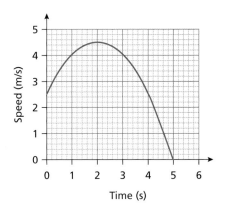

2 The speed–time graph for a model plane is shown.

Does the plane cover a greater distance in the first 4 seconds, or the last 2 seconds? Show clearly how you work out your answer.

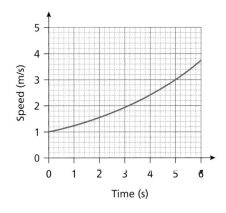

3 A speed–time graph is shown.

a Using two strips of equal width, work out an estimate for the distance travelled in the first 4 seconds.

b Using four strips of equal width, work out another estimate for the distance travelled in the first 4 seconds.

c Which of your answers to parts **a** and **b** is more accurate? Explain your answer.

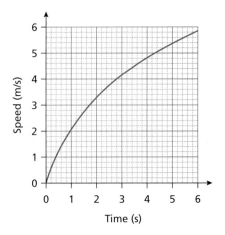

4 The graph of $y = 3x - x^2$ is shown.

a Work out an estimate for the area under the graph between $x = 0$ and $x = 3$

b Given that the exact area is 4.5 square units, work out the percentage error in your answer to part **a**.

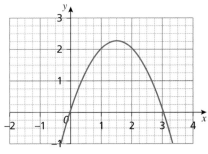

Stretch – can you deepen your learning?

1 The speed–time graph shows information about an ice skater's motion over 20 seconds.

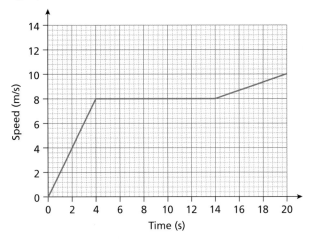

a Work out the total distance travelled by the skater in the 20 seconds.

b How long does it take the skater to cover 100 m?

2 The speed–time graph shows the motion of a car and a motorbike.

At a given time after 60 seconds, the car overtakes the motorbike.

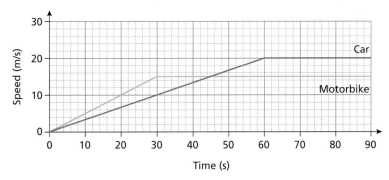

a Work out the distance travelled after 60 seconds by:

　i the motorbike 　　　**ii** the car.

b What can you say about the distance covered by both vehicles at the point of overtaking?

c Work out the time when the car overtakes the motorbike, and the distance covered at that point.

1 Here is the graph of $y = \sin x$ for values of x from 0° to 360°

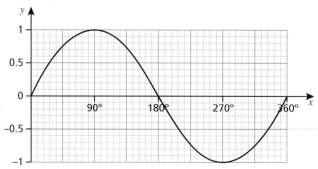

Use the graph to estimate the solutions to the equation $\sin x = -0.3$
in the range 0° to 360° **[2 marks]**

2 Here is the graph of $y = f(x)$

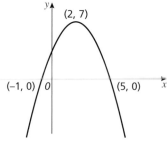

(a) Write down the coordinates of the points of intersection of
the graph of $y = f(x - 2)$ and the x-axis. **[2 marks]**

(b) Write down the coordinates of the turning point of the graph of
$y = f(-x)$ **[1 mark]**

3 Here is the graph of $y = 30 - x^2$

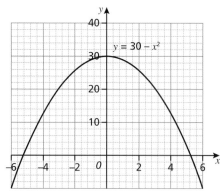

(a) Work out an estimate of the gradient of the graph of $y = 30 - x^2$
when $x = -2$ **[2 marks]**

(b) Use four strips of equal width to estimate the area under the
graph of $y = 30 - x^2$ between $x = 0$ and $x = 4$ **[3 marks]**

Algebra: exam practice

1 Solve the simultaneous equations $2x + y = 5$

$3x + 4y = 5$ **[4 marks]**

2 Work out the nth term of the sequence that starts

   2   8   14   20   26   **[3 marks]**

3 Solve $x^2 - 6x + 8 = 0$ **[2 marks]**

4 A line with gradient -3 passes through the point $(2, 4)$.

Work out the equation of the line.   **[2 marks]**

5 Rearrange $C = 2\pi r$ to make r the subject.   **[2 marks]**

4–6

6 $f(x) = \dfrac{x}{2} + 5$ and $g(x) = x^3$

    (a) Work out the value of $gf(4)$ **[2 marks]**

    (b) Work out an expression for $f^{-1}(x)$ **[2 marks]**

---

7 Solve $2x^2 + x - 3 = 0$ **[3 marks]**

---

8 Simplify $\dfrac{x^2 - x - 12}{x^2 - 9}$ **[3 marks]**

---

9 Work out the equation of a circle of radius $2\sqrt{3}$ and centre $(0, 0)$.   **[2 marks]**

---

10 Sketch the graph of $y = \cos x$ for values of x from 0° to 360°, using a copy of the axes.

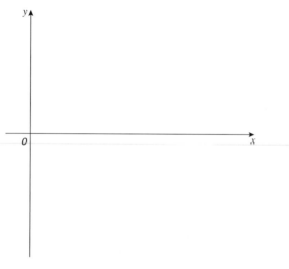

                                                        [2 marks]

---

11 The turning point of the graph of $y = f(x)$ is at the point $(-3, 5)$.

    Write down the coordinates of the turning points of the graphs of:

    (a) $y = f(x + 3)$ **[1 mark]**

    (b) $y = f(x) + 3$ **[1 mark]**

---

12 (a) Write $x^2 - 4x + 9$ in the form $(x + a)^2 + b$, where a and b are integers. **[2 marks]**

    (b) Explain why $x^2 - 4x + 9$ is always positive.   **[2 marks]**

16 Ratio

In this block, we will cover...

16.1 Basic ratio

Example

Zach and Kath share some money in the ratio 3 :

a If they share £120, work out how much mone

b If instead Zach gets £120, how much does Ka

c If instead Kath gets £120 more than Zach, ho

Method

Solution	Cor
a 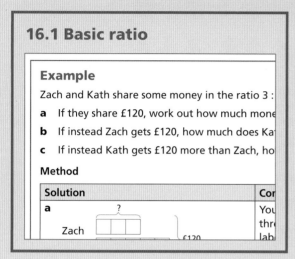	You thro lab

16.2 Fractions and equal ratios

Practice

1 The ratio of part-time to full-time staff in
   What fraction of the staff work part-time?

2 Sven, Junaid and Zach share some money i
   What fraction of the money does Zach rec

3 Marta has some green, purple and yellow
   a What fraction of the pens are yellow?
   b What fraction of the pens are **not** purp

16.3 Ratio and algebra

Consolidate – do you need more

1 $j : k = 6 : 5$ and $j : l = 8 : 5$

   Write the ratio $j : k : l$.

2 Abdullah and Amina share blackberries in
   Abdullah gives three blackberries to Amin
   The ratio of Abdullah's blackberries to Am

   How many blackberries did they share?

Are you ready? (A)

1 Write the ratio of:

    a blue circles to blue triangles

    b circles to triangles

    c green shapes to blue shapes

    d blue shapes to green triangles to green circles.

2 The ratio of benches to trees in a park is 1 : 4

    There are 40 trees. How many benches are there?

3 A food van sells only beef burgers and veggie burgers.

    The ratio of the number of beef burgers sold to the total number of burgers sold is 8 : 11

    Write the ratio of the number of veggie burgers sold to the number of beef burgers sold.

4 Work out:

    a $\frac{1}{3}$ of 42   **b** $\frac{1}{4}$ of 42   **c** $\frac{3}{4}$ of 42   **d** $\frac{3}{4}$ of 84   **e** $\frac{3}{5}$ of 90

Example

Zach and Kath share some money in the ratio 3 : 5

a If they share £120, work out how much money they each get.

b If instead Zach gets £120, how much does Kath get?

c If instead Kath gets £120 more than Zach, how much money do they share in total?

Method

Solution	Commentary
a 3 + 5 = 8 parts altogether £120 ÷ 8 = £15	You can draw a comparison bar model with three boxes for Zach and five boxes for Kath, labelling the total as £120 Label what you want to work out as '?' You can see the total number of parts is 3 + 5 = 8 Now work out each equal part by dividing the total by 8
Zach 15 | 15 | 15 Kath 15 | 15 | 15 | 15 | 15 Zach gets 3 × £15 = £45 Kath gets 5 × £15 = £75	Add this information to each of the boxes representing each equal part. You can now work out how much money each person gets.

b £120 Zach [ ][ ][ ] Kath [ ][ ][ ][ ][ ] ? £120 ÷ 3 = £40	Start with the same comparison bar model, but this time labelling Zach's share as £120 This time, three boxes represent £120, so you can divide £120 by 3 to work out the value of one part.
£120 Zach [£40][£40][£40] Kath [£40][£40][£40][£40][£40] ? Kath gets 5 × £40 = £200	Add this information to each of the boxes representing each equal part. You can now work out how much money Kath gets.
c Zach [ ][ ][ ] ←£120→ }? Kath [ ][ ][ ][ ][ ] £120 ÷ 2 = £60	Again, use the same comparison bar model, but this time label the difference between the bars as £120. Label what you want to work out as '?', which is the total. This time, two boxes represent £120, so you can work out each equal part by dividing £120 by 2
Zach [£60][£60][£60] }? Kath [£60][£60][£60][£60][£60] The total is 8 × £60 = £480	Add this information to each of the boxes representing each equal part. You can now work out how much money they share in total.

Practice (A)

1 Share:

    a £360 in the ratio 5 : 7   **b** 35 bananas in the ratio 4 : 3

    c 42 cm in the ratio 1 : 6 : 7   **d** 700 ml in the ratio 2 : 3

2 Ali and Flo share money in the ratio 2 : 7

    a If they share £63, how much do they each get?

    b If Flo gets £63, how much does Ali get?

    c If Ali gets £63, how much do they share in total?

    d If Flo gets £63 more than Ali, how much do they share in total?

3 A recipe to make a pie crust needs flour and butter in the ratio 3 : 2

Abdullah has 40 g of butter.

Work out how much flour Abdullah will need to make the pie crust.

4 Seb and Bev share pieces of a chocolate bar in the ratio 1 : 3

Seb eats six fewer pieces than Bev.

How many pieces were there altogether?

5 A truck carries 5000 kg of timber, sand and steel in the ratio 2 : 1 : 5

Work out the mass of each material.

6 Share 1.25 kg in the ratio 12 : 13, giving your answer in grams.

7 A festival sells adult tickets and child tickets only.

FESTIVAL TICKETS	
Adult	£20
Child	£12.50

The total number of tickets sold is 3000

The ratio of adult tickets sold to child tickets sold is 5 : 3

Work out the total amount of money the festival receives from the ticket sales.

What do you think? (A)

1 A florist has 540 flowers.

They only have sunflowers, daffodils, tulips and roses.

$\frac{2}{9}$ of the flowers are sunflowers. 50% of the flowers are daffodils.

The ratio of the number of tulips to the number of roses is 3 : 2

Work out the number of tulips the florist has.

2 Filipo and Amina share 60 sweets in the ratio 3 : x, where $x > 0$
   a Write three different possible values of x.
   b Amina says, "x can be any positive integer."
      Explain why she is wrong.

3 Samira, Lida and Rhys share £144

The amount that Samira and Rhys each get is in the ratio 1 : 2

Lida's amount is equal to the mean of Samira and Rhys's amounts.

Work out the amount of money that Rhys gets.

4 The angles in a triangle are in the ratio 1 : 5 : 6

Show that the triangle is right-angled.

Are you ready? (B)

1 Write the highest common factor of: **a** 12 and 30 **b** 40 and 64

2 Write each fraction in its simplest form.

     a $\dfrac{4}{12}$ **b** $\dfrac{15}{20}$ **c** $\dfrac{16}{28}$ **d** $\dfrac{18}{60}$

3 Write each ratio in its simplest form.

     a 15 : 3   **b** 25 : 30   **c** 24 : 40   **d** 84 : 36

4 Convert:

     a 3.5 kilograms to grams

     b 1.42 metres to centimetres

     c 6.07 litres to millilitres.

Example

Ali has £3 and Faith has 90p.

Work out the ratio of the amount of money Ali has to the amount of money Faith has.

Method

Solution	Commentary
£3 : 90p	Firstly, write the amounts as a ratio.
300p : 90p	To compare them, they need to be in the same units so change £3 to 300p.
$\overset{\div 10}{\underset{\div 3}{}}\ \ 300 : 90 \ \ \overset{\div 10}{}$ 30 : 9 10 : 3	Now they are in the same form, you do not need the units.
	10 is a factor of both 300 and 90, so divide both numbers by 10
	You can divide both numbers by 3 to simplify the ratio.

You could do this more quickly by looking for the highest common factor.

The highest common factor of 300 and 90 is 30, so you can divide both numbers by 30

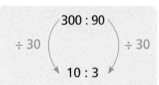

Practice (B)

1 Simplify the ratios.

     a £2 : 50p   **b** £1.50 : £4   **c** 20p : £3   **d** 80p : £2.40   **e** £5 : £3.75

2 Simplify the ratios.

     a 1 kg : 200 g   **b** 2.5 kg : 1000 g

     c 0.5 kg : 750 g : 1 kg   **d** 375 g : 3.75 kg

3 Simplify the ratios.

  a 300 ml : 1 litre

  b 0.7 litres : 350 ml

  c 2500 ml : $\frac{1}{4}$ litre

  d $4\frac{1}{2}$ litres : 900 ml

4 Simplify the ratios.

  a 5.6 km : 8000 m

  b 17 500 m : 5 km

  c 35 m : 35 cm

  d 750 m : $\frac{3}{8}$ km

What do you think? (B)

1 Jakub saves 60% of his monthly pocket money and spends the rest.

Work out the ratio of money Jakub saves each month to how much he spends each month. Give your answer in its simplest form.

2 Beca has 10 coins in her purse.

There are twice as many 10p coins as 20p coins.

The number of 50p coins she has sums to £3.50

Write the ratio of 10p coins to 20p coins to 50p coins.

3 Rhys says the ratio 1 m : 2 cm : 5 mm cannot be simplified.

Show that he is wrong.

4 The ages of Ed and Faith sum to 25

The ages of Faith and Bev sum to 20

Ed is twice as old as Bev.

Write the ages of Ed, Faith and Bev as a ratio in its simplest form.

5 Ed and Marta share counters in the ratio 3 : 5

Marta gives Ed two counters.

They now share the counters in the ratio 2 : 3

Work out how many counters they share in total.

Consolidate – do you need more?

 1 Share:

  a £400 in the ratio 3 : 7

  b 64 strawberries in the ratio 5 : 3

  c 126 mm in the ratio 1 : 2 : 6

  d 1.2 kg in the ratio 1 : 5, giving your answer in grams.

2 Sven and Huda share money in the ratio 3 : 5

    a If they share £240, how much do they each get?

    b If Sven gets £240, how much does Huda get?

    c If Huda gets £240, how much do they share in total?

    d If Huda gets £240 more than Sven, how much do they share in total?

3 Simplify the ratios.

    a £3 : 50p   **b** £2.50 : £6   **c** 80p : £3   **d** 60p : £2.40   **e** £7 : £4.20

4 Simplify the ratios.

    a 1 kg : 400 g   **b** 2.5 kg : 200 g   **c** 0.5 kg : 250 g : 3 kg   **d** $\frac{5}{8}$ kg : 1875 g

5 Simplify the ratios.

    a 700 ml : 1 litre   **b** 0.6 litres : 360 ml   **c** 4900 ml : 0.35 litres   **d** $1\frac{3}{4}$ litres : 750 ml

6 Simplify the ratios.

    a 4 km : 800 m   **b** 1500 m : $\frac{7}{8}$ km   **c** 0.45 m : 225 cm   **d** 1.2 km : 120 000 cm

Stretch – can you deepen your learning?

1 A box contains milk chocolates and dark chocolates.

The chocolates are either in the shape of a square or a triangle.

The ratio of milk chocolates to dark chocolates is 7 : 3

The ratio of square chocolates to triangle chocolates is 5 : 7

    a Work out the least possible number of chocolates in the box.

There are fewer than 150 chocolates in the box.

    b Work out the greatest possible number of square milk chocolates.

2 The average speed of three runners, A, B and C, is in the ratio 16 : 15 : 20

Runner C can run a distance of 5 km in 30 minutes.

Work out by how many seconds Runner A would run 1 km faster than Runner B.

3 a and b are two numbers such that $a > b$.

When you subtract 3 from both a and b, the answers are in the ratio 3 : 1

When you add 12 to both a and b, the answers are in the ratio 3 : 2

Work out the ratio $a : b$, giving your answer in its simplest form.

Are you ready?

1 Write each ratio in its simplest form.

   a 15 : 3   **b** 25 : 30   **c** 24 : 40

2 The ratio of flats to houses in a town is 2 : 5

   Write the ratio of flats to houses in the form: **a** 1 : n **b** n : 1

3 A bar model shows the ratio of electric cars to non-electric cars in a car park.

   a Write the ratio of electric cars to non-electric cars.

   b What fraction of all the cars are electric?

   c What fraction of all the cars are non-electric?

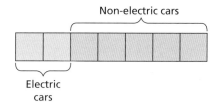

Non-electric cars

Electric cars

Example 1

The ratio of yellow counters to blue counters in a bag is 4 : 3

a What fraction of the counters are blue?

b Bev says, "The number of yellow counters is one-third greater than the number of blue counters."

   Is Bev correct? Explain why.

Method

Solution	Commentary
a Y B $\frac{3}{7}$ of the counters are blue.	Drawing a single bar model and labelling 4 parts yellow and 3 parts blue can help show why the total number of parts is 7
b Y B $4 = \frac{4}{3} \times 3 = 3 + \left(\frac{1}{3} \times 3\right)$ Yes, Bev is correct.	By using a comparison bar model, you can see there is 1 part more representing the yellow counters than the blue counters. So you can say that 1 part is equal to $\frac{1}{3}$ of the total number of blue counters. Alternatively, you could say 1 part is equal to $\frac{1}{4}$ of the yellow counters.

Example 2

$\frac{3}{10}$ of the animals at a farm are cows and the rest are sheep.

Write the ratio of cows to sheep.

Method

Solution	Commentary
Cows Sheep The ratio of cows to sheep is 3 : 7	The number of parts representing cows is 3 The number of parts representing sheep is 7 So the ratio of cows to sheep is 3 : 7

Practice

1. The ratio of part-time to full-time staff in an office is 1 : 4

   What fraction of the staff work part-time?

2. Sven, Junaid and Zach share some money in the ratio 7 : 2 : 6

   What fraction of the money does Zach receive? Write your answer in its simplest form.

3. Marta has some green, purple and yellow felt-tip pens in the ratio 5 : 2 : 3

   a   What fraction of the pens are yellow?

   b   What fraction of the pens are **not** purple?

4. $\frac{2}{5}$ of Abdullah's class are girls.

   He says that the ratio of boys to girls is 5 : 2

   a   Explain why Abdullah is wrong.

   b   Write the correct ratio of boys to girls.

5. Three-quarters of a drink is water and the rest is cordial.

   a   Write the ratio of water to cordial.

   b   Write the ratio of water to the whole drink.

6. Ed has a starter, a main meal and a dessert at a café.

   The cost of his main meal is 60% of the total bill.

   The cost of his starter and dessert are in the ratio 2 : 3

   What fraction of Ed's bill is the cost of his dessert?

What do you think? 💭

1 The ratio of blue tiles to white tiles in a shop is $x : y$.

What fraction of the tiles are blue?

2 Amina, Beca and Chloe paint a room together.

Amina paints $\frac{1}{3}$ of the room.

The rest of the room is painted by Beca and Chloe in the ratio 5 : 4

What proportion of the room does Beca paint?

Consolidate – do you need more?

1 The ratio of mammals to reptiles in a zoo is 5 : 3

What fraction of the animals at the zoo are reptiles?

2 Rhys, Faith and Lida share some toffees in the ratio 4 : 7 : 3

What fraction of the toffees does Rhys receive? Write your answer in its simplest form.

3 A supermarket stocks dairy, soya and oat milk in the ratio 9 : 2 : 3

   a What fraction of the supermarket milk is oat?

   b What fraction of the supermarket milk is dairy-free?

4 At the gym, Kath either runs on the treadmill or attends a class.

Kath spends 55% of her time at the gym on the treadmill.

Write the ratio of the time Kath spends on the treadmill to the time she spends attending a gym class.

5 Four-fifths of the stalls at a market hall sell food. The rest are clothing stalls.

   a Write the ratio of food stalls to clothing stalls.

   b Write the ratio of clothing stalls to all stalls.

6 A skyscraper is made of steel, concrete and glass.

40% of the skyscraper is made of glass.

The ratio of steel to concrete is 11 : 4

What fraction of the skyscraper is made of steel?

Stretch – can you deepen your learning?

1 The ratio of $(a - b) : 2a$ is equivalent to $1 : k$.

Write an expression for b in terms of a and k.

2 $\dfrac{1}{4} : \dfrac{2}{3} = m : 1$

Work out the value of m.

3

The area of triangle E is 10% less than the area of triangle F.

The base length of triangle E is $\dfrac{3}{4}$ of the base length of triangle F.

Work out the ratio $x : y$.

4 The diagram shows a cylindrical tank containing oil and water.

The tank's height and diameter are 3 m and 1.4 m respectively.

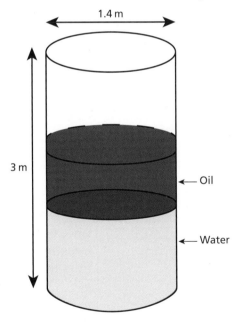

The tank is 40% empty.

The ratio of the volume of oil to the volume of water in the tank is 2 : 3

Given that 1 litre = 1000 cm³, work out the number of litres of oil in the tank.
Give your answer to 3 significant figures.

Are you ready? (A)

1 Work out the lowest common multiple of 24 and 16

2 $15 : 18 = a : 30$

Work out the value of a.

3 The ratio of horses to sheep on a farm is 1 : 3

The ratio of sheep to cows on the same farm is 3 : 2

Write the ratio of horses to cows on the farm.

One way of viewing a ratio is to think of each part as a fraction of the other part.

If Ed and Marta share some money in the ratio 2 : 3, then:

- Ed has $\frac{2}{3}$ of Marta's amount
- and Marta has $\frac{3}{2}$ $\left(\text{or } 1\frac{1}{2}\right)$ lots of Ed's amount.

Ed

Marta

This can be seen in the bar model.

When working with **equivalent ratios**   if $a : b \equiv c : d$ then $\frac{a}{b} \equiv \frac{c}{d}$

Example 1

A park has elm, ash and oak trees.

The ratio of elm trees to ash trees in the park is 3 : 2

The ratio of ash trees to oak trees in the park is 8 : 11

Write the ratio of elm to ash to oak trees.

Method

Solution	Commentary
Elm [ ] Elm [ ][ ][ ][ ][ ][ ][ ][ ][ ][ ][ ][ ] Ash [ ] Ash [ ][ ][ ][ ][ ][ ][ ][ ] Ash [ ][ ][ ][ ][ ][ ][ ][ ] Oak [ ][ ][ ][ ][ ][ ][ ][ ][ ][ ][ ]	Since both the given ratios mention ash trees, make the size of the parts for ash trees the same in both ratios.
$\quad\quad\quad\quad\quad \times 4$ E : A $\quad$ 3 : 2 $\quad \to \quad$ 12 : 8 $\quad\quad\quad\quad \times 4$ A : O $\quad$ 8 : 11 $\quad\quad\quad$ 8 : 11	You can do this by writing the ratio of elm to ash trees as 12 : 8
E : A : O 12 : 8 : 11	Now write the ratio of elm to ash to oak.

Example 2

$12 : x = x - 1 : 1$ where $x > 0$

Work out the value of x.

Method

Solution	Commentary
$\dfrac{12}{x} = \dfrac{x-1}{1}$	Write the the equivalent ratios as an equation involving fractions.
$12 = x(x-1)$ $12 = x^2 - x$	Cross-multiply to form a quadratic equation.
$0 = x^2 - x - 12$	Rearrange the quadratic equation so it is equal to zero and factorise.
$0 = (x-4)(x+3)$ $x = 4$ since $x > 0$	$x = 4$ or $x = -3$ But $x > 0$, therefore $x = 4$

Practice (A)

1 $x : y = 2 : 5$ and $y : z = 10 : 3$

Write the ratio $x : y : z$.

2   The ratio of red pens to green pens is $5 : g$.

The ratio of red pens to blue pens is $3 : b$.

Work out the ratio of red to blue to green pens.

3 $20 - m : 4m = 1 : 1$

Work out the value of m.

4 $y - 8 : 1 = 20 : y$ where $y > 0$

Work out the value of y.

5   A supermarket stocks bottles of skimmed, semi-skimmed and whole milk.

The number of bottles of skimmed milk to whole milk is in the ratio $4 : 5$

The number of bottles of semi-skimmed milk is three times the number of bottles of skimmed milk.

Work out the minimum possible number of bottles of milk at the supermarket.

6   The ratio of students in Year 10 to Year 11 is $7 : 8$

The ratio of the total number of students in Years 10 and 11 to the total number of students in Years 7, 8 and 9 is $4 : 3$

Write the fraction of the total number of students that are in Year 11.

What do you think? (A)

1 In the ratio $a : b : c : d$,

b and d are in the ratio $3 : 4$

b is 20% greater than a.

The ratio $a + d : b + c = 1 : 1$

Work out the ratio $a : b : c : d$ in its simplest form.

2 w, x, y and z are all integers.

The ratio $w : x = x : y = y : z = 5 : 6$

The sum of w, x, y and z is $11k$ where k is an integer.

Work out the value of k.

Are you ready? (B)

1 Simplify $\dfrac{4a^2}{10a}$

2 Multiply out $x(2 - x)$

3 Rearrange to make k the subject of the formula $x = \dfrac{3}{2k}$

4 The ratio of yellow buttons to red buttons in a box is $y : r$.

What fraction of the buttons are red?

Example 1

A box contains only apples and oranges.

The ratio of apples to oranges in the box is $5 : 2$ and there are 30 apples in the box.

Some oranges are added to the box so the ratio of apples to oranges is now $10 : 9$

How many oranges are added to the box?

Method

Solution	Commentary
$30 \div 5 = 6$, so each part of the ratio represents six pieces of fruit. $6 \times 2 = 12$, so there were 12 oranges to start with.	There are 30 apples, so you can work out the number of pieces of fruit represented by each part of the ratio. Write down how many there were initially.
$30 : 12 + n = 10 : 9$ $\dfrac{30}{12 + n} = \dfrac{10}{9}$	Let the number of oranges added be n. Write the ratio as an equation involving fractions.
$270 = 120 + 10n$ $150 = 10n$ $15 = n$, so 15 oranges are added.	Solve the equation.

Example 2

The ratio of $p : q = 3 : 8$

The ratio of $p + 30 : q - 30 = 5 : 6$

Work out the ratio of $p - 5 : \dfrac{q}{3}$

Method

Solution	Commentary
$\dfrac{p}{q} = \dfrac{3}{8}$ and $\dfrac{p+30}{q-30} = \dfrac{5}{6} \Rightarrow 6p + 180 = 5q - 150$	Write each of the ratios as equations involving fractions.
$8p - 3q = 0$ (A) $6p - 5q = -330$ (B)	Rearrange the equations to form simultaneous equations.
$24p - 9q = 0$ (C) $24p - 20q = -1320$ (D) $11q = 1320$ $q = 120$	Make the coefficients of p the same by multiplying equation (A) by 3 and (B) by 4 Label the new equations (C) and (D) respectively. Subtract equation (D) from (C). Solve the equation to find the value of q.
$8p - 360 = 0$ $p = 45$	Substitute the value of q into (A) and find the value of p.
$p - 5 : \dfrac{q}{3} = 45 - 5 : \dfrac{120}{3} = 40 : 40 = 1 : 1$	Use the values of p and q to find the ratio $p - 5 : \dfrac{q}{3}$

Practice (B)

1 The ratio of Flo's age to Jackson's age is 5 : 4

Two years ago their ages were in the ratio 9 : 7

How old is Flo?

2 At the start of the day, the ratio of lemons to limes in a shop is 11 : 9

By the end of the day, 75% of the lemons have been sold and 14 limes have been sold.

The ratio of lemons to limes is now 1 : 2

Work out the total number of lemons and limes in the shop at the start of the day.

3 Filipo and Lida share raspberries in the ratio 5 : 4

Filipo gives two raspberries to Lida.

The ratio of Filipo's raspberries to Lida's raspberries is now 1 : 1

How many raspberries did they share?

4 On Monday, Class A has $3f$ chairs and Class B has $4f$ chairs.

On Tuesday, one chair is added to Class A and four chairs are taken from Class B.

The number of chairs in Class A to the number of chairs in Class B on Tuesday is in the ratio 7 : 8

How many chairs were in each class on Monday?

5 Given that $y^2 : (y - 4) = 16 : 1$, work out the value of y.

6 Given that $w^2 + 3 : 3w - 3 = 7 : 3$, work out the value of w.

7 $x - 1 : x - 3 = 3x + 4 : 3x - 4$

Work out the value of x.

8 The ratio of $t - 3 : u - 3 = 4 : 3$

The ratio of $t + 5 : u + 5 = 6 : 5$

Work out the ratio of $t : u$.

What do you think? (B) 💡

1 $m = \sqrt{5} + \sqrt{p}$ and $n = \sqrt{20} + \sqrt{q}$ where p and q are positive integers.

Given that $p : q = 1 : 4$, work out the ratio $m : n$ in its simplest form.

2 The ratio of $g : h = 2 : 5$

Work out the ratio of $2h : 5g - h$ in its simplest form.

3 $(p - 2)^2 + q : p^2 + (q - 1)^2 = 4 : 9$

$pq : \dfrac{p}{q} = 16 : 1$ where p and q are positive integers.

Work out the values of p and q.

4 $(9 - k)^2 : 2(k^2 + 1) = 1 : 25$ where k is a positive integer.

Work out the value of k.

Consolidate – do you need more?

1 $j : k = 6 : 5$ and $j : l = 8 : 5$

Write the ratio $j : k : l$.

2 Abdullah and Amina share blackberries in the ratio 13 : 9

Abdullah gives three blackberries to Amina.

The ratio of Abdullah's blackberries to Amina's blackberries is now 6 : 5

How many blackberries did they share?

3 The ratio of Chloe's age to Jakub's age is 4 : 3

Four years ago their ages were in the ratio 3 : 2

How old is Chloe?

4 On Sunday, the number of trains and buses that arrived in Halifax was $5c$ and $8c$ respectively.

On Monday, the number of trains that arrived in Halifax increased by 80% and the number of buses that arrived in Halifax increased by 16

The number of trains that arrived in Halifax to the number of buses that arrived in Halifax on Monday was 9 : 10

How many trains arrived in Halifax on Monday?

5 The ratio of $c - 1 : d - 2 = 9 : 5$

The ratio of $c + 2 : d + 2 = 4 : 3$

Work out the ratio of $c : d$.

Stretch – can you deepen your learning?

1 In the ratio $p : q : r : s$,

$q : r$ is in the ratio 4 : 9

s is 50% greater than q

The ratio $qr : ps = 2 : 1$

Work out the ratio $p : q : r : s$ in its simplest form.

2 e, f, g and h are all integers.

The ratio $e : f = f : g = g : h = 2 : 3$

The mean of e, f, g and h is 65

Work out the range of e, f, g and h.

3 The ratio of $v : w = 3 : 5$ and the ratio of $w : x = 6 : 1$

Work out the ratio of $2v + w : 2w + x$ in its simplest form.

4 a and b are positive integers.

$ab : (a + b) = 3 : 2$

Work out the three possible pairs of values for a and b.

Ratio: exam practice

4–6

1 There are 80 doctors and 210 nurses working in a hospital.

Write the ratio of doctors to nurses in the form $1 : n$ **[2 marks]**

---

2 Seb and his sister share some sweets in the ratio 2 : 3

What fraction of the sweets does Seb receive? **[1 mark]**

---

3 A sum of money is shared in the ratio 3 : 4 : 5

The smallest share is £420

How much money is shared? **[2 marks]**

---

4 The scale of a map is 1 : 50 000

Calculate the actual distance between two villages that are 12.3 cm apart on the map.

Give your answer in kilometres. **[3 marks]**

7–9

---

5 Given that $\dfrac{a}{b} = \dfrac{2}{5}$ and $\dfrac{b}{c} = \dfrac{3}{7}$,

work out $a : b : c$ **[3 marks]**

---

6 $x : y = 2 : 3$

A graph is drawn showing y as a function of x.

Write down the equation of the graph. **[2 marks]**

---

7 Benji and Samira share some money in the ratio 4 : 3

Benji gives Samira £200

The ratio of the amount that Benji has to the amount that Samira has is now 8 : 13

Work out the amount of money they originally shared. **[3 marks]**

17 Proportion

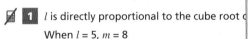
White Rose MATHS

In this block, we will cover...

17.1 Direct proportion

Example 1

The table shows the number of hours worked an employee who is paid at a constant hourly rate.

Hours worked	1	5	10	16
Pay (£)		80		

a   Complete the table.

b   Work out how much pay an employee would

c   Work out how many whole hours an employ

d   Write an expression for the amount of pay an

17.2 Inverse proportion

Practice (A)

1   The time it takes to travel 30 miles depend

Speed (mph)	5	10	20	30	4
Time (h)					

a   Copy and complete the table.

b   What happens when the speed is doub

c   Plot a graph of speed against time for
    horizontal axis and speed on the vertic

2   Three pipes flowing at a constant rate can

17.3 Proportion with powers

Consolidate – do you need more

1   l is directly proportional to the cube root c
    When $l = 5$, $m = 8$

    a   Work out the value of l when $m = 64$

    b   Work out the value of m when $l = 20$

2   d is inversely proportional to the square ro
    When $d = 5$, $f = 2$

    a   Work out the value of d when $f = 8$

    b   Work out the value of f when $d = 20$

Are you ready? (A)

1 Four tins of custard cost £4.86

Work out the cost of:

a 12 tins of custard   **b** 18 tins of custard   **c** 9 tins of custard.

2 Here is a recipe to make 8 flatbreads:

a How much yoghurt and flour are needed to make 20 flatbreads?

b Ed made some flatbreads and used 1575 ml of yoghurt.

Work out how many flatbreads Ed made.

> **Flatbread** (makes 8)
> 450 ml yoghurt
> 125 g flour

3 The exchange rate of British pounds (£) to Norwegian krone (kr) is £1 = 13.05 kr

a Convert £40 into Norwegian krone (kr).

b Convert 500 kr into British pounds (£).

When two quantities are in **direct proportion**, as one quantity increases or decreases, the other quantity increases or decreases at the same rate.

Double number lines can show the multiplicative relationship between quantities in direct proportion, such as exchange rates.

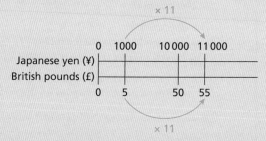

When one of the quantities is multiplied or divided by an amount, the other quantity is also multiplied or divided by the same amount.

Proportion diagrams

These can be used to show different multiplicative relationships and are helpful when problem-solving.

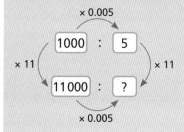

Conversion graphs

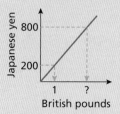

These are useful for converting from one currency to another and for converting between different units of measurement.

A graph that shows two variables X and Y in direct proportion is a straight line that goes through the origin, (0, 0).

When two variables are directly proportional, the ratio of the variables stays constant.

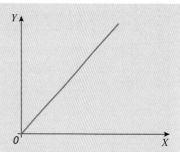

The table below shows the relationship between the values of a variable Y when another variable, X, varies.

X	0	8	12	30	45
Y	0	6	9	22.5	33.75

Excluding when $X = 0$, the ratio $X : Y = 4 : 3$ for each pair of values, or alternatively $\frac{X}{Y} = 1.\dot{3}$

Example 1

The table shows the number of hours worked and the amount of pay earned by an employee who is paid at a constant hourly rate.

Hours worked	1	5	10	16
Pay (£)		80		

a Complete the table.

b Work out how much pay an employee would earn for 7 hours of work.

c Work out how many whole hours an employee would need to work to earn over £200

d Write an expression for the amount of pay an employee would earn for working x hours.

Method

Solution	Commentary
a $\div 5$ $\times 2$ 	Divide by 5 to work out how much an employee earns for 1 hour's work (hourly rate).
Hours worked: 1, 5, 10, 16 / Pay (£): 16, 80, 160 $\div 5$ $\times 2$	Since 10 is double 5, you can double £80
Hours worked: 1, 5, 10, 16 / Pay (£): 16, 80, 160, 256	
$1 + 5 + 10 = 16$ £16 + £80 + £160 = £256	16 can be partitioned into 10, 5 and 1 so you can find the sum of the pay for each of these hours worked.
b ? £16 £16 £16 £16 £16 £16 £16 $7 \times £16 = £112$	Multiply the hourly pay by the number of hours worked.

337

c £200 ÷ £16 = 12.5 hours They would need to work for 13 hours.	Work out the number of hours by dividing the total pay by the hourly rate, and then rounding.
d 16x	For each hour, the employee earns £16. So for x hours worked, multiply the hourly rate by the number of hours worked.

Example 2

The graph can be used to convert between British pounds and Canadian dollars.

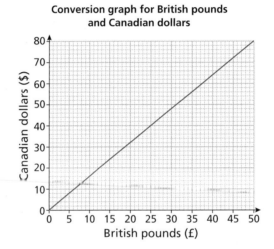

Conversion graph for British pounds and Canadian dollars

Use the graph to convert:

a £40 into Canadian dollars   **b** $48 into pounds   **c** £350 into Canadian dollars.

Method

Solution	Commentary
a 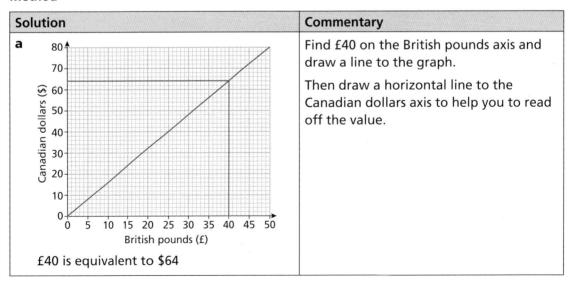 £40 is equivalent to $64	Find £40 on the British pounds axis and draw a line to the graph. Then draw a horizontal line to the Canadian dollars axis to help you to read off the value.

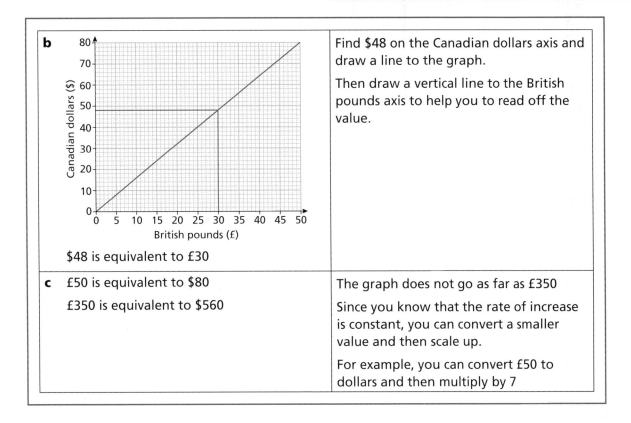

b

$48 is equivalent to £30

Find $48 on the Canadian dollars axis and draw a line to the graph.

Then draw a vertical line to the British pounds axis to help you to read off the value.

c   £50 is equivalent to $80

£350 is equivalent to $560

The graph does not go as far as £350

Since you know that the rate of increase is constant, you can convert a smaller value and then scale up.

For example, you can convert £50 to dollars and then multiply by 7

Practice (A)

1 The amount of honey produced in a beehive is directly proportional to the number of bees in the hive.

A hive with 40 000 bees produces 50 litres of honey.

a Copy and complete the table.

Number of bees	10 000	20 000	40 000	50 000
Number of litres of honey			50	

b Work out the number of bees required to produce 60 litres of honey.

c Assuming that all bees work at the same rate, work out the amount of honey produced by a single bee. Give your answer in millilitres.

2 The cost of soil from a garden supplier is directly proportional to its volume. 40 litres of soil costs £16

a Copy and complete the table to show the costs of different volumes of soil.

Volume of soil, V (litres)	0	10	20	30	40	50	60
Cost of soil, C (£)	0				16		

b Draw the graph of the cost of soil, C, against volume of soil, V.

c Use your graph to work out:

    i the cost of 25 litres of soil   **ii** how much soil you can buy for £50

3 Which of the tables represent two quantities that are directly proportional?

Table 1

A	0	6	12	18	30
B	0	20	40	60	100

Table 2

C	5	45	85	125	205
D	0	20	40	60	100

Table 3

E	0	50	200	1800	5000
F	0	20	40	60	100

4 Which of the graphs represent two quantities that are directly proportional to each other?

A	**B**	**C**	**D**
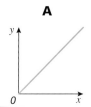		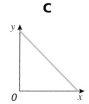	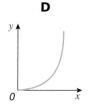

5 The graph shows how to convert between euros (€) and Turkish lira (₺).

Use the graph to convert:

a €15 into Turkish lira

b ₺60 into euros

c €50 into Turkish lira.

d The exchange rate between US dollars and euros is $1 = €0.90

Write the exchange rate for US dollars to Turkish lira.

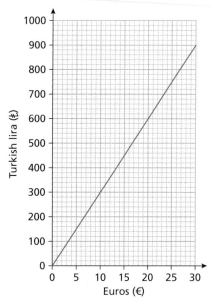

6 It takes a robot 110 minutes to cut a grass field of area 5400 m². Assume that the robot works at a constant rate.

a Work out the area of grass that the robot can cut in 1 hour. Give your answer to 3 significant figures.

b Work out how long it would take the robot to cut the grass in a field with area 100 m². Give your answer to the nearest minute.

7 A train travels at a constant speed. In 1 hour 40 minutes, it covers a distance of 160 miles.

How far did the train travel in 25 minutes?

> You will explore speed in more detail in Chapter 18.1

8 Graph 1 can be used to convert between the radius of a circle and the area of the same circle. Graph 2 shows how to convert between temperatures in Celsius and Fahrenheit.

Graph 1

Conversion graph for the radius and area of a circle

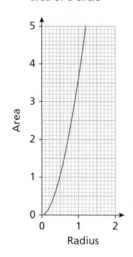

Graph 2

Conversion graph for temperatures in Celsius and Fahrenheit

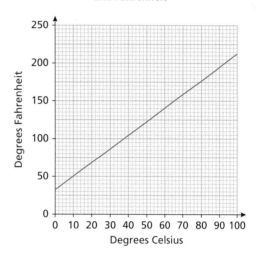

Filipo says, "Both graphs show that as one value increases, the other increases too, so both graphs are in direct proportion."

For each graph, give a reason why Filipo is incorrect.

What do you think? (A)

1 P is directly proportional to Q.

When P is 400, Q is 25

a Work out the ratio of P to Q.

Q is directly proportional to R.

$Q : R = 2 : 3$

b Work out the value of P when R is 6

2 Huda makes 60 peppermint creams using m kg of chocolate.

Write an expression for the amount of chocolate required for Huda to make 150 peppermint creams.

3 The total mass of 3 small boxes and 5 large boxes is 2125 g.

The total mass of 8 small boxes is 1 kg.

Write an expression for the total mass of a small boxes and b large boxes.

Are you ready? (B)

1 Solve the equation $60 = 4j$

2 Which of the following points lie on the line $y = 5x$?

(2, 10)   (7, 12)   (20, 4)   (0.7, 3.5) $(c, 5c)$

If y is directly proportional to x, it can be written as:

$y \propto x$ where $\propto$ means 'directly proportional to'.

This can also be written as:

$y = kx$ where k is the **constant of proportionality**.

In Example 1 on page 337, the constant of proportionality is 16 because pay (£p) is equal to the number of hours worked (h) multiplied by 16, that is $p = 16h$.

The graph of a proportional relationship $y = kx$ will be a straight line going through the origin (0, 0).

Equations of straight lines are covered in Chapter 7.1

Example 1

The total amount of animal feed, in kilograms, eaten at a farm is directly proportional to the number of animals on the farm.

294 kg of animal feed are eaten when 84 animals are on the farm.

a Work out the amount of animal feed eaten when there are 105 animals on the farm.

b Work out the number of animals that would eat exactly 21 kg of animal feed.

Method

Solution	Commentary
a Let f = amount of feed and a = number of animals	Define your variables so it is clear what you are working out at any point.
$f \propto a$ $f = ka$ $294 = 84k$ $k = 3.5$	Form and solve an equation to find the constant of proportionality using the values given in the question. You could write the value of k as an exact fraction, $k = \dfrac{7}{2}$
$f = 3.5a$ $f = 3.5 \times 105$ $f = 367.5$ kg	Use your value of k to form an equation connecting f and a. Substitute in the number of animals (a) to find the total amount of animal feed eaten.
b $f = 3.5a$ $21 = 3.5a$ $a = \dfrac{21}{3.5} = 6$	Start with the formula you worked out in part **a**. Substitute $f = 21$ Rearrange to find the value of a.

Example 2

The graph shows how y varies with x.

a Work out the equation of the straight line.

b Work out the value of y when $x = 60$

c Show that the point (544, 680) lies on the line.

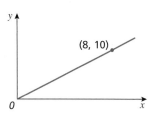

Method

Solution	Commentary
a $y = mx$ $m = \dfrac{\text{change in } y}{\text{change in } x} = \dfrac{10}{8} = \dfrac{5}{4}$ $y = \dfrac{5}{4}x$	The graph shows a straight line through the origin, which means that y is directly proportional to x. The equation of a straight line through the origin is of the form $y = mx$. Therefore, the gradient represents the constant of proportionality. Work out the gradient of the straight line to find the equation connecting y and x.
b $y = \dfrac{5}{4} \times 60$ $y = 75$	Substitute in the value of x to find the value of y.
c $y = \dfrac{5}{4}x$ $680 = \dfrac{5}{4}x$ $x = 544$	Use the equation found in part **a**, this time substituting either x or y to find the value of y or x respectively.

Practice (B)

  1 a is directly proportional to b.

a is given by the formula $a = 0.9b$

a Work out the value of a when $b = 45$

b Work out the value of b when $a = 45$

 2 $c \propto d$

Which of the following graphs could represent the relationship of c and d?

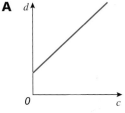

A

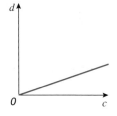

B

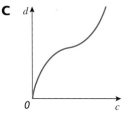

C

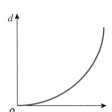

D

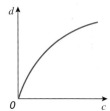

E

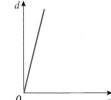

F

3 w is directly proportional to z. When $w = 804$, $z = 134$

   a Write a formula connecting w and z.

   b Work out the value of w when $z = 51$

   c Work out the value of z when $w = 222$

4 The force applied to an object is directly proportional to the acceleration of the object.

   When the force on the object is $37.5\,N$, the acceleration is $50\,m/s^2$.

   a Work out the force applied to an object when the acceleration is $21\,m/s^2$.

   b Work out the acceleration of the object when the force applied is $2\,N$.

5 d is directly proportional to f. When $d = 0.2$, $f = 11$

   a Work out the value of d when $f = 121$ **b** Work out the value of f when $d = 13.8$

6 j is directly proportional to l. When $j = \dfrac{3}{4}$, $l = \dfrac{2}{3}$

   a Work out the value of j when $l = 32$ **b** Work out the value of l when $j = 621$

7 The graph shows how y varies with x.

   a Work out the value of y when $x = 35$

   b Show that the point $(2\dfrac{2}{3}, 8)$ lies on the line.

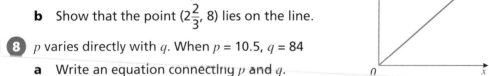

8 p varies directly with q. When $p = 10.5$, $q = 84$

   a Write an equation connecting p and q.

   b Work out the value of p when $q = 14$, giving your answer as a mixed number.

   c Work out the value of q when $p = 122$

What do you think? (B)

1 t varies directly with u. The ratio of $t : u = 2 : 3$

   a Write a formula for t in terms of u.

   b Work out the value of t when $u = \dfrac{3}{4}$

   c Work out the value of u when $t = \dfrac{26}{27}$, giving your answer as a mixed number.

2 x is directly proportional to y. When x is 60, y is 48

   y is directly proportional to z. When $y = 1.44$, $z = 0.12$

   Write a formula for x in terms of z.

3 The table shows pairs of values for x and y.

x	400	$r + 6$
y	$3r$	21

   a Work out the value of r when x is directly proportional to y.

   b Write a formula for y in terms of x.

   c Write a formula for x in terms of y.

Consolidate – do you need more?

1 The mass of CO_2 (kg) absorbed by a garden in a year is directly proportional to the number of trees in the garden. A garden of 300 trees absorbs 6600 kg.

   a Copy and complete the table.

Number of trees	100	200	300	500
Mass of CO_2 absorbed (kg)			6600	

   b Work out how many kilograms of CO_2 a single tree absorbs.

   c Work out the number of trees required to absorb at least 5000 kg of CO_2.

2 The graph shows how to convert between British pounds (£) and Swiss francs (CHF).

Use the graph to copy and complete the table.

British pounds (£)	0				36		
Swiss francs (CHF)	0	10	20	30	40	50	60

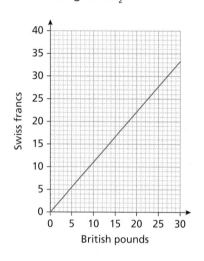

3 e is directly proportional to f.

   e is given by the formula $e = 0.7f$

   a Work out the value of e when $f = 42$

   b Work out the value of f when $e = 42$

4 g is directly proportional to h. When $g = 2.5$, $h = 10$

   a Work out the value of g when $h = 72$

   b Work out the value of h when $g = 13.5$

5 w varies directly with x. When $w = 12$, $x = 64$

   a Write an equation connecting w and x.

   b Work out the value of w when $x = 160$

   c Work out the value of x when $w = 123$

Stretch – can you deepen your learning?

1 The table shows expressions for x and y in terms of q, where $q \neq 1$

x	$q - 1$	$5(q - 1)$
y	$2(q - 5)$	$2(3q - 1)$

Form and solve an equation to find the value of q when x is directly proportional to y.

2 The fuel tank of a car holds n litres of petrol.

The car can travel 400 miles on a full tank of petrol.

After driving n miles, there are 43.75 litres of petrol left in the tank.

Stating any assumptions you have made, work out the two possible values of n.

Hint: You will need to form and solve a quadratic equation. See Block 11.

Are you ready? (A)

1 The area of a rectangle is 24 cm².

Work out the width if the length of the rectangle is:

    a 8 cm   **b** 16 cm   **c** 1.6 cm   **d** 5 cm   **e** $\frac{1}{5}$ cm

2 Work out the average speed of a car, in miles per hour, if it travels:

    a 80 miles in 2.5 hours   **b** 25 miles in 80 minutes.

If two quantities are in **inverse proportion**, when one quantity increases, the other decreases at the same rate.

For example, a factory uses machines to make milk bottles.

It takes one machine 4 working days to produce 1000 bottles.

However, if there are *more* of the same machines, it will take *less* time. If there are two identical machines, the 1000 milk bottles can be produced in half the time, 2 days.

Number of machines	Number of working days
1	4
2	2

$\times 2$ (on machines), $\div 2$ (on days)

If there are three machines, it will take $1\frac{1}{3}$ days to produce 1000 milk bottles.

Number of machines	Number of working days
1	4
3	$1\frac{1}{3}$

$\times 3$ (on machines), $\div 3$ (on days)

This is an example of inverse proportion. When you increase the number of machines, the time taken decreases at the same rate.

Number of machines	Number of days	
1	4	$1 \times 4 = 4$
2	2	$2 \times 2 = 4$
6	$\frac{2}{3}$	$6 \times \frac{2}{3} = 4$

$\times 2$, $\div 2$; $\times 3$, $\div 3$

Notice that the product of the number of machines and the number of days is always 4. This is because 4 days' worth of production time are being shared between the machines.

Example 1

A swimming pool is being filled with water by hosepipes. It takes 5 hosepipes 4 hours to fill the pool.

Assuming all hosepipes work at the same rate:

a   how long will it take 2 hosepipes to fill the swimming pool

b   how many hosepipes will it take to fill the swimming pool in 30 minutes?

Method

Solution	Commentary
a $\div 5$ (5 hosepipes 4 hours) $\times 5$ $\times 2$ (1 hosepipe 20 hours) $\div 2$ 2 hosepipes 10 hours So 2 hosepipes will take 10 hours.	Start by working out how long one hosepipe will take. Since you have five times fewer hosepipes, it will take five times as long to do the work.
	If you now double the number of hosepipes, it will take half the time.
b $\times 8$ (5 hosepipes 4 hours) $\div 8$ 40 hosepipes 0.5 hours So it will take 40 hosepipes to do the work in 30 minutes.	Five hosepipes take 4 hours.
	30 minutes = 0.5 hours
	To finish in 0.5 hours would mean dividing the time by 8, so eight times as many hosepipes are needed.

Example 2

The variables x and y are inversely proportional to each other.

x	10	20	30		
y	24			6	4

a   Complete the table.

b   Express this relationship in the form $xy = k$, where k is a constant to be found.

c   Using each pair of x and y values, plot the graph of $xy = k$.

Method

Solution	Commentary
a <table><tr><td>x</td><td>10</td><td>20</td><td>30</td><td>40</td><td>60</td></tr><tr><td>y</td><td>24</td><td>12</td><td>8</td><td>6</td><td>4</td></tr></table> $20 = 10 \times 2$, so when $x = 20$, $y = 24 \div 2 = 12$ $30 = 10 \times 3$, so when $x = 30$, $y = 24 \div 3 = 8$ $6 = 12 \div 2$, so when $y = 6$, $x = 20 \times 2 = 40$ $4 = 24 \div 6$, so when $y = 4$, $x = 10 \times 6 = 60$	Since x and y are inversely proportional, as x decreases at a rate, y increases at the same rate. So if x is halved then y is doubled.
b $xy = 240$	The product of x and y is a constant, in this case 240. This is always true of two variables that are inversely proportional to each other.

c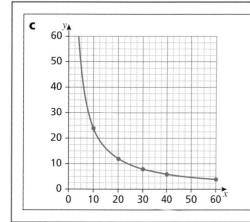

Draw a coordinate grid, then plot the points and join them with a smooth curve.

Notice that the graph approaches both the x- and y-axes but never reaches either.

Practice (A)

1 The time it takes to travel 30 miles depends on the speed at which you travel.

Speed (mph)	5	10	20	30	45	60
Time (h)						

   a Copy and complete the table.

   b What happens when the speed is doubled?

   c Plot a graph of speed against time for the data in the table. Plot time on the horizontal axis and speed on the vertical axis.

2 Three pipes flowing at a constant rate can fill a tank with oil in 1 hour.

   a How long will it take 6 pipes that flow at the same rate to fill the tank?

   b How long will it take a single pipe flowing at the same rate to fill the tank?

3 Six kitchen workers can peel 10 kg of potatoes in 20 minutes.

   Assume all the kitchen workers peel at the same rate.

   a Which is the correct statement?

      A The number of kitchen workers is inversely proportional to the mass of the potatoes.

      B The number of kitchen workers is inversely proportional to the time taken.

      C The mass of the potatoes is inversely proportional to the time taken.

   b How long would it take 15 kitchen workers to peel 10 kg of potatoes?

   c How many kitchen workers are needed to peel 10 kg of potatoes in 20 seconds?

4 A car travelling at a constant speed of 48 mph takes 4 minutes to complete one lap of a race track.

   a How long will it take a car travelling at an average speed of 60 mph to complete one lap of the race track?

   b Work out the average speed of the car needed to complete one lap of the race track in 90 seconds.

5 Five people take 1 hour and 15 minutes to unload a van.

Assume that all people work at the same rate.

a How many minutes would it take 10 people to unload the van?

b Filipo says, "One person contributes 15 minutes. So it would take 1 hour with four people."

Show that Filipo is incorrect.

6 It takes 30 builders 12 months to construct a block of flats.

Assume that all the builders work at the same rate.

a How many builders are required to construct the block of flats in 3 months?

b How many months would it take 20 builders to complete the work?

c Discuss any other assumptions you have made as a class.

7 The variables x and y are inversely proportional.

a Copy and complete the table.

x	4		0.8	
y	6	1.5		360

b Express this relationship in the form $xy = k$, where k is a constant to be found.

8 Five cleaners can clean 12 offices in 1 hour.

a How long would it take:

i two cleaners to clean 6 offices   **ii** three cleaners to clean 3 offices?

b How many cleaners would be needed to clean 48 offices in 2 hours?

c Discuss any assumptions made about your answers to parts **a** and **b**.

What do you think? (A)

1 Twenty people can construct a small house in 15 days.

For the first 6 days, five of the workers are not available.

For the remaining days, all 20 people work on the construction of the house.

a How long will it take to build the house?

b Discuss any assumptions you have made about your answer for part **a**.

2 State whether each table shows variables that are **directly proportional**, **inversely proportional** or **neither**.

a

a	5	16	0.8
b	8	2.5	50

b

c	9	2.25	36
d	6	1.5	24

c

e	4	4.5	10
f	1	1.5	7

d

g	6	24	72
h	11	44	132

Are you ready? (B)

1 Solve the equation $18.9 = \dfrac{k}{44.1}$

2 Solve the equation $18.9 = \dfrac{44.1}{x}$

Recall from Chapter 17.1 that if y is directly proportional to x, this can be written as $y \propto x$ or $y = kx$.

If y is inversely proportional to x, this is the same as y being directly proportional to $\dfrac{1}{x}$

Therefore, this can be written as $y \propto \dfrac{1}{x}$ or $y = \dfrac{k}{x}$

The graph of $y = \dfrac{k}{x}$ is a **reciprocal** curve.

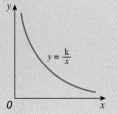

$y = \dfrac{k}{x}$ can be rearranged to $xy = k$, which shows that the product of the two variables is always constant.

Rearranging formulae is covered in more detail in Chapter 8.2

Example

g is inversely proportional to h.

When $g = 175$, $h = 1.4$

Work out the value of h when $g = 70$ and $h \geqslant 0$

Method

Solution	Commentary
$g \propto \dfrac{1}{h}$ $g = \dfrac{k}{h}$	Form and solve an equation to find the constant of proportionality.
$175 = \dfrac{k}{1.4}$ $1.4 \times 175 = k$ $k = 245$	Work out the value of k by substituting in the values of g and h.
$g = \dfrac{245}{h}$ $70 = \dfrac{245}{h}$ $h = \dfrac{245}{70}$ $h = 3.5$	Rewrite the equation with the value of k that you found. Substitute in $g = 70$ Rearrange the equation to work out the value of h.

Practice (B)

1 w is inversely proportional to v.

When $w = 64$, $v = 3.75$

   a Show that when $w = 32$, $v = 7.5$

   b Work out the value of v when $w = 400$ and $v > 0$

   c Work out the value of w when $v = 0.8$

2 d is inversely proportional to e.

When $d = 12.6$, $e = 0.\dot{3}$

   a Work out the value of e when $d = 0.125$ and $e > 0$

   b Work out the value of d when $e = 6$

3 t is inversely proportional to r.

When $t = \dfrac{2}{3}$, $r = \dfrac{4}{5}$

   a Work out the value of r when $t = \dfrac{2}{9}$ and $r > 0$

   b Work out the value of t when $r = 2\dfrac{2}{9}$

4 The value of a variable f changes with the value of another variable, j.

When $f = 36$, $j = 24$

Work out the value of j if $f = 90$ and if:

   a f is directly proportional to j **b** f is inversely proportional to j.

5 The temperature in degrees Celsius (t) of an oven when cooking a jacket potato is inversely proportional to the time taken in minutes (m) to cook the potato.

When the oven is 180°C, it takes 45 minutes to cook a jacket potato.

   a Work out the temperature required to cook the potato for 1 hour.

   b Work out the time required to cook the potato if the oven temperature is 400°C.

   c Comment on how realistic your answers are.

6 A satellite orbits Earth at a constant speed of 930 mph and takes 24 hours.

Work out how long the satellite would take to orbit Earth if the satellite was travelling at a constant speed of 1395 mph.

7 b is inversely proportional to c.

When $b = 2.3 \times 10^4$, $c = 8.05 \times 10^{-2}$

Work out the value of c when $b = 0.5$

What do you think? (B) 💭

1 x is inversely proportional to y.

y is directly proportional to z.

When $x = 16$, $y = 44$ and $z = 396$

Work out the value of x if $z = 72$

2 n is inversely proportional to m.

m is inversely proportional to p.

When n is 1.4, m is 0.5 and p is 4.8

a Work out n and p when $m = 4$

b Copy and complete the sentence. n is _____ proportional to p.

3 Kath conjectures, "If the adjacent side length is fixed, the hypotenuse and the side length opposite angle θ are inversely proportional."

Use a counterexample to disprove Kath's conjecture.

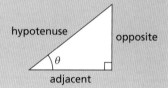

Consolidate – do you need more?

1 It takes 8 workers 20 days to complete a project.

Assume that all the workers work at the same rate.

a How many workers are required to complete the project in 4 days?

b How many days would 10 workers take to complete the project?

2 Four people can make 6 pizzas in 10 minutes.

Assume everyone works at the same rate.

a How long would it take 16 people to make 6 pizzas?

b How many people are needed to make 6 pizzas in 8 minutes?

3 p is inversely proportional to q.

When $p = 75$, $q = 1.6$

a Work out the value of q when $p = 180$ and $q > 0$

b Work out the value of p when $q = 0.48$

4 b is inversely proportional to d.

When $b = 12.5$, $d = \frac{2}{5}$

a Work out the value of d when $b = 37.5$ and $d > 0$

b Work out the value of b when $d = \frac{5}{7}$

Stretch – can you deepen your learning?

1 x is directly proportional to z.

x is inversely proportional to y.

When x is 40, y is 1.125 and z is $\dfrac{5}{36}$

   a   Work out x and y when $z = 10$

   b   Copy and complete the sentence. y is _____ proportional to z.

   c   Write a formula for y in terms of z.

2   Chloe conjectures, "The hypotenuse and the cosine of the angle are inversely proportional if the adjacent side length is fixed."

Investigate Chloe's conjecture.

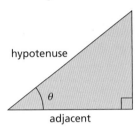

3 a is inversely proportional to b.

When $a = 9$, $b = \dfrac{1}{6}$

   a   Express the relationship of a and b in the form $a = \dfrac{k}{b}$ where k is a constant to be found.

   b   Work out the values of a and b that satisfy the equation $a + b = \dfrac{21}{4ab}$

Are you ready?

1. Evaluate $3p^2$ when $p = 4$

2. Solve $13824 = 125b^3$

3. Solve $3.5 = \dfrac{\sqrt{v}}{4}$

If y is **directly proportional** to x then $y = kx$, where k is the **constant of proportionality**.

In the same way, if y is directly proportional to x^2 then $y = kx^2$

Similarly, if y is inversely proportional to x, this can be written as $y = \dfrac{k}{x}$

Then if y is inversely proportional to the cube of x, this means that $y = \dfrac{k}{x^3}$

Example 1

a is directly proportional to the cube of b.

When $a = 32$, $b = 12$

Work out the value of a when $b = 9$

Method

Solution	Commentary
$a \propto b^3$ $a = kb^3$ $32 = 1728k$ $k = \dfrac{1}{54}$	Form and solve an equation to find the constant of proportionality, writing 'the cube of b' as b^3. Work out the value of k by substituting in the values of a and b.
$a = \dfrac{b^3}{54}$ $a = \dfrac{9^3}{54}$ $a = \dfrac{729}{54}$ $a = 13.5$	Rewrite the equation with the value of k you found to give a formula connecting a and b. Substitute $b = 9$ into your formula. Solve to find the value of a.

Example 2

c is inversely proportional to the square root of d.

When $c = 10$, $d = \dfrac{9}{4}$

Work out the value of d when $c = 20$. Write your answer as a fraction.

Method

Solution	Commentary
$c \propto \dfrac{1}{\sqrt{d}}$ $c = \dfrac{k}{\sqrt{d}}$	Form and solve an equation to find the constant of proportionality, writing 'the square root of d' as $\sqrt{d}$.
$10 = \dfrac{k}{\sqrt{\frac{9}{4}}}$ $15 = k$	Work out the value of k by substituting in the values of c and d.
$c = \dfrac{15}{\sqrt{d}}$ $20 = \dfrac{15}{\sqrt{d}}$ $\sqrt{d} = \dfrac{15}{20} = \dfrac{3}{4}$	Rewrite the equation with the value of k you found to give a formula connecting c and d. Substitute $c = 20$ into your formula.
$d = \dfrac{9}{16}$	Rearrange to work out the value of d, leaving your answer as a fraction.

Practice

 1 z is directly proportional to the square root of y.

When $z = 4$, $y = 25$

    a Write a formula for z in terms of y.   **b** Work out the value of z when $y = 64$

 2 r is inversely proportional to the square of t.

When $r = 1.5$, $t = 6$

Benji says, "The constant of proportionality is 9"

    a Explain why Benji is incorrect.

    b Work out the value of t when $r = 6$

    c Work out the value of r when $t = 9$. Give your answer as a fraction in its simplest form.

3 y varies proportionally to x^2.

Which graph cannot represent the relationship between y and x?

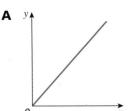 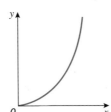 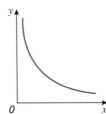

🔢 **4** h is inversely proportional to the cube root of g.

When $g = 4096$, $h = 500$

a Work out the value of g when $h = 1000$

b Work out the value of h when $g = 15\,625$

🔢 **5** q is directly proportional to p.

p is directly proportional to the square of r.

When $q = 25.6$, $p = 32$

When $p = 58.8$, $r = 7$

Work out the value of q when $r = 25$

6 Here are four graphs, A, B, C and D:

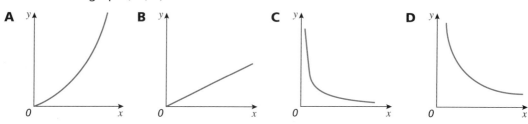

Match each graph with a statement in the table below.

Proportionality relationship	Graph (letter)
y is directly proportional to x	
y is inversely proportional to x	
y is directly proportional to x^2	
y is inversely proportional to x^2	

What do you think? 💡

1 Decide whether each statement is **true** or **false** as the given lengths vary.

a The circumference of a circle is directly proportional to the radius of the circle.

b The surface area of a cube is directly proportional to the length of its edge.

c The area of a circle is directly proportional to the square of the radius of the circle.

d The length of a rectangle with an area of k square units is inversely proportional to its width.

2 The table shows corresponding pairs of values for x and y.

x	p	pq		
y	q^2		p^2q^2	1

a Copy and complete the table if x and y are in direct proportion.

b Copy and complete the table if x and y are in inverse proportion.

Consolidate – do you need more?

 1 l is directly proportional to the cube root of m.

When $l = 5$, $m = 8$

   a Work out the value of l when $m = 64$

   b Work out the value of m when $l = 20$

 2 d is inversely proportional to the square root of f.

When $d = 5$, $f = 2$

   a Work out the value of d when $f = 8$

   b Work out the value of f when $d = 20$

3 Match the graphs to the correct statements.

    A 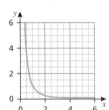   **B** 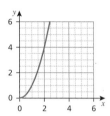   **C** 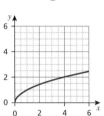   **D**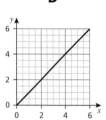

y is directly proportional to x	y is inversely proportional to x^2	y is directly proportional to x^2	y is directly proportional to $\sqrt{x}$

Stretch – can you deepen your learning?

1 x^2 and y^3 are directly proportional.

When $x = 4a$, $y = a$

Work out an expression for y when $x = 3$ in the form $\left(\dfrac{pa}{q}\right)^{\frac{1}{3}}$ where p and q are integers to be found.

2 r and t^2 are inversely proportional to each other.

When $r = x$, $t = \sqrt{x - y}$

Work out an expression for r in terms of x and y when $t = 1$

3 The table shows pairs of values for p and q that vary proportionally where $x \neq 0$

p	$x - 1$		$2x^2 - 2$	
q	$x + 1$	1		$x - \dfrac{1}{x}$

   a Copy and complete the table if p and q are inversely proportional.

   b Copy and complete the table if p and q are directly proportional.

Proportion: exam practice

White Rose
MATHS

1 15 pens cost £3.60

Work out the cost of 25 pens. **[2 marks]**

4–6

2 0.6 kg of potatoes cost 72p.

Work out the cost of 1 kg of potatoes. **[2 marks]**

3 m is directly proportional to n.

When $m = 8$, $n = 0.4$

Work out the value of n when $m = 60$ **[3 marks]**

4 The table shows some values of a and b.

a	7.5	45	300
b	3	21	12

Is a proportional to b? Justify your answer. **[2 marks]**

5 p is inversely proportional to q.

Given that $p = 72$ when $q = 3$, express p in terms of q. **[3 marks]**

6 A new printer prints three times as many sheets as an old printer in the same amount of time.

It took 30 minutes to print 500 sheets on the old printer.

How long will it take to print 800 sheets using the new printer? **[3 marks]**

7–9

7 18 workers can complete a job in 5 days.

Stating any assumptions you make, work out how many more workers are needed to complete the job in 3 days. **[3 marks]**

8 x is inversely proportional to the square of y.

When $x = 6$, $y = 4$

Work out the value of x when $y = 10$ **[3 marks]**

18 Rates

White Rose
MATHS

In this block, we will cover...

18.1 Speed

Example 1

Junaid travels 160 miles by train from Swansea to

The journey from Swansea to Bristol is 60 miles, a
of 40 mph.

The journey from Bristol to London takes 1 hour

a   Work out how long the journey from Swanse
hours and minutes.

b   Work out the average speed of the journey fr
correct to 3 significant figures.

Method

18.2 Density

Practice

1   250 cm³ of brass has a mass of 2130 g.

Use this information to work out the dens

2   A piece of oak has a mass of 3.25 g and a v

a   Use this information to find the density

b   Work out the mass of a piece of oak if

18.3 Other rates

Consolidate – do you need more

1   Huda can chop an onion in 45 seconds.

a   How many onions can Huda chop in 12

b   How long will it take Huda to chop 30

c   What assumptions have you made?

2   A machine can slice 48 000 carrots per hou

a   Work out the number of carrots sliced

b   At the same rate, how many carrots ca

c   How long will it take the machine to sl

Are you ready?

1 Write 140 minutes in hours and minutes.

 **2** Solve $3k = 75$

3 Rearrange these formulae to make r the subject.

    a $d = \dfrac{r}{f}$ **b** $rd = f$ **c** $d = \dfrac{f}{r}$

4 Round each number to 3 significant figures.

    a 14.076   **b** 10 079.1106   **c** 0.029 42

Using your calculator

On some calculators, you can use the [·, ,,] button to convert between hours, minutes and seconds.

To convert 2 hours and 36 minutes into hours, type [2] [·, ,,] [3] [6] [·, ,,] [0] [·, ,,] [=]

```
2˚36˚0˚
          2° 36'0"
```

, then press the [·, ,,] button to show 2.6 hours.

You can also convert minutes into hours.

To convert 164 minutes into hours, type [0] [·, ,,] [1] [6] [4] [·, ,,] [0] [·, ,,] [=] and the display shows

```
0˚164˚0˚
          2° 44'0"
```

, which you can read as 2 hours and 44 minutes.

Pressing the [S⟺D] button shows that this is equivalent to 2.73 hours.

Speed is the rate of change of distance with time, or the rate at which an object is moving.

You can work out speed using this formula:

$$\text{speed} = \frac{\text{distance}}{\text{time}} \text{ or } s = \frac{d}{t}$$

Common units of speed include miles per hour (mph), kilometres per hour (km/h) and metres per second (m/s).

On a graph, the gradient of a line segment is calculated by $\dfrac{\text{change in } y}{\text{change in } x}$

In a **distance–time graph**, the **gradient** of a line segment is calculated by $\dfrac{\text{change in } \textbf{distance}}{\text{change in } \textbf{time}}$
It therefore represents the speed of that part of the journey.

Here is a distance–time graph:

The speed at each section is:

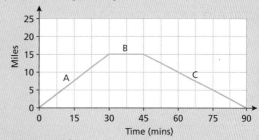

A: $\dfrac{15 \text{ miles}}{0.5 \text{ hours}} = 30\,\text{mph}$

B: $\dfrac{0 \text{ miles}}{0.25 \text{ hours}} = 0\,\text{mph}$ (a horizontal line represents being stationary)

C: $\dfrac{-15 \text{ miles}}{0.75 \text{ hours}} = -20\,\text{mph}$

A negative gradient represents speed in an opposite direction, so the speed for C is 20 mph.

> **Speed** is the rate of change of distance with time, whereas **velocity** is the rate of change of distance with time in a particular direction.

Example 1

Junaid travels 160 miles by train from Swansea to London via Bristol.

The journey from Swansea to Bristol is 60 miles, and the train travels at an average speed of 40 mph.

The journey from Bristol to London takes 1 hour and 15 minutes.

a Work out how long the journey from Swansea to Bristol takes. Give your answer in hours and minutes.

b Work out the average speed of the journey from Swansea to London. Give your answer correct to 3 significant figures.

Method

Solution	Commentary
a $t = \dfrac{d}{s}$ $t = \dfrac{60 \text{ miles}}{40 \text{ mph}}$ $t = 1.5$ hours (1 hour and 30 minutes)	Rearrange the formula $s = \dfrac{d}{t}$ to make t the subject.
b Total time = 1 hour 30 minutes + 1 hour 15 minutes = 2 hours and 45 minutes = 2.75 hours	Write the time in hours.
$s = \dfrac{160}{2.75} = 58.18\ldots$ mph The average speed is 58.2 mph.	To find the average speed for the whole journey, calculate 'total distance' ÷ 'total time'. Remember to give your answer correct to 3 significant figures.

Example 2 🖩

Ali's 21 km journey to work takes exactly 36 minutes.

a   Show that Ali's average speed is 35 km/h.

b   A greyhound's top speed is 9.8 metres per second.

Show that Ali's average speed on his journey to work is slower than the top speed of a greyhound.

Method

Solution	Commentary
a 36 minutes = 0.6 hours $s = \dfrac{21}{0.6} = 35$ km/h	Write the number of minutes in terms of hours ($36 \div 60 = 0.6$). Use the formula $s = \dfrac{d}{t}$ to show the average speed is 35 km/h.
b **Method A** $9.8 \times 60 = 588$ 9.8 m/s = 588 m/min	Multiply 9.8 by 60 to change metres per second into metres per minute.
$588 \times 60 = 35\,280$ 588 m/min = 35 280 m/h	Multiply by 60 again to change metres per minute into metres per hour.
35 280 m/h = 35.28 km/h	Divide by 1000 to change metres per hour into kilometres per hour.
$35 < 35.28$	Compare the speeds to show that Ali's average speed to work is slower than the top speed of a greyhound.
Method B $35 \div 60 = 0.583\ldots$ 35 km/h = 0.583… km/min	Alternatively, you could start with Ali's speed and convert this into metres per second. First, divide 35 by 60 to change km/h into km/min.
$0.583 \div 60 = 0.009\,72\ldots$ 0.583… km/min = 0.009 72 km/s	Divide by 60 again to change km/min into km/s.
0.009 72 km/s = 9.72 m/s	Multiply by 1000 to change km/s into m/s.
$9.72 < 9.8$	Compare the speeds to show that Ali's average speed to work is slower than the top speed of a greyhound.

Example 3

The graph shows Jakub's cycle ride to the cinema and the amount of time he stays there.

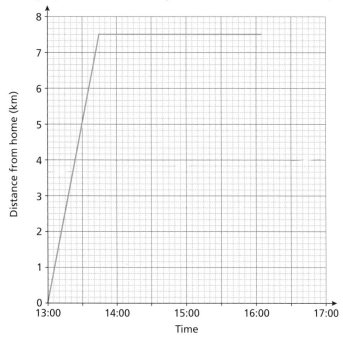

a How long does Jakub stay at the cinema?

b For his journey home, Jakub takes the exact same route and cycles at the same average speed as his journey to the cinema.

What time does Jakub arrive home?

c Amina was also at the cinema. She lives the same distance as Jakub from the cinema and leaves for home at the same time he does. Amina arrives home 5 minutes before Jakub.

Work out the average speed of Amina's journey home from the cinema.

Method

Solution	Commentary
a + 15 min + 2 hours + 5 min 13:45 14:00 16:00 16:05 2 hours and 20 minutes	The horizontal line indicates the time Jakub is at the cinema. Find the time between 13:45 and 16:05 by splitting into easy sections.
b 16:05 + 45 minutes 16:50	On the return journey, Jakub must have covered the same distance in the same amount of time. It took him 45 minutes to cycle to the cinema.
c $\dfrac{7.5}{0.\dot{6}} = 11.25$ 11.25 km/h	Amina arrives home at 16:45, which means it takes her 40 minutes to get home. 40 minutes = 40 ÷ 60 = 0.$\dot{6}$ hours To calculate her average speed, divide the distance she travels by 0.$\dot{6}$

Practice

1 Calculate the average speed of each journey. Write your answers using appropriate units.

    a 350 miles travelled in 4 hours

    b 174 kilometres travelled in 1.5 hours

    c 1161 miles travelled in $4\frac{1}{2}$ hours

    d 320 metres travelled in 50 seconds

2 Calculate the distance travelled in each journey. Write your answers using appropriate units.

    a Average speed of 45 mph for 2.6 hours

    b Average speed of 154 km/h for 1.15 hours

    c Average speed of 210 mph for 180 minutes

    d Average speed of 23 m/s for 1 hour

3 Calculate the time taken for each journey. Write your answers using appropriate units.

    a 120 miles at an average speed of 48 mph

    b 372 kilometres at an average speed of 74.4 km/h

    c 135 metres at an average speed of 5.4 m/s

    d 2500 metres at an average speed of 50 km/h

4 Amina went on a day trip to a museum.

    Here is a description of her journey:

        She left home at 10 am.

        She arrived at the museum, 15 km away, at 10.30 am.

        She stayed at the museum for 2 hours 40 minutes.

        She went straight home, arriving back at 1.45 pm.

    a Represent this information on a distance–time graph.

    b Work out Amina's average speed for her journey home.

5 The distance of Mario's journey home from work is 17 miles.

    He travels the first 5 miles in 10 minutes. He then stops for 5 minutes at a petrol station before completing his journey home at an average speed of 40 mph.

    a Represent this information on a distance–time graph.

    b Work out the average speed of Mario's journey home. Give your answer to 3 significant figures.

6 This distance–time graph of a car journey has no numbers.

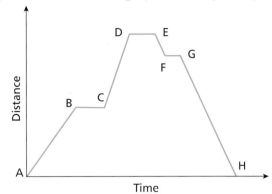

Between which two points was the car travelling slowest?

7 The graph shows Bobbie's journeys to work on Monday and Tuesday.

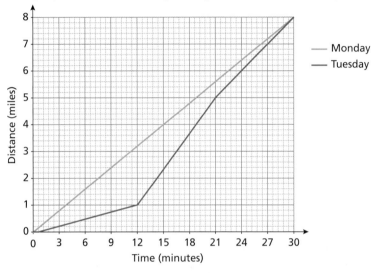

Filipo says, "Bobbie's average speed was greater on Tuesday."

Explain why Filipo is incorrect.

8 A team of four runners run a 4 × 100 m relay race.

Two of the runners run at an average speed of 8 m/s.

The two other runners each take 10.5 seconds to run 100 metres.

Work out the average speed of the team. Give your answer correct to 3 significant figures.

9 Calculate the average speed of each journey in miles per hour.

    a   36 miles in 25 minutes   **b**   27 miles in 36 minutes

    c   37.5 miles in 75 minutes   **d**   209 miles in 38 minutes

10 Rob walks 7.2 km on a mountain climb.

He takes $2\frac{1}{2}$ hours to reach the top of the mountain.

He then walks back down the mountain the same way he went up.

His average speed walking down the mountain is 3.5 km/h.

Calculate the average speed of Rob's entire journey up and down the mountain.

11 A cheetah has a top speed of 33.15 m/s.

Show that the cheetah's top speed is greater than 115 km/h.

What do you think?

1 Bev is travelling 210 miles from Liverpool to Dundee.

She drives the first 100 miles in 1 hour 40 minutes. The rest of her journey takes 2 hours.

a Work out how much quicker Bev's journey would have been if she had stayed at the same average speed as the first 100 miles of her journey.

b Work out the percentage change of the speed from the first 100 miles to the rest of the journey.

Consolidate – do you need more?

1 Calculate the average speed of each journey. Write your answers using appropriate units.

a 225 miles travelled in 3 hours

b 240 kilometres travelled in 2.4 hours

c 1800 miles travelled in $4\frac{1}{2}$ hours

d 424 metres travelled in 53 seconds

2 Calculate the distance travelled in each journey. Write your answers using appropriate units.

a Average speed of 12.6 mph for 0.7 hours

b Average speed of 700 km/h for $1\frac{1}{4}$ hours

c Average speed of 45.7 mph for 100 minutes

d Average speed of 100 km/h for 10 seconds

3 Calculate the time taken for each journey. Write your answers using appropriate units.

a 120 miles at an average speed of 80 mph

b 96 kilometres at an average speed of 32 km/h

c 288 metres at an average speed of 7.2 m/s

d 250 metres at an average speed of 1 km/h

4 Huda went for a hike.

Here is a description of her journey:

She left home at 9 am.

She arrived at Coniston, 10 km away, at 11:30 am.

She rested for 20 minutes before continuing another 4 km until she arrived at Levers Water, 1 hour and 10 minutes later.

She stopped for another rest for 30 minutes.

She then walked back home, resting for 15 minutes exactly halfway.

She arrived home at 5 pm.

a Represent this information on a distance–time graph.

b Work out Huda's average speed for the part of her journey from Coniston to Levers Water.

Stretch – can you deepen your learning?

1 A train travels at an average speed of 60 mph for half of its total journey.

For the remaining half of the journey, the train travels at an average speed of 30 mph.

Zach says, "The train's average speed for the whole journey must be 45 mph."

Show that Zach is wrong.

2 A car travels at 25 m/s for 4 seconds. It then accelerates for 2 seconds to 30 m/s until it slows down and comes to rest 4 seconds later.

The speed–time graph represents this information.

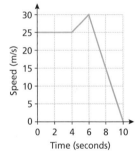

a Work out the distance travelled in the first 4 seconds.

b What does the area under the graph represent?

c Work out the total distance travelled by the car.

> See Block 15 for details about the relationship between speed–time graphs and distance.

 3 A van travels the first third of its journey at an average speed of a mph.

The second third of the journey is travelled twice as fast as the first third.

The final third is travelled twice as fast as the second third.

Show that the average speed for the entire journey is $\dfrac{12a}{7}$ mph.

Are you ready?

1 Convert: **a** 1750 g into kilograms **b** 17.5 kg into grams.

2 Work out the volume of:

a a cuboid with dimensions 4.5 cm by 6 cm by 8 cm

b a cube with side length 0.5 m.

3 Solve each equation.

a $14a = 224$ **b** $12 = \dfrac{b}{30}$ **c** $40 = \dfrac{274}{c}$

Density is a measure of mass per unit of volume. For example, 1 cm³ of aluminium has a mass of 2.7 g, but 1 cm³ of tungsten has a mass of 19.28 g. So tungsten has a greater density than aluminium.

The density of an object can be found by dividing its mass by its volume.

$$\text{Density} = \frac{\text{Mass}}{\text{Volume}}$$

$$\text{or } D = \frac{M}{V}$$

Common units of density are g/cm³ or kg/m³.

Example 1 🖩

A 200 cm³ piece of copper has a mass of 1.792 kg.

An 80 cm³ piece of zinc has a mass of 328 g.

a Which metal has the greater density, copper or zinc? Justify your answer.

b The density of iridium is 22.56 g/cm³.

i Work out the mass of iridium that has the same volume as the piece of copper.

ii Work out the volume of iridium that has the same mass as the piece of zinc.
Give your answer to 3 significant figures.

Method

Solution	Commentary
a Copper: $\dfrac{1792}{200} = 8.96 \text{ g/cm}^3$ Zinc: $\dfrac{328}{80} = 4.1 \text{ g/cm}^3$	Write the mass of copper in grams so that the units of density will both be g/cm³. Use Density = $\dfrac{\text{Mass}}{\text{Volume}}$
8.96 > 4.1 so copper has a greater density than zinc.	Justify your conclusion.

b i $D = \dfrac{M}{V}$ $22.56 = \dfrac{M}{200}$	Substitute the information into the formula for density.
$M = 22.56 \times 200$ $M = 4512\,g = 4.512\,kg$	Rearrange to find the mass. You can write the mass in kilograms instead of grams.
ii $D = \dfrac{M}{V}$ $22.56 = \dfrac{328}{V}$	Substitute the information into the formula for density.
$V = \dfrac{328}{22.56}$ $V = 14.539\ldots$	Rearrange to find the volume.
$V = 14.5\,cm^3$ (3 s.f.)	Give your answer correct to 3 significant figures.

Example 2 🖩

80 ml of coffee and 280 ml of milk are mixed to make a latté.

Coffee has a density of 0.55 g/ml.

Milk has a density of 1.035 g/ml.

Work out the density of the latté.

Method

Solution	Commentary
Coffee **Milk** $M = ?$ $M = ?$ $V = 80\,ml$ $V = 280\,ml$ $D = 0.55\,g/ml$ $D = 1.035\,g/ml$ Mass of coffee = 80 × 0.55 = 44 g Mass of milk = 280 × 1.035 = 289.8 g	To find the density of the latté, you need the total mass and volume. Use the information given to find the mass of the coffee and of the milk.
Density of latté = $\dfrac{\text{Total mass of the latté}}{\text{Total volume of the latté}}$ $= \dfrac{333.8}{360} = 0.927\,g/ml$	Substitute into the formula to find the density of the latté. Remember to provide units with your answer.

Practice

🖩 **1** 250 cm³ of brass has a mass of 2130 g.

Use this information to work out the density of brass.

🖩 **2** A piece of oak has a mass of 3.25 g and a volume of 5.2 cm³.

a Use this information to work out the density of oak.

b Work out the mass of a piece of oak if it has a volume of 3.6 cm³.

3 The table shows information about the mass, volume and density of different objects.

Copy and complete the table.

Object	Mass (g)	Volume (cm³)	Density (g/cm³)
A	49.6	6.4	
B	9.408		4.8
C		4	0.68
D	72.05		13.1

4 The density of an A4 piece of paper is 250 kg/m³.

Kath cuts an A4 piece of paper in half.

She says, "The density of each piece of paper is 125 kg/m³."

Explain why Kath is incorrect.

5 The density of olive oil is 0.917 kg/l.

Work out the volume, in litres, of olive oil that has a mass of:

a 2 kg **b** 0.5 kg **c** 1250 g.

6 Diamond has a density of 3.51 g/cm³.

Work out the mass, in grams, of diamond that has a volume of:

a 3 cm³ **b** 4.8 cm³ **c** 1 m³.

7 Each of the shapes below have a mass of 8.4 kg.

Work out the density of each shape. Give your answers correct to 3 decimal places and include units.

a

2.5 cm
2.5 cm 2.5 cm

b

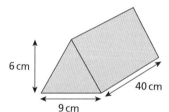

6 cm
9 cm
40 cm

c

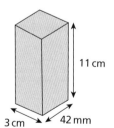

11 cm
3 cm 42 mm

d

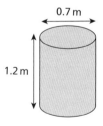

0.7 m
1.2 m

8 A bracelet made of platinum has a mass of 51 grams and a volume of 2.3 cm³.

Work out the density of platinum, giving your answer correct to 3 significant figures and in kg/m³.

9 Here are the volumes and masses of three spheres made of different materials:

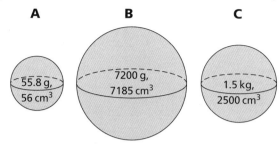

A   B   C

55.8 g, 56 cm³

7200 g, 7185 cm³

1.5 kg, 2500 cm³

An object will float on a liquid if its density is less than the density of the liquid. The density of water is 0.997 g/cm³.

Work out which of the spheres will float on water.

10 Metal A and metal B are melted down and combined to produce an alloy C.

Metal A has a volume of 0.4 m³ and a density of 320 kg/m³.

Metal B has a mass of 356 kg and a density of 712 kg/m³.

Work out the density of alloy C, giving your answer correct to 3 significant figures.

Consolidate – do you need more?

1 A titanium object has a mass of 90 g and a volume of 20 cm³.

Use this information to work out the density of titanium.

2 The table shows the mass and volume of four pieces of rosewood.

Copy and complete the table. Give your answers to 3 significant figures.

Mass (g)	Volume (cm³)
205	250
200	
	400
125	

3 A piece of cedar wood has a volume of 2.72 cm³ and a density of 0.55 g/cm³.

The density of mahogany is 0.85 g/cm³.

Work out the volume of a piece of mahogany that has the same mass as the piece of cedar wood.

4 A cuboid with dimensions 3 cm by 6 cm by 8 cm has a mass of 288 g.

Work out the density of the cuboid.

5 Liquid A and liquid B are combined to make liquid C.

Liquid A has a volume of 0.4 litres and a density of 32 g/l.

Liquid B has a mass of 250 g and a density of 18 g/l.

Work out the density of liquid C.

Stretch – can you deepen your learning?

1 The diagram shows the uniform cross-section of a bench made of a type of wood.

The length of the bench is 1.8 m and its mass is 50 kg.

Calculate the density of the wood, giving your answer in kg/m³.

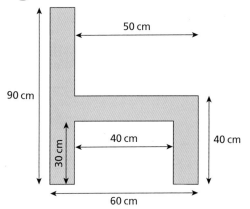

2 Magnesium has a density of 1.74 g/cm³.

30 cm³ of magnesium and some copper are melted to produce an alloy.

The alloy has a mass of 192.2 g and a density of 4.21 g/cm³.

Use this information to work out the density of copper.

3 A fruit punch is made of lemonade and fruit juice in the ratio 4 : 5

The density of the lemonade is 0.92 g/cm³.

The density of the fruit juice is 1.01 g/cm³.

1.6 kg of fruit juice is used to make the fruit punch.

Work out the density of the fruit punch. Give your answer to 3 significant figures

4 The volume of object A is three times the volume of object B.

The mass of object A is four times the mass of object B.

Express the density of object B as a percentage of the density of object A.

5 The mass of the prism shown is $3x^2$ grams.

Show that the density of the prism is $\dfrac{3}{(x+1)}$ g/cm³.

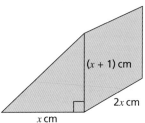

6 Hemisphere A is made of lead and hemisphere B is made of tin.

A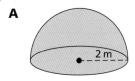

B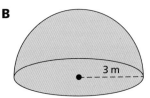

Lead has a density of 11 300 kg/m³.

Tin has a density of 7260 kg/m³.

Volume of a sphere = $\dfrac{4}{3}\pi r^3$

Which hemisphere has the greater mass? You must show how you get your answer.

Are you ready?

1 £1 can buy 1.15 euros.

    a How many euros can be bought with £80?

    b How many pounds are needed to buy €575?

2 Two workers can build five cabinets in 10 hours.

How long would it take four workers to build five cabinets if they were working at the same rate?

3 A machine can peel 15 potatoes per minute.

    a How many potatoes can be peeled if the machine runs for 3 hours?

    b How long will it take the machine to peel 9000 potatoes?

The container shown in the diagram is full of water.

Water is then emptied through a hole at the bottom at a constant rate.

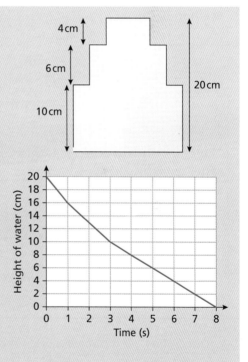

The graph shows the height of the water in the container as it is emptied over time.

The **gradient** of each line segment represents the rate of change of the height of water for each part of the container. The gradient can be calculated by working out $\dfrac{\text{change in } y}{\text{change in } x}$

Between 0 and 1 second, the gradient is:

$$\frac{\text{change in } y}{\text{change in } x} = \frac{20 - 16}{0 - 1} = \frac{4}{-1} = -4$$

The rate of change is 4 cm/s. The gradient is negative as the height is decreasing over time.

Between 1 and 3 seconds, the gradient is:

$$\frac{16 - 10}{1 - 3} = \frac{6}{-2} = -3$$

The rate of change is 3 cm/s.

Between 3 and 8 seconds, the gradient is:

$$\frac{10 - 0}{3 - 8} = \frac{10}{-5} = -2$$

The rate of change is 2 cm/s.

Owing to the different widths, the height of water in the container is **not** changing at a constant rate.

The **average rate of change** of the height of water can be found by calculating the gradient between when the container was full (0, 20) and when it was empty (8, 0).

Gradient $= \dfrac{20 - 0}{0 - 8} = -2.5$ The average rate of change is 2.5 cm/s.

Example 1

A soft drinks machine can fill a 500 ml cup in 12 seconds.

A water fountain machine takes 1 minute to fill a 2-litre bottle.

Which machine flows at a faster rate?

Method

Solution	Commentary
Method A Soft drinks machine 500 ml = 12 seconds × 4 ( ) × 4 2000 ml = 48 seconds 1 minute = 60 seconds Therefore the soft drinks machine flows faster as 48 seconds < 60 seconds.	Scale up the rate of change of the soft drinks machine to find how long it will take to fill 2 litres. Compare the rates of change. Justify your conclusion.
Method B $\dfrac{500}{12} = 41.\dot{6}$ ml/s $\dfrac{2000}{60} = 33.\dot{3}$ ml/s $41.\dot{6}$ ml/s > $33.\dot{3}$ ml/s so the soft drinks machine has a faster flow rate.	Find the rate of change in millilitres per second by dividing the volume of liquid by the time taken. Remember to change litres to millilitres and minutes to seconds. State your conclusion clearly, and justify it.

Example 2

Mario takes 1 hour to wash four cars. Emily takes 30 minutes to wash four cars of the same size.

Assume they both work at a constant rate.

a   How long will it take Mario and Emily to wash four cars if they work together?

b   What percentage of a car will get washed in 1 minute if they work together?

Method

Solution	Commentary
a In 1 hour: • Mario washes 4 cars • Emily washes 8 cars. 1 hour = 60 minutes  ? 60 ÷ 3 = 20 It will take them 20 minutes to wash four cars together.	Find the number of cars cleaned in a single unit of time. If Emily washes four cars in 30 minutes, working at the same rate she will wash eight cars in 1 hour. Therefore, Mario and Emily can wash 12 cars every hour. Using a bar model can help to see how long Mario and Emily need to wash four cars together. Write the number of hours in minutes. Write a statement to conclude your workings.
b 12 cars are washed every 60 minutes so $\frac{12}{60}$ will be washed every minute. $\frac{12}{60} = \frac{1}{5} = 20\%$ of one car will be washed in 1 minute if they work together.	Find the number of cars washed every 60 minutes and divide this by 60 Remember to write your answer as a percentage.

Example 3

Water is poured into four different containers, A, B, C and D, at a constant rate.

A	B	C	D

a Match the containers to the graph that shows how the depth of water changes over time.

E	F	G	H
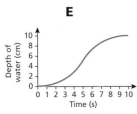	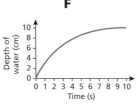	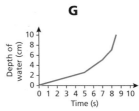	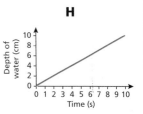

b Between which times was the rate of flow the fastest in Graph G?

Method

Solution	Commentary
a <table><tr><th>Container</th><th>Graph</th></tr><tr><td>A</td><td>H</td></tr><tr><td>B</td><td>F</td></tr><tr><td>C</td><td>E</td></tr><tr><td>D</td><td>G</td></tr></table>	Containers with completely vertical sides fill at a constant rate so container A matches to graph H. The sides of container B get wider as the water fills, therefore it will fill quickly then slow down. Container C fills quicker as it approaches halfway and then more slowly until it reaches the top. Container D has vertical sides with increasingly narrower widths. As the container fills, the rate of change of the height of water suddenly quickens as it reaches a new section.
b 8 and $8\frac{1}{2}$ minutes	Graph G has its steepest line segment between these times.

Practice

1 Abdullah can fold a shirt in 4 seconds.

    a How many shirts can Abdullah fold in 5 minutes?

    b How long will it take him to fold 100 shirts?

2 Lida packs 30 boxes in 10 minutes.

Rob packs 40 boxes in 12 minutes.

Work out who packs boxes faster.

3 The graph shows the water flow out of two containers, X and Y.

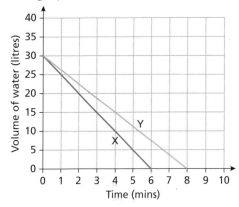

Work out the rate at which water is leaving each container.

4 A machine can make 11 400 aluminium cans per hour.

    a Work out the number of cans made per minute.

    b At the same rate, how many cans can the machine make in two weeks?

    c How long will the machine take to make 1 000 000 cans? Give your answer to the nearest minute.

5 Ed is typing an essay. For the first two minutes, he types 60 words per minute.

For the next four minutes, he types 40 words per minute.

a Draw a graph to represent this information.

b Ed says, "Over the whole six minutes, my average rate was 50 words per minute."

Show that Ed is wrong.

6 At an airport, a security officer checks 8 passports every 10 minutes.

a Work out the average rate, in passports per minute, at which the security officer works.

b How many passports will the security officer check in 45 minutes?

c How long will it take the security officer to check 100 passports?

d Write any assumptions about your answers to parts **b** and **c**.

7 £1 is equivalent to 34 Turkish lira. £1 is also equivalent to 100 Indian rupees.

a How many Indian rupees are equivalent to 50 Turkish lira? Give your answer to the nearest rupee.

b Write an expression for how many Turkish lira are equivalent to y Indian rupees.

8 A container is in the shape of a cuboid. It has dimensions 30 cm by 40 cm by 60 cm. The container is initially empty. Liquid is poured into the container at a rate of 250 ml per minute.

How long will it take to completely fill the container?

Consolidate – do you need more?

1 Huda can chop an onion in 45 seconds.

a How many onions can Huda chop in 12 minutes?

b How long will it take Huda to chop 30 onions?

c What assumptions have you made?

2 A machine can slice 48 000 carrots per hour.

a Work out the number of carrots sliced per minute.

b At the same rate, how many carrots can the machine slice in 3 days?

c How long will it take the machine to slice 100 carrots?

3 The graph shows a jar being filled at different rates.

a Between which times was the rate of flow the fastest?

b Work out the rate at which the jar was being filled:

    i in the first 2 minutes

    ii between 2 and 3 minutes

    iii between 3 and 5 minutes.

c Work out the average rate at which the jar was being filled over the five-minute period.

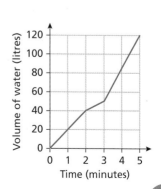

4 A cylindrical tank has a diameter of 5 metres and a height of 3 metres.

The tank is initially empty. Water flows into the tank from a tap at a constant rate of 750 litres per minute.

$1\,m^3 = 1000$ litres

How long will it take to completely fill the tank?
Give your answer to the nearest minute.

Stretch – can you deepen your learning?

1 A container has a capacity of 15 000 litres.

It takes 1 hour to fill the container using tap X.

It takes 20% less time using taps X and Y together.

How long would it take to fill the container using tap Y only?

2 The diagram shows a water tank.

The tank is initially empty. Water flows into the tank at a rate of 270 litres per minute.

Work out how long it will take the tank to be filled.
Give your answer in hours and minutes to the nearest minute.

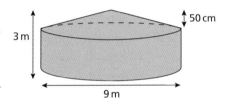

3 m · 50 cm · 9 m

3 It takes Bev c hours to complete a job.

It takes Flo d hours to complete the same job.

Write an expression for how long it will take Bev and Flo to complete the job if they work together, stating any assumptions that you have made.

4–6

1 Ed drives at an average speed of 30 mph for two-and-a-half hours to attend a meeting.

His return journey along the same route takes three hours.

Work out Ed's average speed for the return journey. **[3 marks]**

2 The density of aluminium is 2.7 g/cm³.

A block of aluminium has mass 945 g.

Work out the volume of the block. **[2 marks]**

3 It takes 4 hours to empty an oil tank using 15 pumps.

How long will it take to empty the tank using 25 of the same pumps? **[2 marks]**

4 Rhys runs for 20 minutes at an average speed of 9 miles per hour.

He then runs for 40 minutes at an average speed of 6 miles per hour.

Benji runs the same distance as Rhys in 42 minutes.

Work out Benji's average speed in miles per hour. **[2 marks]**

5 100 Argentinian pesos are worth 0.12 US dollars.

100 Argentinian pesos are worth 17.87 Japanese yen.

Complete the statement: "1 US dollar is worth _____ Japanese yen."

Give your answer to the nearest yen. **[2 marks]**

7–9

6 A car travels 90 m in 3 seconds.

Use the fact that 5 miles ≈ 8 km to estimate the car's speed in miles per hour. **[3 marks]**

7 The diagram shows the pattern for making a hat, consisting of two identical semicircles and a rectangle.

One square metre of the fabric needed to make the hat has mass 380 g.

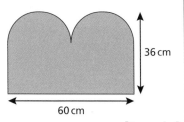

36 cm

60 cm

Calculate, to the nearest gram, the mass of fabric needed to make the hat. **[4 marks]**

8 200 ml of liquid A and 300 ml of liquid B are mixed together to make liquid C.

Liquid A has a density of 1.2 g/ml.

Liquid B has a density of 0.8 g/ml.

Work out the density of liquid C. **[4 marks]**

Ratio, proportion and rates of change: exam practice

4–6

1 A force of 50 Newtons acts on an area of 15 cm².

The force stays the same but the area increases.

What happens to the pressure?

$$\text{pressure} = \frac{\text{force}}{\text{area}}$$

[1 mark]

2 Flo and Seb share some money in the ratio 2 : 5

Seb gets £42 more than Flo.

How much money did they share?   **[2 marks]**

3 A cuboid is 10 cm long, 8 cm wide and 6 cm high.

The mass of the cuboid is 1.2 kg.

Work out the density of the cuboid, giving your answer in g/cm³.   **[3 marks]**

4 The value of x is 40% more than the value of y.

Write the ratio $x : y$ in its simplest integer form.   **[2 marks]**

5 ABC is a straight line.

AB : BC = 2 : 5

The length of BC is 35 cm.

A ———— B ——————————— C

Work out the length of AC.   **[3 marks]**

6 Marta has three times as many coins as Faith.

Jackson has four fewer coins than Marta.

Marta, Faith and Jackson have 52 coins in total.

Write the ratio of the number of coins Faith has to the number of coins Jackson has in the form 1 : n.   **[3 marks]**

7 It takes $2\frac{1}{4}$ hours to fill a cylindrical water tank that is 30 cm tall.

(a) Calculate the time it takes to fill $\frac{3}{5}$ of the tank.   **[2 marks]**

(b) What is the depth of water in the tank after 27 minutes?   **[2 marks]**

8 Bev, Seb and Rhys share £228

The amounts Bev and Seb get are in the ratio 6 : 5

The amounts Seb and Rhys get are in the ratio 3 : 1

How much does Bev get?   **[4 marks]**

9 A bus journey is 13 km long.

The bus travels at an average speed of 9 km/h for 20 minutes.

The bus must complete the whole journey in exactly 1 hour to be on time.

At what average speed must the bus travel for the remainder of the journey to be on time? **[3 marks]**

10 24 workers can clear a 50-acre site in 18 days.

What is the minimum number of workers needed to clear a 70-acre site in 15 days? **[3 marks]**

11 p is three-quarters of q.

$2r = 3q$

Write the ratio $p : q : r$ in its simplest integer form. **[3 marks]**

12 k is directly proportional to the square of v.

$k = 63$ when $v = 3$

Work out the value of k when $v = 4$ **[3 marks]**

13 A and B are in the ratio 5 : 3

C and D are in the ratio 2 : 7

$3A + 4B = 4(C + D)$

Work out A : B : C : D

Give your answer in simplest integer form. **[4 marks]**

Glossary

Acceleration – the rate at which an object's speed is changing

Algebraic fraction – a fraction whose numerator and/or denominator are algebraic expressions

Annual – covers a period of one year

Axis – a line on a graph that you can read values from

Balance – an amount of money in an account

Base – the number that gets multiplied when using a power/index

Best buy (or **best value**) – the item which is cheapest when equal-sized amounts of different items are compared

Bill – shows how much money is owed for goods or services

Binomial – an expression with two terms

Coefficient – a number in front of a variable, e.g. for $4x$ the coefficient of x is 4

Collect like terms – put like terms in an expression together as a single term

Common denominator – two or more fractions have a common denominator when their denominators are the same

Commutative – when an operation can be performed in any order

Conjecture – a statement that might be true that has not yet been proved

Consecutive – following on, e.g. 14, 15, 16… or n, $n + 1$, $n + 2$…

Constant – not changing

Convert – change from one form to another, e.g. a percentage to a decimal

Coordinate – an ordered pair used to describe the position of a point

Counterexample – an example that disproves a statement

Credit – an amount of money paid into an account

Cube root – the cube root of a number is a value that, when multiplied by itself three times, gives that number

Curve – a line on a graph showing how one quantity varies with respect to another

Debit – an amount of money taken out of an account

Decimal – a number with digits to the right of the decimal point

Decimal places – the number of digits to the right of the decimal point in a number

Decrease – make something smaller

Decreasing (or descending) sequence – a sequence where every term is smaller than the previous term

Degree of accuracy – how precise a number is

Denominator – the bottom number in a fraction; it shows how many equal parts one whole has been divided into

Density – the ratio of the mass of an object to its volume; the mass per unit volume

Deposit – an amount of money paid into a bank account

Depreciate – reduce or decrease in value

Descending – decreasing in size

Difference – in arithmetic, the result of subtracting a smaller number from a larger number; in sequences, the gap between numbers in a sequence

Digits – the numerals used to form a number

Direct proportion – two quantities are in direct proportion when as one increases or decreases, the other increases or decreases at the same rate

Directed numbers – numbers that can be negative or positive

Divide in a ratio – share a quantity into two or more parts so that the shares are in a given ratio

Dividend – the amount you are dividing

Division – the process of splitting a number into equal parts

Divisor – the number you are dividing by

Double number line – two lines used to represent ratio problems

Equal – having the same value; the sign = is used between numbers and calculations that are equal in value, and the sign ≠ is used when they are not

Equation – a statement with an equals sign, which states that two expressions are equal in value

Equivalent – numbers or expressions that are written differently but are always equal in value

Error interval – the range of values a number could have taken before being rounded

Estimate – give an approximate answer

Evaluate – work out the numerical value of

Even number – a number that is divisible by 2

Exchange rate – the rate at which the currency of one country is exchanged for that of another

Expand – multiply to remove brackets from an expression

Exponential – containing, involving or expressed as a power (exponent); an exponential function is of the form $y = k^x$

Express – write, often in a different form

Expression – a collection of terms involving mathematical operations

Factor – a positive integer that divides exactly into another positive integer

Factor pair – a pair of numbers that multiply together to give a number

Factorise – find the factors you need to multiply to make an expression

Fibonacci sequence – the next term in a Fibonacci sequence is found by adding the previous two terms together

Find – work out the value of

Formula (plural: **formulae**) – a rule connecting variables written with mathematical symbols

Fraction – a number that compares equal parts of a whole

Function – a relationship with an input and an output

General term – an expression that generates the terms of a sequence

Geometric sequence – a sequence is geometric if the value of each successive term is found by multiplying or dividing the previous term by the same number

Gradient – the steepness of a line

Graph – a diagram showing how values change

Graphical – using a graph

Highest common factor (HCF) – the greatest number that is a factor of every one of a set of numbers

Identity – a statement that is true no matter what the values of the variables are

Improper fraction – a fraction in which the numerator is greater than the denominator

Inclusive – including the end points of a list

Increase – make something larger

Increasing (or ascending) sequence – a sequence where every term is greater than the previous term

Index (plural: **indices**) – an index number (or power) tells you how many times to multiply a number by itself

Inequality – a comparison between two quantities that are not equal to each other

Inequality symbol – a symbol comparing values showing which is greater and which is smaller

Integer – a whole number

Intercept – the point at which a graph crosses, or intersects, a coordinate axis

Interest – a percentage fee paid when borrowing money or a percentage earned when you deposit money into a savings account

Inverse – the opposite of a mathematical operation; it reverses the process

Inverse function – the inverse function of a function f is a function that undoes the operation of f

Inverse proportion – if two quantities are in inverse proportion, when one quantity increases, the other decreases at the same rate

Irrational number – a number that cannot be written in the form $\frac{a}{b}$ where a and b are integers

Is equivalent to $\equiv$ – equal to for all values of the variables in an expression

Iteration – repeating a process

Kilo- – one thousand

Like terms – terms whose variables are the same, e.g. $7x$ and $12x$

Linear – forming a straight line

Linear equation – an equation with a simple unknown like a, b, or x, i.e. there are no squared terms, cubed terms, etc.

Linear sequence – a sequence whose terms are increasing or decreasing by a constant difference

Loss – if you buy something and then sell it for a smaller amount; loss = amount paid – amount received

Lower bound – the bottom limit of a rounded number

Lowest common multiple (LCM) – the smallest number that is a multiple of every one of a set of numbers

Maximum point – the point on a graph where a function has its greatest value

Mental strategy – a method that enables you to work out the answer in your head

Midpoint – the point halfway between two others

Minimum point – the point on a graph where a function has its least value

Mixed number – a number presented as an integer and a proper fraction

Multiple – the result of multiplying a number by a positive integer

Multiplier – a number you multiply by

Negative numbers – numbers less than zero

Non-linear – not forming a straight line

Non-linear sequence – a sequence whose terms are not increasing or decreasing by a constant difference

Non-unit fraction – a fraction with a numerator that is not 1

Number line – a line on which numbers are marked at intervals

Numerator – the top number in a fraction that shows the number of parts

Odd number – a number that when divided by 2 gives a remainder of 1, e.g. 1, 17, 83

Operation – a mathematical process such as addition, subtraction, multiplication or division

Order of magnitude – size of a number in powers of 10

Order of operations – the rules that tell you the order in which to perform each part of a calculation

Origin – the point where the x-axis and y-axis meet

Original value – a value before a change takes place

Parabola – a type of curve that is approximately U-shaped and has a line of symmetry

Parallel – in the same direction; parallel lines have the same gradient

Per annum – every year

Per cent – parts per hundred

Percentage – the number of parts per hundred

Perpendicular – at right angles to

Piece-wise graph – a graph that consists of more than one straight line

Point of intersection – the point where two graphs cross each other

Position-to-term rule – the rule that links the position of the term to the value of the term

Power (or exponent) – this is written as a small number to the right and above the base number, indicating how many times to use the number in a multiplication, e.g. the 5 in 2^5

Powers of 10 – the result of multiplying 10 by itself a number of times to give a value such as 10, 100, 1000, 10 000, and so on

Prime factor decomposition – writing numbers as a product of their prime factors

Prime number – a positive integer with exactly two factors, 1 and itself

Priority – a measure of the importance of something

Product – the result of a multiplication

Profit – if you buy something and then sell it for a higher amount; profit = amount received – amount paid

Proof – an argument that shows that a statement is true

Proportion – a part, share, or number considered in relation to a whole

Prove – show that something is always true

Quadratic – of the form $ax^2 + bx + c$

Quadratic formula – the formula that gives the roots of a quadratic equation

Quotient – the result of a division

Ratio – a ratio compares the sizes of two or more values

Rational number – a number that can be written in the form $\frac{a}{b}$ where a and b are integers

Rationalise the denominator – rewrite a fraction with a surd denominator as an equivalent fraction with a rational denominator

Real number – all positive and negative numbers, including decimals and fractions

Reciprocal – the result of dividing 1 by a given number; the product of a number and its reciprocal is always 1

Recurring decimal – a recurring decimal has digits that are in a repeating pattern like 0.3333… or 0.171 717…

Reduce – make something smaller

Reflection – a type of geometrical transformation, where an object is flipped to create a mirror image

Remainder – the amount left over after dividing one integer by another

Repeated percentage change – when an amount is changed by one percentage followed by another

Reverse percentage – a problem where you work out the original value

Root – the nth root of a number x is the number that is equal to x when multiplied by itself n times

Round – give an approximate value of a number that is easier to use

Satisfy – make an equation or inequality true

Sector – a part of a circle formed by two radii and a fraction of the circumference

Sequence – a list of items in a given order, usually following a rule

Significant figures – the most important digits in a number that give you an idea of its size

Simplify – rewrite in a simpler form, e.g. rewrite $8 \times h$ as $8h$

Simultaneous – at the same time

Solution – a value you can substitute in place of the unknown in an equation to make it true

Solution set – a range of values for which a statement is true

Solve – find a value that makes an equation true

Speed – the rate at which an object is moving

Square number – a positive integer that is the result of an integer multiplied by itself

Square root – a square root of a number is a value that, when multiplied by itself, gives the number

Standard form – a number written in the form $A \times 10^n$ where A is at least 1 and less than 10, and n is an integer

Subject – the variable in a formula that is expressed in terms of the other variables

Substitute – replace letters with numerical values

Successive – coming after another term in a sequence

Surd – a root that cannot be written as an integer

Tangent (to a circle) – a straight line that touches the circumference of a circle at one point only

Term – in algebra, a single number or variable, or a number and variable combined by multiplication or division; in sequences, one of the members of a sequence

Term-to-term rule – a rule that describes how you get from one term of a sequence to the next

Terminating decimal – a decimal fraction with a finite number of digits, e.g. 0.75

Translation – a type of geometrical transformation, where an object is moved left or right and/or up or down

Trial and improvement – a method of finding a solution to a mathematical problem where you make a guess (a trial), see if it works in the problem, and then refine it to get closer to the actual answer (improvement)

Triangular number – a positive integer that is the sum of consecutive positive integers starting from 1

Truncate – remove digits from a decimal number

Turning point – the point at which a quadratic graph changes direction

Unit fraction – a fraction with a numerator of 1

Unknown – a variable (letter) whose value is not yet known

Unlike terms – terms whose variables are not exactly the same, e.g. $7x$ and 12 or $5a$ and $5a^2$

Upper bound – the top limit of a rounded number

Variable – a numerical quantity that might change, often denoted by a letter, e.g. x or t

y-intercept – the point at which a graph crosses or intersects the y-axis

This page has deliberately been left blank

Answers

Block 1 Number review

Chapter 1.1

Are you ready? (A)

1 $57 + 32$ $50 + 7 + 30 + 2$ $60 + 32 - 3$
2 $84 - 40 + 4$ $84 - 30 - 6$

Practice (A)

1 a 577   **b** 1817   **c** 93.3
   d 2.13   **e** 574   **f** £12.01
2 a 5926   **b** 55.96   **c** 716
   d 0.327   **e** £3.51   **f** £32.66
3 a -7 **b** 1   **c** -1 **d** -1
   e -2 **f** 2   **g** -16 **h** 2
   i -33 **j** -159 **k** 33   **l** -33
4 a 16   **b** 38.4   **c** -39
   d 22.4   **e** -0.6 **f** 16.6
5 a 128, 150   **b** 7.5, 6.6   **c** $-8, -24$ **d** $-0.1, 0.5$

What do you think? (A)

1 a 623   **b** 618   **c** 6.21
   d 821   **e** 621   **f** 625

2 a

$$\begin{array}{r} 7\,5\,7 \\ +\ 3\,6\,4 \\ \hline 1\,1\,2\,1 \\ {}_{1}\ {}_{1} \end{array}$$

b

$$\begin{array}{r} 6\,5\,.\,3 \\ 5\,3\,.\,0\,3 \\ +\ 3\,0\,.\,7\,4 \\ \hline 1\,4\,9\,.\,0\,7 \\ {}_{1} \end{array}$$

Are you ready? (B)

1 a 60.8   **b** 0.0608   **c** 608 000   **d** 0.608
   e 19   **f** 32   **g** 1.9   **h** 3200
2 a 3.6   **b** 36   **c** 8.5   **d** 30.6

Practice (B)

1 a 192   **b** 414   **c** 882   **d** 2247
   e 1216   **f** 7968   **g** 8988   **h** 44 296
2 a 32.4   **b** 56.4   **c** 141.12   **d** 67.032
   e 14   **f** 1.44   **g** 0.21   **h** 0.04088
3 a 86   **b** 163   **c** 1050.75   **d** 1093.5
   e $53.\dot{6}$ **f** 257.4   **g** 120.675   **h** 751.375
4 a 10   **b** 50   **c** 30   **d** 62.5
   e 4   **f** 5   **g** 0.2   **h** 0.04
5 Area of the rectangle = 23.56 mm²
   $x = 9.424$ mm
6 a 162, 486   **b** 10, 5   **c** 78.125, 195.3125
   d 6.4, 1.28

What do you think? (B)

1 a 207   **b** 3465   **c** 5662.8
2 The final digit of the product is the same as the final digit of the product of the final digits in the multiplication.

Are you ready? (C)

1 a 49   **b** 16   **c** 27   **d** 16
2 a 11   **b** 13   **c** 24   **d** 36
3 a 9   **b** 12   **c** 12   **d** 27

Practice (C)

1 a 19   **b** 17   **c** 79   **d** 41
2 a 23   **b** 188   **c** 16   **d** 42
   e 2   **f** 100   **g** 21   **h** 60
3 a 21   **b** -3 **c** 9   **d** -18
4 a $(2 + 5) \times 3 - 1 = 20$ **b** $6^2 - 4 \times (3 + 6) = 0$
   c $(5 - 3) \times (8 \div 2) = 8$ **d** $18 \times 4^2 \div (4 \times 8) = 9$

5 a 9   **b** 27   **c** 81   **d** 36
6 a -15 **b** -1 **c** -1 **d** -6

What do you think? (C)

1 a $4 \times 6^2 + 7 \times 3 = 165$
   b Compare answers as a class.

Consolidate

1 a 1479   **b** 38 962   **c** 1276.35   **d** 0.3
2 a 689   **b** 11 918   **c** 3167.34   **d** 0.008 59
3 a 17 784   **b** 23 664   **c** 21.488   **d** 0.043 68
4 a 683   **b** 6834.8   **c** 70   **d** 65
5 a -11 **b** 24   **c** 11   **d** -7
   e -11 **f** -144 **g** -507 **h** 27
6 a 31   **b** 28   **c** -29 **d** 17
   e 369   **f** 60   **g** 3   **h** 576
7 a 243, 729   **b** $-65, -58$ **c** 0.45, 0.28   **d** 24, 4.8

Stretch

1 a $975 \times 86 = 83\,850$
   b Adding a 3-digit and a 2-digit number gives a greatest sum of $987 + 65 = 1052$. Adding a 4-digit and a 1-digit number gives a greatest sum of $9876 + 5 = 9881$
2 Compare answers as a class.

Chapter 1.2

Are you ready? (A)

1 a $\frac{3}{4}$ **b** $\frac{9}{10}$ **c** $\frac{2}{3}$ **d** $\frac{5}{6}$

2 a $\frac{4}{7}$ **b** $\frac{6}{11}$ **c** $4\frac{5}{7}$ **d** $2\frac{1}{4}$

3 a $14\frac{2}{5}$ **b** $7\frac{3}{10}$ **c** $8\frac{1}{3}$ **d** $13\frac{2}{3}$

Practice (A)

1 a $\frac{7}{8}$ **b** $\frac{14}{15}$ **c** $\frac{13}{16}$ **d** $\frac{11}{24}$

   e $\frac{31}{40}$ **f** $1\frac{11}{56}$ **g** $\frac{5}{6}$ **h** $1\frac{37}{80}$

2 a $-\frac{1}{12}$ **b** $-\frac{3}{16}$ **c** $-\frac{1}{4}$ **d** $\frac{1}{18}$

   e $-\frac{17}{63}$ **f** $\frac{1}{12}$ **g** $\frac{1}{18}$ **h** $-\frac{5}{24}$

3 a $\frac{1}{3}$ **b** $\frac{31}{90}$ **c** $\frac{19}{24}$ **d** $-\frac{38}{63}$

4 a $1\frac{37}{72}$ **b** $2\frac{11}{24}$ **c** $\frac{13}{24}$ **d** $6\frac{37}{56}$

   e $-\frac{31}{36}$ **f** $-5\frac{41}{55}$ **g** $8\frac{2}{9}$ **h** $32\frac{14}{45}$

5 a $1\frac{61}{80}$ cm   **b** $2\frac{2}{3}$ m   **c** $3\frac{5}{12}$ m   **d** $16\frac{2}{3}$ cm

6 a $5\frac{43}{72}$ m   **b** $\frac{23}{36}$ m

7 a $r = \frac{2}{45}$ **b** $p = 6\frac{37}{56}$

8 Check answers using calculator.

What do you think? (A)

1 a Possible example is $\frac{4}{7} + \frac{5}{6}$
    Possible counterexample is $\frac{1}{2} + \frac{1}{3}$
   b Possible example is $\frac{5}{6} - \frac{1}{2}$
    Possible counterexample is $\frac{35}{8} - \frac{19}{7}$

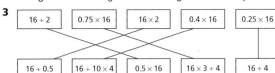

Answers

2 Compare methods as a class.

Are you ready? (B)

1 a $\frac{23}{7}$ **b** $\frac{23}{9}$ **c** $\frac{93}{10}$ **d** $\frac{47}{8}$

2 a $2\frac{3}{8}$ **b** $2\frac{5}{6}$ **c** $7\frac{7}{8}$ **d** $7\frac{5}{7}$

3

$16 \div 2$	0.75×16	16×2	0.4×16	0.25×16

$16 \div 0.5$	$16 \div 10 \times 4$	0.5×16	$16 \times 3 \div 4$	$16 \div 4$

Practice (B)

1 a $\frac{1}{48}$ **b** $\frac{3}{35}$ **c** $\frac{3}{16}$ **d** $\frac{1}{4}$

 e $\frac{9}{121}$ **f** $\frac{5}{18}$ **g** $\frac{3}{28}$ **h** $\frac{1}{9}$

2 a $\frac{3}{5}$ **b** $\frac{3}{4}$ **c** $2\frac{2}{5}$ **d** $4\frac{4}{5}$

3 a 5 **b** 7 **c** $\frac{4}{3}$ or $1\frac{1}{3}$

 d $\frac{9}{5}$ or $1\frac{4}{5}$ **e** $\frac{1}{6}$ **f** $-\frac{1}{8}$

 g $\frac{5}{3}$ or $1\frac{2}{3}$ **h** $\frac{8}{7}$ or $1\frac{1}{7}$

4 a $\frac{3}{4}$ **b** $\frac{7}{15}$ **c** $\frac{5}{6}$ **d** $\frac{3}{4}$

 e 30 **f** $\frac{9}{80}$ **g** $1\frac{2}{3}$ **h** $\frac{4}{15}$

5 a $\frac{7}{40}$ **b** 2 **c** $4\frac{13}{35}$ **d** 6

6 a $3\frac{11}{15}$ **b** $1\frac{11}{19}$ **c** $1\frac{71}{153}$ **d** $1\frac{2}{7}$

7 a $\frac{1}{4}$ **b** $5\frac{11}{16}$ **c** $\frac{16}{147}$ **d** $\frac{64}{189}$

8 a $\frac{26}{45}$ cm² **b** $\frac{7}{15}$ m² **c** $1\frac{113}{192}$ m² **d** $\frac{19}{32}$ cm²

9 a $1\frac{1}{14}$ **b** $-\frac{68}{105}$ **c** $1\frac{31}{55}$ **d** $2\frac{351}{392}$

10 Check answers using calculator.

What do you think? (B)

1 $\boxed{8}\frac{\boxed{3}}{\boxed{4}} \times \boxed{7}\frac{\boxed{5}}{\boxed{6}} = 68\frac{13}{24}$

2 Compare methods as a class.

Consolidate

1 a $\frac{75}{77}$ **b** $\frac{17}{45}$ **c** $\frac{13}{24}$ **d** $-\frac{1}{24}$

 e $\frac{13}{60}$ **f** $\frac{37}{50}$ **g** $1\frac{5}{52}$ **h** $-\frac{37}{112}$

2 a $3\frac{37}{45}$ **b** $1\frac{1}{18}$ **c** $2\frac{1}{15}$ **d** $1\frac{11}{40}$

 e $5\frac{83}{120}$ **f** $2\frac{43}{72}$ **g** $8\frac{29}{30}$ **h** $2\frac{29}{30}$

3 a $\frac{6}{35}$ **b** $\frac{1}{6}$ **c** $6\frac{4}{11}$ **d** $\frac{9}{64}$

 e $\frac{34}{35}$ **f** $7\frac{11}{15}$ **g** $4\frac{1}{14}$ **h** $6\frac{19}{25}$

4 a $\frac{6}{11}$ **b** $1\frac{3}{14}$ **c** $1\frac{3}{32}$ **d** $\frac{3}{22}$

 e $4\frac{2}{7}$ **f** $2\frac{19}{63}$ **g** $\frac{49}{72}$ **h** $\frac{172}{185}$

Stretch

1 $3\frac{3}{8}$ cm²

2 Example answers:

 a $\frac{1}{4} + \frac{2}{8} + \frac{3}{6} = 1$ **b** $\frac{3}{4} \times \frac{4}{5} \times 1\frac{2}{3}$

3 a $0.6 \times 1.25 = \frac{3}{5} \times 1\frac{1}{4} = \frac{3}{5} \times \frac{5}{4} = \frac{3}{4}$

 b Compare answers as a class.

Chapter 1.3

Are you ready? (A)

1 2, 5, 19, 23, 29

2 a 3, 6, 15, 18, 108, 180 **b** 10, 15, 35, 180

 c 15, 180

Practice (A)

1 a $2^3 \times 5$ **b** $2^4 \times 3$

 c $2^2 \times 3 \times 7$ **d** $2 \times 3 \times 5^2$

 e $2^2 \times 7^2$ **f** 5×7^2

 g $2 \times 3 \times 5^2 \times 7$ **h** 3^6

2 a $2^3 \times 3 \times 5$ **b** $2 \times 3 \times 5$

 c $2^4 \times 3 \times 5$ **d** $2^4 \times 3 \times 5^3$

3 a $3^3 \times 5^3 \times 7^3$ **b** $2^3 \times 3^2 \times 5^3 \times 7^3$

 c $2 \times 3^4 \times 5^3 \times 7^3$ **d** $2^2 \times 3^2 \times 5^5 \times 7^3$

4 a $2 \times 3^2 \times 5^2$ **b** 2

5 15

6 2

7 a Yes, 5×7 **b** Yes, $2^2 \times 5$

 c No, $24 = 2^3 \times 3$ **d** Yes, $2 \times 3^2 \times 5$

 e No, $50 = 2 \times 5^2$ **f** No, $135 = 3^3 \times 5$

What do you think? (A)

1 a x **b** xy **c** Compare answers as a class.

Are you ready? (B)

1 a 1, 2, 4, 5, 10, 20

 b 1, 2, 3, 4, 6, 8, 12, 24

 c 1, 2, 3, 4, 6, 9, 12, 18, 36

 d 1, 2, 3, 4, 6, 7, 12, 14, 21, 28, 42, 84

2 a 8, 16, 24, 32 **b** 12, 24, 36, 48

 c 20, 40, 60, 80 **d** 24, 48, 72, 96

3 a $2^2 \times 5$ **b** $2^3 \times 3^2$

 c $2^2 \times 5^2$ **d** $2^3 \times 3 \times 5$

Practice (B)

1 a HCF = 20, LCM = 120 **b** HCF = 12, LCM = 240

 c HCF = 18, LCM = 180 **d** HCF = 6, LCM = 180

2 a $60 = 2^2 \times 3 \times 5$, $84 = 2^2 \times 3 \times 7$

 b ξ

Prime factors of 60		Prime factors of 84
5	2, 2	7
	3	

 c i HCF = $2^2 \times 3 = 12$ **ii** LCM = $2^2 \times 3 \times 5 \times 7 = 420$

3 a HCF = 12, LCM = 840 **b** HCF = 8, LCM = 1200

 c HCF = 20, LCM = 5040 **d** HCF = 65, LCM = 975

4 HCF = 30, LCM = 1800

5 a $2^2 \times 7$ **b** $2^3 \times 3^2 \times 5^2 \times 7$

6 a 5^2 **b** 5

 c $2^2 \times 3^2 \times 5^3 \times 7^2$ **d** $2^3 \times 3^2 \times 5^3 \times 7^2 \times 11$

7 Example answers: 24 and 180, 60 and 72, 36 and 120

8 10 and 30, 15 and 30, 6 and 30
9 9000

What do you think? (B)

1 a i B cannot be 25 because 720 is not a multiple of 25
  ii B cannot be 36 because the LCM would be 72
  iii B could be 180 because 720 is a multiple of both 72 and 180
  b 240

Consolidate

1 a 2×5^2 **b** $2 \times 3^2 \times 5$
  c $2 \times 5 \times 13$ **d** $2^2 \times 3 \times 5^2$
2 a $576 = 2^6 \times 3^2$ **b** $1296 = 2^4 \times 3^4$
  c $5832 = 2^3 \times 3^6$
3 a 16   **b** 30   **c** 21   **d** 48
4 a 72   **b** 48   **c** 54   **d** 144
5 a $2^2 \times 5$ **b** $2^3 \times 3 \times 5^2 \times 7$ **c** 15
6 a $2 \times 3 \times 7 \times 11$ **b** $2^3 \times 3^3 \times 7^2 \times 11$ **c** 7623

Stretch

1 8:44 am
2 3 m
3 a If b is the LCM then $\frac{120}{1} = 120$. Any other multiple of 120 can be used.
  b 1×120

Number review: exam practice

1 $2^3 \times 3 \times 5^2$
2 a $5\frac{17}{20}$ **b** $\frac{7}{9}$
3 0.002 04
4 Example answer: $a = 35$, $b = 21$
5 400 000
6 1080
7 a $1\frac{2}{3}$ **b** $\frac{7}{1000}$
8 46.8
9 $\frac{3}{5}$
10 $\frac{13}{20}$
11 5, 2, 7
12 $3375 = 3^3 \times 5^3 = (3 \times 5)^3$ (or $15 \times 15 \times 15 = 3375$)

Block 2 Indices

Chapter 2.1

Are you ready? (A)

1 a 2^4 **b** 0.9^3 **c** 6^6 **d** $4^2 \times 5^2$
2 a 9, 64, 100, 1   **b** 27, 64, 125, 1, 216
3 a 1, 4, 9, 16, 25, 36, 49, 64, 81, 100
  b 1, 8, 27, 64, 125, 216, 343, 512, 729, 1000

Practice (A)

1 a 81   **b** 64   **c** 512   **d** 625
  e 64   **f** 256   **g** 6.25   **h** 3.375
2 a 2187   **b** 3125   **c** 3375
  d 1 728 000   **e** 0.512   **f** 104.8576
  g 71 639 296   **h** 12 812.904
3 a 17   **b** 19   **c** 47   **d** 83
  e 6.8   **f** 92.8   **g** 0.05   **h** 0.82
4 a 6   **b** 9   **c** 8.2   **d** 6.5
5 a 100   **b** 1 000 000   **c** 81   **d** 128
6 Accept answers that differ by ±0.1 from these:
  a 5.2   **b** 12.2   **c** 14.1   **d** 19.9
  e 3.1   **f** 5.3   **g** 2.1   **h** 0.9

What do you think? (A)

1 a $\boxed{2}^6 = 64$ $\boxed{4}^3 = 64$ $\boxed{8}^2 = 64$
  b $\boxed{3}^4 = 81$ $\boxed{9}^2 = 81$

2 ≈ 14 or 15
  Discuss different possibilities depending on degree of rounding.
3 Compare answers as a class.

Are you ready? (B)

1 a 81   **b** 27   **c** 9   **d** 3
2 a $\frac{1}{8}$ **b** $\frac{1}{15}$ **c** 4
  d $\frac{7}{5}$ **e** $\frac{9}{19}$ **f** 4

Practice (B)

1 a $\frac{1}{5}$ **b** 1   **c** $\frac{1}{8}$ **d** $\frac{1}{636}$
  e $\frac{1}{8}$ **f** $\frac{1}{16}$ **g** $\frac{1}{49}$ **h** $\frac{1}{81}$
2 a 7^{-1} **b** 6^{-1} **c** 2^{-3} **d** 8^{-4}
3 a 2^{-1} **b** 2^{-4} **c** 2^{-6} **d** 2^{-3}
4 a 25   **b** $\frac{125}{27} = 4\frac{17}{27}$ **c** $\frac{81}{16} = 5\frac{1}{16}$ **d** $\frac{64}{121}$
5 a $\frac{2}{5}$ **b** $\frac{3}{32}$ **c** $2\frac{1}{2}$ **d** $\frac{1}{16}$

What do you think? (B)

1 No, $5^2 \times 5^{-2} = 5^0 = 1$
2 <
3 $\frac{1}{16}$

Consolidate

1 a 81   **b** 343   **c** 125
  d 729   **e** 256   **f** 2197
2 a 11   **b** 14   **c** 4
  d 8   **e** 5   **f** 10
3 Accept answers that differ by ±0.1 from these:
  a 7.1   **b** 8.9   **c** 2.2
  d 12.6   **e** 4.6   **f** 4.4
4 a $\frac{1}{3}$ **b** $\frac{1}{5}$ **c** $\frac{5}{4} = 1\frac{1}{4}$
  d $\frac{1}{125}$ **e** $\frac{1}{36}$ **f** $\frac{9}{25}$
5 a 2^{-8} 6^{-2} 5^{-2} 5^0 2^2
  b 2×4^{-1} $2 - 4^{-1}$ 4×2^{-1} $2 + 4^{-1}$ $2 \div 4^{-1}$

Stretch

1 $\boxed{0}^5 = 0$ $\left(\boxed{\frac{1}{6}}\right)^3 = \frac{1}{216}$ $\boxed{2}^{-2} = \frac{1}{4}$ $\left(\boxed{\frac{3}{4}}\right)^{-3} = 2\frac{10}{27}$
2 $ab > b^a$ when a is greater than 1 and b is a positive decimal less than 1
  $a^b > b^a$ when a is a positive odd integer and b is negative

Chapter 2.2

Are you ready? (A)

1 a 10 000   **b** 100   **c** 100 000   **d** 0.1
2 a 4000   **b** 63 000   **c** 3030   **d** 363 000
3 a 3^7 **b** p^{-3} **c** 10^6 **d** w^3

Practice (A)

1 a 8×10^3 **b** 2×10^5
  c 3.2×10^4 **d** 5.6×10^3
  e 9.8×10^5 **f** 5.84×10^5
  g 9.52×10^2 **h** 1.651×10^6
  i 1×10^6 **j** 7×10^5
  k 4×10^4 **l** 1.25×10^5
2 a 3×10^{-2} **b** 8×10^{-5}
  c 3.8×10^{-3} **d** 6.08×10^{-3}
  e 6.34×10^{-4} **f** 3.684×10^{-6}
  g 3.065×10^{-3} **h** 5.38×10^{-3}
  i 5×10^{-1} **j** 3×10^{-2}
  k 1.8×10^{-1} **l** 8.75×10^{-2}

3 a 30 000   **b** 600 000
c 92 000   **d** 635 500
e 68 050 000   **f** 3 840 000
g 126 500   **h** 305 400 000
4 a 0.03   **b** 0.0006
c 0.0038   **d** 0.000 682
e 0.000 050 8   **f** 0.000 000 398
g 0.000 002 865   **h** 0.000 208 05
5 a 2.9×10^4 **b** 8.65×10^{-2}
c 6.38×10^{-1} **d** 8.5
e 3.84×10^2 **f** 3.68×10^{-2}
g 8.468×10^5 **h** 5.8×10^{-7}
6 a 3.5×10^3 6×10^3 3.5×10^4 3×10^6
b 3.2×10^3 1.23×10^4 1.32×10^4 3.21×10^4
c 2.08×10^{-3} 8.2×10^{-3} 2.8×10^{-2} 8.02×10^{-2}
d 3.3×10^{-7} 2.3×10^{-6} 3.3×10^{-6} 3.2×10^{-2}

What do you think? (A)

1 0.35×10^2 is the greatest, as all the others are less than 1
2 All three statements are sometimes true.

Are you ready? (B)

1 a 3.2×10^4 **b** 8.6×10^{-2}
c 8.462×10^5 **d** 3.54×10^{-4}
2 a 10^7 **b** 10^4 **c** 10^2 **d** 10^{-2}
3 a 8.4   **b** 4   **c** 16.686   **d** 0.5
4 a 3.2×10^4 **b** 6×10^2 **c** 1.92×10^5 **d** 3.2×10^{-4}

Practice (B)

1 a 6×10^6 **b** 5×10^8 **c** 7.5×10^5
d 6.4×10^2 **e** 7.8×10^2 **f** 9.3×10^{-2}
2 a 3×10^3 **b** 2×10^2 **c** 3.5×10^6
d 3.25×10^{-4} **e** 3.1×10^7 **f** 1.5×10^{-1}
3 a 1.8×10^9 **b** 3.5×10^4
c 1.08×10^8 **d** 2.072×10^{12}
e 5.184×10^9 **f** $1.176\,49 \times 10^{14}$
g 4.2×10^2 **h** 2.08×10^3
i 2.015×10^{-6} **j** 5.476×10^{-7}
4 a 7.5×10^2 **b** 6.25×10^3 **c** 5×10^2
d 6×10^{-3} **e** 4×10^{-2} **f** 7×10^2
5 a 6×10^4 **b** 9×10^7 **c** 6×10^8 **d** 8.75×10^{-8}
6 a 9.62×10^{11} **b** 6.44×10^{-6}
c 7.84×10^8 **d** 3.375×10^{-9}
e 4×10^7 **f** 3.2×10^3
g 5.75×10^2 **h** 1.7×10^{-8}
7 90 times bigger
8 1.169×10^{-21} grams
9 23.2 hours

What do you think? (B)

1 $(3 \times 10^{-3}) \times (x \times 10^2) = 7.5 \times 10^y$ $x = 2.5,\ y = -1$

$\dfrac{6.8 \times 10^x}{1.7 \times 10^y} = z \times 10^6$ x and y could have many values,
e.g. 9 and 3, 10 and 4,
and $z = 4$

$\dfrac{x \times 10^{-2}}{y \times 10^z} = 6 \times 10^4$ x and y could have many values,
e.g. 9 and 1.5, 8.4 and 1.4,
and $z = -6$

$(x \times 10^4) \times (y \times 10^z) = 8.4 \times 10^{24}$ x and y could have many values, e.g. 2.1 and 4, 1.2 and 7, and $z = 20$

Are you ready? (C)

1 a 7.2×10^4 **b** 3.4×10^{-2}
c 9.84×10^4 **d** 5.8×10^{-5}
2 a 8.5   **b** 2.7
c 3.856   **d** 4.252
3 a 0.32×10^4 **b** 68.9×10^3
c 0.0083×10^{-1} **d** 350×10^{-4}

Practice (C)

1 a 8.6×10^3 **b** 3.6×10^{-7} **c** 8.94×10^{-5}
d 2.83×10^4 **e** 2.4×10^4 **f** 9.9×10^{-2}

2 a 2.72×10^4 **b** 3.28×10^{-3} **c** 6.63×10^{-4}
d 8.14×10^4 **e** 8.766×10^4 **f** 3.841×10^{-2}
3 a 5.67×10^{-2} **b** 8.8×10^3 **c** 8.7×10^{-5}
d 8.054×10^6 **e** 1.6×10^4 **f** 8.03×10^{-3}
4 a 4.7×10^{-3} **b** 6.8×10^5
c 9.4×10^3 **d** 2.082×10^5

What do you think? (C)

1 a $\left(\boxed{6} \times 10^{\boxed{8}}\right) + \left(\boxed{5} \times 10^{\boxed{7}}\right)$ **b** $\left(\boxed{7} \times 10^{\boxed{8}}\right) - \left(\boxed{6} \times 10^{\boxed{5}}\right)$

Consolidate

1 a i 3×10^3 **ii** 8×10^5 **iii** 5.2×10^4
iv 6.85×10^4 **v** 2.034×10^5
b i 3000   **ii** 84 000   **iii** 68 400 000
iv 6080   **v** 50 450
2 a i 6×10^{-2} **ii** 8.4×10^{-4} **iii** 2.508×10^{-3}
iv 3.052×10^{-3} **v** 5.46×10^{-4}
b i 0.0005   **ii** 0.0023   **iii** 0.000 352
iv 0.000 003 05   **v** 0.000 435 2
3 a 8×10^7 **b** 3.5×10^2 **c** 7.8×10^4 **d** 9×10^{-7}
e 3.2×10^{-6} **f** 2.1×10^2 **g** 2.4×10^7 **h** 3×10^{-7}
i 3×10^{-3} **j** 6.724×10^7
4 a 6×10^4 **b** 4.6×10^5 **c** 1.28×10^3
d 1.23×10^{-3} **e** 9.3×10^{-3} **f** 8.44×10^{-3}
g 1.022×10^5 **h** 5.3×10^{-4} **i** 1.034×10^{-1}
j 8.36×10^{-3}
5 a 7.5×10^7 **b** 3.25×10^4
c 6.25×10^6 **d** 7.4949×10^7
6 7.1832×10^4

Stretch

1 $(4.2688 \times 10^{12})\,\mathrm{cm}^3$
2 $(5.236 \times 10^6)\,\mathrm{m}^2$
3 a $p = 3,\ q = 7$ **b** $n = 6.1$ **c** $y = 3.5$
4 a $p^2 = (9 \times 10^{4x})^2 = 8.1 \times 10^{8x+1}$
b $a \times 10^{-7} = c \times 10^{-5} - b \times 10^{-5}$

$a = \dfrac{c \times 10^{-5} - b \times 10^{-5}}{10^{-7}} = \dfrac{c - b}{10^{-2}} = (c - b) \times 10^2$

Chapter 2.3

Are you ready? (A)

1 a 9   **b** 25   **c** 8   **d** 64
2 a x^6 **b** y^{12} **c** x **d** y
3 a $\dfrac{1}{5}$ **b** $\dfrac{1}{9}$ **c** 7   **d** $\dfrac{25}{9} = 2\dfrac{7}{9}$

Practice (A)

1 a 6   **b** 9   **c** 12   **d** 15   **e** 5
f 6   **g** 3   **h** 2   **i** 1   **j** 5
2 a $\dfrac{1}{5}$ **b** $\dfrac{1}{2}$ **c** -2 **d** $\dfrac{1}{6}$ **e** $-\dfrac{1}{5}$
3 a $a^{\frac{1}{2}}$ **b** $b^{\frac{1}{2}}$ **c** $c^{\frac{1}{3}}$ **d** $d^{\frac{1}{7}}$ **e** $e^{\frac{1}{n}}$
4 a $3x^2$ **b** $3a^3$ **c** $5b^6$ **d** $6y^3z^2$ **e** $\dfrac{1}{4h^2k^3}$
f $\dfrac{1}{3x^2y}$ **g** $-\dfrac{1}{5d^2e^3f^{-1}} = -\dfrac{f}{5d^2e^3}$ **h** $-\dfrac{1}{8x^2y^3}$

What do you think? (A)

1 $a^{\frac{1}{2}} > a^{\frac{1}{4}}$ Sometimes true, when $a > 1$

$\left(a^{\frac{1}{4}}\right)^{\frac{1}{4}} = a^{\frac{1}{8}}$ Sometimes true, e.g. when $a = 1$

$a < a^{\frac{1}{3}}$ Sometimes true, when a is a negative integer, or $0 < a < 1$

2 a 16^2 **b** $16^{\frac{1}{2}}$ **c** $16^{\frac{1}{4}}$
3 a 2×10^2 **b** 4×10^2 **c** 4×10^3

Are you ready? (B)

1 a a^{21} **b** b^{-6} **c** c **d** $d^{\frac{5}{2}}$
2 a 4   **b** 8   **c** 4   **d** 6
3 a 27   **b** 125   **c** 4   **d** 625
4 a $\dfrac{1}{5}$ **b** $\dfrac{1}{16}$ **c** $\dfrac{1}{32}$ **d** $\dfrac{16}{9} = 1\dfrac{7}{9}$

Practice (B)

1 a 27   **b** 16   **c** 64   **d** 243   **e** 8
  f 8   **g** 25   **h** 729   **i** 128   **j** 256

2 a $x^{\frac{1}{2}}$ **b** $x^{\frac{3}{2}}$ **c** $x^{\frac{4}{3}}$ **d** $x^{\frac{5}{6}}$ **e** $x^{\frac{n}{m}}$

3 a $64x^9$ **b** $9m^6$ **c** $32w^{10}$ **d** $625x^8y^{12}$

4 a $\frac{1}{16}$ **b** $\frac{1}{64}$ **c** $\frac{1}{16}$ **d** $\frac{1}{4}$

  e $\frac{27}{8}=3\frac{3}{8}$ **f** $\frac{27}{64}$ **g** $\frac{4}{9}$ **h** 81

What do you think? (B)

1 a $8^{\frac{2}{3}}=(-8)^{\frac{2}{3}}$ True, both 4
  b $8^{\frac{2}{3}}\times8^{\frac{2}{3}}=8^{\frac{4}{3}}$ True, add the powers
  c $16^{-\frac{3}{2}}\times4=2^{-3}$ False, 2^{-4}

2 a i $16^{\frac{1}{2}}$ **ii** $16^{\frac{3}{4}}$ **iii** $16^{\frac{5}{4}}$ **iv** $16^{\frac{7}{4}}$ **v** $16^{\frac{3}{2}}$
  b i $8^{\frac{2}{3}}$ **ii** 8^1 **iii** $8^{\frac{5}{3}}$ **iv** $8^{\frac{7}{3}}$ **v** 8^2
  Compare answers as a class. Students should notice that it is not generally the case that the powers in part **b** are double those in part **a**.

Consolidate

1 a 10   **b** 2   **c** 8   **d** 2
  e 5   **f** 4   **g** $\frac{2}{3}$ **h** $\frac{3}{2}=1\frac{1}{2}$

2 a 4   **b** 100   **c** 512   **d** 64
  e 8   **f** $\frac{8}{27}$ **g** $\frac{64}{125}$ **h** $\frac{243}{32}=7\frac{19}{32}$

3 a $\frac{1}{2}$ **b** $\frac{1}{16}$ **c** $\frac{1}{125}$ **d** $\frac{1}{81}$
  e 2   **f** $\frac{256}{81}=3\frac{13}{81}$ **g** $\frac{125}{729}$ **h** $\frac{243}{1024}$

Stretch

1 128
2 a 2   **b** $4^{\frac{1}{2}}$
3 a $x=\frac{3}{5}$ **b** $x=20$ **c** $x=\frac{17}{2}$

Indices: exam practice

1 a 0.00157   **b** 6.4×10^5 **c** 7.2×10^4
2 49
3 a 1   **b** 36   **c** $\frac{1}{8}$ **d** 4
4 0.000519 51.9×10^{-3} 5.19×10^4 519×10^3
5 3^8
6 $\frac{7}{2}$
7 3000
8 2^{16}
9 $5^{(6-x)}$
10 8^{25}
11 $\frac{125}{27}$
12 $\frac{7}{4}$

Block 3 Surds

Chapter 3.1

Are you ready? (A)

1 1, 4, 9, 16, 25, 36, 49, 64, 81, 100
2 a 4   **b** 9   **c** 49   **d** 25   **e** 25
3 a 4   **b** 9   **c** 12

Practice (A)

1 a Yes   **b** Yes   **c** No   **d** Yes
  e No   **f** No   **g** No   **h** Yes
2 a $2\sqrt{3}$ **b** $2\sqrt{2}$ **c** $6\sqrt{5}$ **d** $3\sqrt{5}$
  e $4\sqrt{2}$ **f** $3\sqrt{7}$ **g** $5\sqrt{10}$ **h** $20\sqrt{2}$

3 a $12\sqrt{3}$ **b** $12\sqrt{6}$ **c** $30\sqrt{2}$ **d** $25\sqrt{3}$
  e $8\sqrt{30}$ **f** $36\sqrt{10}$ **g** $60\sqrt{6}$ **h** $16\sqrt{5}$
4 a $2\sqrt{10}$ **b** $8\sqrt{2}$ **c** $4\sqrt{3}$ **d** $6\sqrt{2}$

What do you think? (A)

1 $\frac{\sqrt{80x}}{\sqrt{5x}}=\frac{4\sqrt{5x}}{\sqrt{5x}}=4$
2 a $\sqrt{45}$ **b** $\sqrt{96}$ **c** $\sqrt{1344}$ **d** $\sqrt{1050}$

Are you ready? (B)

1 a $2\sqrt{3}$ **b** $3\sqrt{3}$ **c** $2\sqrt{15}$ **d** $5\sqrt{2}$
2 a $10x$ **b** $2x+7y$ **c** $10x+5x^2$ **d** $4x+9y$

Practice (B)

1 a $4\sqrt{5}$ **b** $2\sqrt{7}$ **c** $3\sqrt{3}$ **d** $15\sqrt{11}$
2 a $3\sqrt{3}$ **b** $\sqrt{3}$ **c** $5\sqrt{2}$ **d** $12\sqrt{6}$
3 a $2\sqrt{2}+2\sqrt{6}$ **b** $\sqrt{2}$ **c** $30\sqrt{3}+12\sqrt{2}$
  d $51\sqrt{3}$ **e** $6\sqrt{3}$ **f** $5\sqrt{5}$
  g $9\sqrt{3}$ **h** $4\sqrt{5}$
4 a $16\sqrt{3}+12\sqrt{2}$ cm   **b** $16\sqrt{2}$ cm   **c** $2\sqrt{15}+20\sqrt{3}$ m

What do you think? (B)

1 $\sqrt{3}$ and $\sqrt{243}$, $\sqrt{12}$ and $\sqrt{192}$, $\sqrt{27}$ and $\sqrt{147}$, $\sqrt{48}$ and $\sqrt{108}$, $\sqrt{75}$ and $\sqrt{75}$

Consolidate

1 a $2\sqrt{3}$ **b** $4\sqrt{3}$ **c** $4\sqrt{2}$ **d** $2\sqrt{10}$
  e $7\sqrt{2}$ **f** $4\sqrt{5}$ **g** $2\sqrt{7}$ **h** $3\sqrt{3}$
2 a $5\sqrt{14}$ **b** $5\sqrt{5}$ **c** $10\sqrt{10}$ **d** 14
  e $12\sqrt{3}$ **f** $2\sqrt{30}$ **g** $10\sqrt{5}$ **h** $18\sqrt{2}$
3 a $15\sqrt{2}$ **b** $16\sqrt{6}$ **c** $12\sqrt{2}$ **d** $48\sqrt{15}$
4 a $5\sqrt{5}$ cm   **b** $2\sqrt{14}$ cm
5 a $24\sqrt{3}$ **b** $24\sqrt{3}$ **c** $675\sqrt{6}$ **d** 400
6 a $5\sqrt{3}$ **b** $3\sqrt{3}$ **c** $4\sqrt{7}$ **d** $6\sqrt{6}$
  e $4\sqrt{3}$ **f** $\sqrt{2}$ **g** $14\sqrt{2}$ **h** $24\sqrt{5}$

Stretch

1 $\sqrt{km^2}$
2 $xy\sqrt{xz}$
3 $5\sqrt{26}$
4 $6a\sqrt{5}-3a\sqrt{5b}$

Chapter 3.2

Are you ready? (A)

1 a $5\sqrt{2}$ **b** $2\sqrt{15}$ **c** $7\sqrt{2}$ **d** $10\sqrt{3}$
2 a $4x$ **b** $27x$ **c** $45y^2$ **d** $4n$

Practice (A)

1 a $\sqrt{30}$ **b** $\sqrt{21}$ **c** $6\sqrt{42}$ **d** $27\sqrt{66}$
2 a $\sqrt{10}$ **b** $\sqrt{2}$ **c** $3\sqrt{3}$ **d** $3\sqrt{5}$
3 a $2\sqrt{15}$ **b** $9\sqrt{2}$ **c** 10   **d** 60
  e $147\sqrt{2}$ **f** $6\sqrt{7}$ **g** $36\sqrt{15}$ **h** $8\sqrt{6}$
4 a $48\sqrt{6}$ cm²   **b** $12\sqrt{3}$ m²   **c** $16\sqrt{15}$ cm²
5 $32\sqrt{6}$ cm³
6 108 cm³

What do you think? (A)

1 a $a=1,b=12$ $a=2,b=6$ $a=3,b=4$
  b $c=1,d=20$ $c=2,d=10$ $c=4,d=5$
  c There are the same number of pairs, as 12 and 20 have the same number of factors.
  d Discuss as a class.

Are you ready? (B)

1 a $4\sqrt{5}$ **b** $7\sqrt{2}$ **c** $3\sqrt{7}$ **d** $6\sqrt{5}$

2 a $5\sqrt{6}$ **b** $\sqrt{15}$ **c** $9\sqrt{2}$ **d** $6\sqrt{35}$

3 a $2\sqrt{7}$ **b** $5\sqrt{6}$ **c** $3\sqrt{2}$ **d** $\sqrt{3}$

4 a $6x + 15$ **b** $10x^2 - 15x$

    c $x^2 - 3x - 10$ **d** $x^2 - 12x + 36$

Practice (B)

1 a $3\sqrt{2} + 12$ **b** $3 + 4\sqrt{3}$ **c** $\sqrt{14} - \sqrt{21}$

    d $5\sqrt{7} + 5\sqrt{3}$ **e** $3\sqrt{2} - 2\sqrt{3}$ **f** $4\sqrt{6} - 8$

    g $18\sqrt{2} - 24\sqrt{3}$ **h** $6\sqrt{10} - 10$

2 a $5\sqrt{5} + 11$ **b** $\sqrt{10} + \sqrt{30} + 2\sqrt{3} + 6$

    c $\sqrt{6} - \sqrt{10} + \sqrt{15} - 5$ **d** $-\sqrt{2} - 4$

    e $9 - 2\sqrt{14}$ **f** $15 + 6\sqrt{6}$

    g -11 **h** 5 **i** 51

3 a $8\sqrt{3} - 6\sqrt{6} + 4\sqrt{2} - 6$ **b** $12 + 4\sqrt{3}$

    c $28 + \sqrt{5}$ **d** $42 - 3\sqrt{14} + 4\sqrt{7} - 2\sqrt{2}$

    e $29 + 12\sqrt{5}$ **f** $56 - 16\sqrt{6}$

4 a $\left(3 + 4\sqrt{3}\right)\text{cm}^2$

    b $\left(\sqrt{10} + 4\sqrt{5} + 3\sqrt{2} + 12\right)\text{cm}^2$

    c $\left(40.5 + 16\sqrt{5}\right)\text{m}^2$

5 $3\sqrt{2}$ cm

What do you think? (B)

1 $\left(\sqrt{50} - 3\sqrt{2}\right)^2 = \left(5\sqrt{2} - 3\sqrt{2}\right)^2 = \left(2\sqrt{2}\right)^2 = 8$, which is rational.

2 $\left(3\sqrt{g} + 2g\right)^2 = 9g + 6g\sqrt{g} + 6g\sqrt{g} + 4g^2$

$$= 9g + 12g\sqrt{g} + 4g^2$$

$\left(\sqrt{6g} + 2g\right)^2 = 6g + 2g\sqrt{6g} + 2g\sqrt{6g} + 4g^2$

$$= 6g + 4g\sqrt{6g} + 4g^2$$

Beca is wrong, they are not equivalent.

3 Parts **g**, **h** and **i**. Compare to the difference of two squares.

Consolidate

1 a $\sqrt{10}$ **b** $\sqrt{35}$ **c** 4 **d** $\sqrt{10}$

    e $24\sqrt{3}$ **f** $4\sqrt{3}$ **g** $24\sqrt{6}$ **h** $\dfrac{256\sqrt{3}}{\sqrt{2}}$ or $128\sqrt{6}$

2 a $3\sqrt{3} - 6$ **b** $2\sqrt{5} + \sqrt{15}$ **c** $4\sqrt{10} - 8$ **d** $16\sqrt{3} - 32$

3 a $-5 - 2\sqrt{3}$ **b** $10 + 2\sqrt{3} + 5\sqrt{6} + 3\sqrt{2}$ **c** 4

    d $36 + 2\sqrt{2}$ **e** $15 - 10\sqrt{5} + 9\sqrt{2} - 6\sqrt{10}$ **f** $408 - 112\sqrt{2}$

4 $20\sqrt{3}$ cm

5 a $4\sqrt{5}$ cm   **b** $25\,\text{cm}^2$

Stretch

1 $16\sqrt{5}$

2 $x = 8$ and $y = 3$

3 a $2\sqrt{3x}$ **b** $54x\sqrt{2x}$ **c** $\dfrac{1}{81x^4 y\sqrt{3xy}}$

Chapter 3.3

Are you ready?

1 a $5\sqrt{2}$ **b** $2\sqrt{3}$ **c** $12\sqrt{2}$ **d** $16\sqrt{6}$

2 a $\sqrt{15}$ **b** $6\sqrt{2}$ **c** $63\sqrt{2}$ **d** $32\sqrt{3}$

3 a $3\sqrt{5} - 6$ **b** $\sqrt{10} + 3\sqrt{2}$

    c $2\sqrt{6} + 4\sqrt{3}$ **d** $6\sqrt{10} - 10\sqrt{2}$

Practice

1 a $\dfrac{3\sqrt{5}}{5}$ **b** $\dfrac{5\sqrt{11}}{11}$ **c** $\dfrac{\sqrt{10}}{5}$ **d** $4\sqrt{2}$

    e $\dfrac{\sqrt{3}}{3}$ **f** 2 **g** $\sqrt{2}$ **h** $\dfrac{\sqrt{6}}{4}$

2 a $\dfrac{\sqrt{3}}{2}$ **b** $\dfrac{4\sqrt{2}}{3}$ **c** $\dfrac{\sqrt{5}}{3}$ **d** $\sqrt{3}$

    e $\dfrac{\sqrt{15}}{6}$ **f** $\dfrac{2}{3}$ **g** $\sqrt{5}$ **h** $\dfrac{5\sqrt{2}}{4}$

3 a $\dfrac{\sqrt{15} + \sqrt{5}}{5}$ **b** $\dfrac{4\sqrt{5} - \sqrt{10}}{5}$ **c** $\dfrac{\sqrt{10} + 5\sqrt{5}}{5}$

    d $\sqrt{2} - \sqrt{6}$ **e** $2\sqrt{2} - \sqrt{3}$ **f** 15

    g $\dfrac{9 - 8\sqrt{3}}{6}$ **h** $\dfrac{2 + \sqrt{10}}{2}$

4 a $\dfrac{3\sqrt{2} - 2\sqrt{6}}{2}$ **b** $\dfrac{20 + 25\sqrt{2}}{4}$

    c $2\sqrt{5} - 3\sqrt{15}$ **d** $2\sqrt{2}$

5 a $h = \sqrt{5}$ cm   **b** $h = 2\sqrt{30}$ cm

6 $4\sqrt{10}$

What do you think?

1 When simplified, they will both give the same answer.

$$\dfrac{6 + \sqrt{3}}{2\sqrt{6}} \times \dfrac{2\sqrt{6}}{2\sqrt{6}} = \dfrac{12\sqrt{6} + 6\sqrt{2}}{24} = \dfrac{2\sqrt{6} + \sqrt{2}}{4}$$

$$\dfrac{6 + \sqrt{3}}{2\sqrt{6}} \times \dfrac{\sqrt{6}}{\sqrt{6}} = \dfrac{6\sqrt{6} + 3\sqrt{2}}{12} = \dfrac{2\sqrt{6} + \sqrt{2}}{4}$$

2 $12 + 6\sqrt{5}$, which is irrational.

3 a $\dfrac{\sqrt{6}}{3}$ **b** $\dfrac{\sqrt{6}}{2}$ **c** 1

    d $\dfrac{\sqrt{6}}{6}$ **e** $\dfrac{\sqrt{2}}{6}$ **f** $\dfrac{1}{6}$

Consolidate

1 a $\sqrt{5}$ **b** $2\sqrt{3}$

    c $\dfrac{2\sqrt{6}}{3}$ **d** $\dfrac{2\sqrt{15}}{15}$

2 a $\dfrac{3\sqrt{6} + 2\sqrt{3}}{6}$ **b** $\dfrac{\sqrt{15} - 4\sqrt{5}}{5}$

    c $\dfrac{2\sqrt{30} + 3\sqrt{2}}{6}$ **d** $\dfrac{8\sqrt{3} - 3\sqrt{6}}{2}$

Stretch

1 $\sqrt{a} - 1$

2 $25\sqrt{2}$

Surds: exam practice

1 $4\sqrt{3}$

2 $10\sqrt{2}$

3 $12\sqrt{5}$

4 $8\sqrt{3}$

5 7

6 $7 - 3\sqrt{3}$

7 $2\sqrt{3}$

8 $7 + 4\sqrt{3}$

9 The expression simplifies to $a^2 - b$, which is the difference between two integers.

10 $\dfrac{\sqrt{2}}{8}$

11 $p = 5$, $q = 15$

12 45

13 Rational (46)

14 $\dfrac{5\sqrt{7} + 42}{7}$

Block 4 Percentages

Chapter 4.1

Are you ready? (A)

1 **a** 0.5   **b** 0.4   **c** 0.75
   d 0.$\dot{3}$ **e** 0.625   **f** 0.3
2 **a** 80%   **b** 40%   **c** 3%
   d 84%   **e** 107%   **f** 1.5%

3 **a** $\frac{8}{25}$ **b** $\frac{9}{20}$ **c** $\frac{23}{100}$

   d $\frac{3}{100}$ **e** $\frac{1}{500}$ **f** $\frac{21}{20} = 1\frac{1}{20}$

Practice (A)

1

Fraction	Decimal	Percentage
$\frac{1}{4}$	**0.25**	**25%**
$\frac{7}{10}$	0.7	**70%**
$\frac{63}{100}$	**0.63**	63%
$\frac{1}{5}$	0.2	**20%**
$\frac{3}{8}$	**0.375**	**37.5%**
$\frac{1}{50}$	**0.02**	2%
$\frac{2}{25}$	**0.08**	**8%**
$\frac{1}{200}$	**0.005**	0.5%
$\frac{1}{4000}$	0.000 25	**0.025%**

2 **a i** 0.625   **ii** 62.5%
   b i 0.45   **ii** 45%
   c i 0.36   **ii** 36%
   d i 0.575   **ii** 57.5%
   e i 0.725   **ii** 72.5%
   f i 0.94   **ii** 94%
   g i 0.7875   **ii** 78.75%
   h i 0.425   **ii** 42.5%

3 **a** $\frac{7}{10}$ **b** $\frac{6}{25}$ **c** $\frac{8}{25}$ **d** $\frac{19}{20}$

   e $\frac{1}{25}$ **f** $\frac{3}{1000}$ **g** $\frac{7}{8}$ **h** $\frac{319}{500}$

4 **a** $\frac{1}{4}$ $\frac{1}{3}$ 35% 0.4   **b** 60% $\frac{13}{20}$ 0.66 $\frac{2}{3}$

   c 0.89 $\frac{9}{10}$ 92% $\frac{19}{20}$ **d** 1.01 105% $\frac{11}{10}$ $\frac{10}{9}$

What do you think? (A)

1 **a** 0.0$\dot{1}$4 285$\dot{7}$ **b** 0.$\dot{2}$85 71$\dot{4}$

   c 1.857 14$\dot{2}$ **d** 0.0$\dot{7}$1 428$\dot{5}$

Are you ready? (B)

1 **a** 0.428 57$\dot{1}$ **b** 0.$\dot{6}$ **c** 0.$\dot{5}$ **d** 0.$\dot{7}\dot{2}$
2 **a** 0.555 555   **b** 0.363 636
   c 0.522 222   **d** 0.246 246
3 **a** $\frac{1}{2}$ **b** $\frac{7}{10}$ **c** $\frac{21}{25}$ **d** $\frac{3}{50}$

4 $\frac{3}{7}$ $\frac{5}{6}$ $\frac{5}{11}$ $\frac{8}{15}$

Practice (B)

1 **a** 0.$\dot{6}$ **b** 0.8$\dot{3}$ **c** 0.857 14$\dot{2}$ **d** 0.$\dot{7}\dot{2}$

2 **a** $\frac{5}{9}$ **b** $\frac{8}{9}$ **c** $\frac{16}{33}$ **d** $\frac{241}{333}$

3 **a** $\frac{1}{3}$ **b** $\frac{3}{11}$ **c** $\frac{9}{11}$ **d** $\frac{26}{99}$

   e $\frac{74}{99}$ **f** $\frac{304}{999}$ **g** $\frac{211}{333}$ **h** $\frac{94}{111}$

4 **a** $\frac{7}{15}$ **b** $\frac{23}{90}$ **c** $\frac{37}{90}$ **d** $\frac{2}{55}$

   e $\frac{127}{330}$ **f** $\frac{14}{33}$ **g** $\frac{8171}{9900}$ **h** $\frac{128}{555}$

5 **a** $1\frac{5}{9}$ **b** $2\frac{38}{99}$ **c** $4\frac{73}{90}$ **d** $3\frac{14}{165}$

6 $43\frac{1}{3}$

7 $\frac{5}{6}$

8 $\frac{31}{55}$ $\frac{4}{7}$ 0.5$\dot{7}$ $\frac{3}{5}$

What do you think? (B)

1 $\frac{277}{330}$

2 $\frac{32}{45}$

3 For one digit, the decimal is 0.0n0n0n0n...
   For two digits, the decimal is 0.$ababab$...

Consolidate

1

Fraction	Decimal	Percentage
$\frac{1}{5}$	**0.2**	**20%**
$\frac{9}{10}$	0.9	**90%**
$\frac{27}{50}$	**0.54**	54%
$\frac{13}{100}$	0.13	**13%**
$\frac{3}{20}$	**0.15**	**15%**
$\frac{2}{25}$	**0.08**	8%
$\frac{9}{25}$	**0.36**	**36%**
$\frac{37}{5000}$	**0.0074**	0.74%
$\frac{21}{2500}$	0.0084	**0.84%**

2 **a** 47% $\frac{1}{2}$ $\frac{13}{25}$ 0.55   **b** 14% $\frac{3}{20}$ $\frac{1}{6}$ 0.17

   c 0.09 0.099 $\frac{1}{10}$ $\frac{3}{15}$ **d** $\frac{4}{5}$ 0.84 $\frac{17}{20}$ 86%

3 **a** $\frac{7}{9}$ **b** $\frac{2}{9}$ **c** $\frac{65}{99}$ **d** $\frac{28}{33}$

   e $\frac{19}{33}$ **f** $\frac{268}{333}$ **g** $\frac{146}{333}$ **h** $\frac{5}{27}$

4 **a** $\frac{2}{45}$ **b** $\frac{7}{18}$ **c** $\frac{13}{330}$ **d** $\frac{46}{99}$

   e $\frac{437}{1650}$ **f** $2\frac{1}{3}$ **g** $1\frac{38}{99}$ **h** $7\frac{47}{198}$

Stretch

1 **a** $\frac{10}{33}$ **b** $1\frac{61}{90}$ **c** $\frac{1252}{4455}$

2 $\frac{2}{5} + \frac{x}{10} = \frac{4 + x}{10}$

3 $\frac{1}{3} + \frac{y}{10} = \frac{10 + 3y}{30}$

Chapter 4.2

Are you ready? (A)

1 **a** 8   **b** 36   **c** 6.5   **d** 4.55
2 **a** 6.5   **b** 0.34   **c** 4.8   **d** 0.38

Practice (A)

1 **a** 15.6   **b** 25.5   **c** 66   **d** 13
   e 427.5   **f** 56.1   **g** 17.1   **h** 319.2
2 **a** 11.7   **b** 145.6   **c** 210.8   **d** 147.9
3 **a** 29   **b** 14.5   **c** 7.25   **d** 0.145
4 **a** 38.4   **b** 37.8   **c** 54.56   **d** 267.8
5 £14.40
6 40
7 £552.50
8 168
9 3510
10 £267.80

What do you think? (A)

1 No, $0.8 \times 0.9 = 0.72$, which means she got a discount of 28% overall.
2 **a** 50% of 80 = 80% of 50
     $0.5 \times 80 = 0.8 \times 50 = 8 \times 5 \div 10$
   b 64% of 25 = 25% of 64 = 16
   c Discuss with a partner.

Are you ready? (B)

1 **a** 0.3   **b** 0.12   **c** 0.68   **d** 0.03
2 **a** 249   **b** 305   **c** 20   **d** 168
3 **a** £354.33   **b** £24.23   **c** £1.28   **d** £5684.39

Practice (B)

1 **a** 0.82   **b** 0.63   **c** 0.03   **d** 0.007
2 **a** 15.36   **b** 87.92   **c** 157.08   **d** 319.62
   e 9.12 g   **f** 720.79 g   **g** £3.70   **h** £23.06
3 **a** 1.23   **b** 0.86   **c** 1.28   **d** 1.68
   e 0.48   **f** 0.97   **g** 0.932   **h** 1.092
4 **a** 116.44   **b** 407.16   **c** 314.88   **d** 23.36
   e 39.6   **f** 64.89   **g** 75.62   **h** 24.24
5 **a** 88%   **b** £2068
6 £2550
7 £13 334.80
8 88 109

What do you think? (B)

1 $0.32 \times 84 = 0.84 \times 32 = 84 \times 32 \div 100$ but $1.32 \times 84 \neq 1.84 \times 32$
2 Less than: $1.1 \times 0.9 = 0.99$, which is less than 1

Are you ready? (C)

1 **a** $\frac{3}{20}$ **b** $\frac{3}{4}$ **c** $\frac{37}{50}$ **d** $\frac{13}{20}$
2 **a** 67%   **b** 54%   **c** 85%   **d** 76%
3 **a** 198.39   **b** £6.84   **c** £652.50   **d** 3574.30

Practice (C)

1 **a** 72%   **b** 74%   **c** 70%   **d** 65%
   e 62%   **f** 72%   **g** 37%   **h** 6%
2 **a** 56.25%   **b** 49.23%   **c** 76%   **d** 52.06%
   e 76.43%   **f** 30.72%   **g** 18%   **h** 25%
3 33.3%
4 Maths = 86.7%, English = 87.5%
   Seb did better in his English test.
5 16% loss
6 56% profit
7 5.5% interest per year
8 33.3% profit
9 4% loss
10 8.57% loss
11 16.4%

Consolidate

1 **a** 72.8   **b** 76.8   **c** 32.2   **d** 23.4
   e 28.8   **f** 307.1   **g** 1.59   **h** 9.6
2 **a** 243   **b** 280   **c** 14.4   **d** 44.8
   e 57.2   **f** 153.6   **g** 667.08   **h** 38.6
3 **a** 45.6   **b** 5.2   **c** 26.1   **d** 4.76
4 **a** 44.45   **b** 116.28   **c** 15.36   **d** 78.2
   e 52.2   **f** 16.5   **g** 6.97   **h** 287.844
5 **a** 15%   **b** 72%   **c** 57.5%   **d** 61.5%
   e 24.3%   **f** 12.3%   **g** 18.3%   **h** 3.84%
6 £310

Stretch

1 Lower. $0.75 \times 1.3 = 0.975 = 97.5\%$ of the original price

2 **a** $N \times \left(1 + \dfrac{a}{100}\right)$ **b** $N \times \left(1 - \dfrac{a}{100}\right)$

3 **a** 20%   **b** 56.25%
4 115 raisin cookies = 24%

Chapter 4.3

Are you ready?

1 **a** 27   **b** 73.6   **c** 92   **d** 33.44
2 **a** 1.2   **b** 0.7   **c** 1.12   **d** 0.38
3 **a** 1.56   **b** 0.9975   **c** 0.9775   **d** 0.675

Practice

1 **a** $20 \times 1.1 \times 1.1$ **b** $120 \times 0.9 \times 0.9$
   c $60 \times 0.8 \times 0.75$ **d** $160 \times 1.3 \times 1.15$
   e $40 \times 1.15 \times 1.3$ **f** $85 \times 1.12 \times 1.06$
   g $163 \times 0.77 \times 0.95$ **h** $82 \times 0.95 \times 1.035$
2 £38.48
3 **a** £12 000   **b** £11 520   **c** £10 192.16
4 **a** £546.36   **b** £133.39
5 **a** 250 cm   **b** 7.8125 cm
6 £867.10
7 43.56 cm^2
8 373.248 cm^3
9 £3444.60
10 **a** 8.54 billion   **b** 2048

What do you think?

1 **a** 9 years
   b Various possible answers relating to reliability
2 **a** **i** 38.4% decrease   **ii** 30.2% increase
   b 6.9% decrease overall

Consolidate

1 **a** $75 \times 1.1 \times 1.2$ **b** $150 \times 1.15 \times 1.08$
   c $220 \times 0.88 \times 0.85$ **d** $342 \times 1.2 \times 0.97$
2 **a** 212   **b** 267–268   **c** 809–810
3 £257 397
4 £459
5 **a** £1539   **b** £1620.07   **c** £1938.94
6 2812–2813   3164–3165   3559–3560

Stretch

1 £780
2 15 years
3 2.470%
4 $x = 4.2$

Chapter 4.4

Are you ready?

1 **a** 0.8   **b** 0.65   **c** 0.28   **d** 0.038
2 **a** 1.2   **b** 0.6   **c** 1.32   **d** 0.98
3 **a** $x = 40$ **b** $y = 80$ **c** $a = 45$ **d** $w = 0.7$

Practice

1 **a** 50   **b** 100   **c** 400   **d** 500
2 **a** 48   **b** 144   **c** 24   **d** 240

3 25

4 £23

5 £12.80

6 £375

7 650 g

8 £70 profit

9 £36 200

10 £680

What do you think?

1 512 cm³

2 £150

3 If a 100 g bar costs £1, then increasing the bar size gives 110 g for £1.
Decreasing the cost gives 100 g for 90p, which is equivalent to 111.1 g for £1 and is better value.

Consolidate

1 a 15   **b** 7.5   **c** 3.75   **d** 75

2 £180

3 £15 200

4 80 mm

5 £156.75

6 £49

7 £725.50

8 58 shapes

Stretch

1 187.5 g

2 £250

3 a $\frac{2700}{100 + x} \times 100$ **b** $\frac{5.60}{100 - y} \times 100$

   c $\frac{t}{(100 - a)(100 - b)} \times 10\,000$

4 £1243.91 (to the nearest penny)

Percentages: exam practice

1 125%

2 £9479.20

3 8640

4 64%

5 500%

6 £390

7 926 100

8 0.375, terminating
0.$\dot{5}$, recurring
0.7$\dot{2}$, recurring

9 $\frac{4}{33}$

10 $\frac{17}{45}$

11 $\frac{49}{60}$ 0.81$\dot{6}$

Block 5 Accuracy and estimation

Chapter 5.1

Are you ready?

1 a 600 **b** 20 000 **c** 40 **d** 0.3 **e** 0.0007

2 1020.846   984   1250   1499.999

3 a 12 **b** 45   **c** 36   **d** 30   **e** 28

Practice

1 a 80   **b** 130   **c** 25   **d** 12
  e 10   **f** 9   **g** 55   **h** 13

2 a 80   **b** 7500   **c** 8.3   **d** 4.5
  e 500   **f** 15.75   **g** 800   **h** 29

3 a 100 cm²   **b** 220 m²   **c** 1200 mm²

4 a 150 cm²   **b** 14.1 cm

5 126 cm²

6 a 26.58%   **b** 7.73%   **c** 35.98%

What do you think?

1 a 4 days

  b No, rounding up the numerator increases the estimate, but rounding up the denominator has the opposite effect of reducing the estimate.

2 a Underestimate   **b** Overestimate

  c Underestimate   **d** Cannot tell

Consolidate

1 a 16   **b** 10   **c** 0.04   **d** 1800
  e 803   **f** 6.25   **g** 24.5   **h** 0.5

2 a 40   **b** 112.5   **c** 9.1   **d** 0.8
  e 14 000 **f** 33.25   **g** 12   **h** 15.5

3 £400

4 1500

5 13 hours and 20 minutes

Stretch

1 a 17 days

  b It wouldn't affect it, as 26 and 32 both round to 30 to 1 significant figure.

2 9 cm

3 3000 minutes or 50 hours

4 Many possible examples.
If a number rounds down, the answer will be the same; but if the number rounds up, it will give a different answer.

Chapter 5.2

Are you ready? (A)

1 197   201.58

2 a 84.4   **b** 3.5   **c** 0.3   **d** 99.0

3 a 300   **b** 0.08   **c** 40 000   **d** 400 000

4 a 0.08 **b** 6.849 **c** 0.003   **d** 3.68

Practice (A)

1 a i lower = 250   **ii** upper = 350
   iii $250 \leqslant a < 350$

  b i lower = 235   **ii** upper = 245
   iii $235 \leqslant b < 245$

  c i lower = 45 650   **ii** upper = 45 750
   iii $45\,650 \leqslant c < 45\,750$

  d i lower = 32 495   **ii** upper = 32 505
   iii $32\,495 \leqslant d < 32\,505$

  e i lower = 18 500   **ii** upper = 19 500
   iii $18\,500 \leqslant e < 19\,500$

  f i lower = 349.5   **ii** upper = 350.5
   iii $349.5 \leqslant f < 350.5$

  g i lower = 1995   **ii** upper = 2005
   iii $1995 \leqslant g < 2005$

  h i lower = 455.5   **ii** upper = 456.5
   iii $455.5 \leqslant h < 456.5$

2 a $450 \leqslant a < 550$ **b** $75\,000 \leqslant b < 85\,000$
  c $515 \leqslant c < 525$ **d** $1845 \leqslant d < 1855$

3 a $3.45 \leqslant a < 3.55$ **b** $2.355 \leqslant b < 2.365$
  c $0.5835 \leqslant c < 0.5845$ **d** $320.25 \leqslant d < 320.35$
  e $84.345 \leqslant e < 84.355$ **f** $3.5835 \leqslant f < 3.5845$
  g $3.245 \leqslant g < 3.255$ **h** $0.0835 \leqslant h < 0.0845$

4 a $3.5 \leqslant a < 3.6$ **b** $8.4 \leqslant b < 8.5$
  c $9534 \leqslant c < 9535$ **d** $0.08 \leqslant d < 0.09$

5 $18.8\,\text{cm} \leqslant \text{length} < 21.2\,\text{cm}$

6 $6.53\,\text{cm} \leqslant \text{length} < 6.55\,\text{cm}$

7 $8.45 \leqslant x < 8.46$

8 $11.8125\,\text{mm} \leqslant x < 11.9375\,\text{mm}$

What do you think? (A)

1 a $95 \leqslant x < 150$ **b** $99.5 \leqslant x < 105$
  c $99.95 \leqslant x < 100.5$

2 a $2.9\,\text{cm} \leqslant r < 3.3\,\text{cm}$ **b** $3.07\,\text{cm} \leqslant r < 3.11\,\text{cm}$

3 $25.7 < p \leqslant 25.8$

Are you ready? (B)

1 a $17.5 \leqslant x < 18.5$ **b** $247.5 \leqslant x < 248.5$
  c $345 \leqslant x < 355$ **d** $6450 \leqslant x < 6550$
  e $83\,500 \leqslant x < 84\,500$ **f** $3450 \leqslant x < 3550$

2 a $8.4 \leqslant x < 8.5$ **b** $6.49 \leqslant x < 6.50$
    c $0.84 \leqslant x < 0.85$ **d** $0.09 \leqslant x < 0.10$
    e $84.67 \leqslant x < 84.68$ **f** $945.324 \leqslant x < 945.325$

Practice (B)

1 a $580\,\text{m}$ **b** $17\,625\,\text{m}^2$
2 $1225 \leqslant x < 1275$
3 $180 \leqslant x < 366.6$
4 a $433.5 \leqslant x < 444.5$ **b** $399.15 \leqslant x < 409.25$
    c $759.15 \leqslant x < 769.25$ **d** $793.5 \leqslant x < 804.5$
    e $37\,327.5 \leqslant x < 38\,837.5$ **f** $26\,507.25 \leqslant x < 27\,381.25$
    g $56\,801.25 \leqslant x < 57\,711.25$ **h** $79\,987.5 \leqslant x < 81\,857.5$
5 a $19.85 \leqslant x < 29.95$ **b** $46.155 \leqslant x < 56.165$
    c $6.155 \leqslant x < 16.165$ **d** $59.85 \leqslant x < 69.95$
6 a $35.71 \leqslant x < 54$ **b** $1.30 \leqslant x < 1.83$
    c $3.90 \leqslant x < 4.44$ **d** $14.71 \leqslant x < 18$
7 a $6.91 \leqslant x < 7.94$ **b** $5799.02 \leqslant x < 5825.06$
    c $7.67 \leqslant x < 7.81$ **d** $487.13 \leqslant x < 561.43$
8 a $54.825\,\text{cm}^2$ **b** $3.5\,\text{cm}$
9 $56.3\,\text{mph}$
10 a Upper bound = $18.72\,\text{cm}$ Lower bound = $17.33\,\text{cm}$
     b Upper bound = $14.60\,\text{cm}$ Lower bound = $14.38\,\text{cm}$

What do you think? (B)

1 $2498\,\text{cm}^2$ (to the nearest integer)
    Discuss different possibilities depending on the degree
    of rounding.
2 Compare answers as a class.

Consolidate

1 a $31.5 \leqslant x < 32.5$ **b** $615 \leqslant x < 625$
    c $750 \leqslant x < 850$ **d** $24\,500 \leqslant x < 25\,500$
    e $8950 \leqslant x < 9050$ **f** $8500 \leqslant x < 9500$
    g $3450 \leqslant x < 3550$ **h** $3495 \leqslant x < 3505$
2 a $9.15 \leqslant x < 9.25$ **b** $95.475 \leqslant x < 95.485$
    c $0.00075 \leqslant x < 0.00085$ **d** $984.55 \leqslant x < 984.65$
    e $3.55 \leqslant x < 3.65$ **f** $8.5675 \leqslant x < 8.5685$
    g $0.06475 \leqslant x < 0.06485$ **h** $3.835 \leqslant x < 3.845$
3 a $352 \leqslant x < 353$ **b** $9.50 \leqslant x < 9.60$
    c $6.802 \leqslant x < 6.803$ **d** $0.002 \leqslant x < 0.003$
    e $32.84 \leqslant x < 32.85$ **f** $962.4 \leqslant x < 962.5$
    g $94.6 \leqslant x < 94.7$ **h** $0.008 \leqslant x < 0.009$
4 a $80.75 \leqslant x < 90.85$ **b** $0.688 \leqslant x < 0.802$
    c $248.0625 \leqslant x < 251.2225$ **d** $6077.5 \leqslant x < 7087.5$
    e $77.65 \leqslant x < 78.75$
5 a $37.025 \leqslant x < 37.135$ **b** $20.75 \leqslant x < 21.26$
    c $7487.93 \leqslant x < 7556.81$
    d $99\,402.27 \leqslant x < 100\,392.18$
    e $457.865 \leqslant x < 467.975$

Stretch

1 $75.199\,\text{cm} \leqslant C < 83.135\,\text{cm}$
2 a $0.561 \leqslant a < 0.578$ **b** $-0.90 \leqslant b < -0.84$
3 a $9.079\,\text{m}^2 \leqslant x < 9.294\,\text{m}^2$ **b** $2.572\,\text{m}^3 \leqslant x < 2.664\,\text{m}^3$
4 329

Accuracy and estimation: exam practice

1 a Example answer: $1.052\,208\,839$ **b** 1.1
2 a Example answer: $30 \times 50 = 1500$
    b Example answer: An underestimate as both
      numbers have been rounded down
3 $\approx \dfrac{3^3 + \sqrt{100}}{0.5} = \dfrac{27 + 10}{0.5} = 74$
4 $4.7 \leqslant x < 4.8$
5 $0.635 \leqslant y < 0.645$
6 a Example answer: $3 \times 20^2 = 1200\,\text{cm}^2$
    b Example answer: An underestimate as $\pi > 3$
      and $21 > 20$
7 50
8 57
9 Example answer: It should be $8.5 - 4.5 = 4$
10 £1071

11 11 books
12 4.06
13 2.41

Number: exam practice

1 $80\,000$
2 $\dfrac{7}{30}$
3 9
4 $5\sqrt{2}$
5 24
6 £800
7 Example answer: $9 \times 10 = 90$
8 $2^4 \times 3^5 \times 5^4$
9 £400.23
10 $\dfrac{22}{30}$
11 $\dfrac{81}{16}$
12 48
13 $39.5 \div 2.85 = 13.85\ldots$
    So only 13 would fit in this case.
14 $x = 13$

Block 6 Understanding algebra

Chapter 6.1

Are you ready?

1 a -1 **b** 18 **c** -16
2 a $-2x + 7y$ **b** $2 + 8t$
    c $2w^2 + 12w$ **d** $b - a - 2ab$
3 a false   **b** true   **c** false   **d** true
4 a $k + 2$ **b** $k - 7$ **c** $\dfrac{k}{2}$ or $\dfrac{1}{2}k$
    d $\dfrac{k}{4}$ **e** jk

Practice

1 a expression   **b** formula
    c equation   **d** identity
2 $4x + 8$
3 a $2t^3$ **b** $12x^5$ **c** $6ab^3$
    d $3d^2$ **e** $3fg^2$ **f** $8g^3h^6$
4 $4t + 3s$
5 a $x + 4$ **b** $x + 1$ **c** $3x + 5$
6 $2y + 45x$
7 a 9 **b** -17 **c** 13 **d** -104
8 £123.05

What do you think?

1 a odd   **b** odd   **c** odd   **d** odd
2 a Compare answers as a class.
     Possible answers include:
     i $2p + q$ **ii** $2(p + q)$
    b No

Consolidate

1 a 2 **b** 14 **c** -16 **d** -64
2 a £7   **b** £8.13   **c** Yes
3 a $6efg$ **b** $6ef + 4eg + 12fg$
4 a formula   **b** identity
    c equation   **d** expression

Stretch

1 C
2 $6p + 9$
3 $4 : 3$
4 $a = -4$ and $b = 2$
5 $(x + 3) \times 2 + 4 - 2x = 2x + 6 + 4 - 2x = 10$

Chapter 6.2

Are you ready? (A)

1 a $6a$ **b** $-16y$ **c** $2b^2$ **d** $-6jk$ **e** $-30t^3$
2 a 3 **b** 6 **c** 14 **d** 4

Practice (A)

1 a $6x + 12$ **b** $16 - 8m$
 c $3w + 30$ **d** $-6c - 24$
 e $-9q + 36$ **f** $12x - 24y$
2 a $b^2 - 6b$ **b** $12t - 9t^2$
 c $15g^2 - 5g$ **d** $2h^3 + 2gh$
 e $2w^2y - 3w^3$ **f** $2x^2y - 3xy^3$
3 a $5c + 24$ **b** $7x$
 c $7e - 21$ **d** $13t + 33$
 e $7g^2 - 11g$ **f** $10k - 13jk$
4 a $4(x + 2)$ **b** $6(x + 4)$
 c $12(x - 2)$ **d** $8x(x + 2)$
 e $5x(3x + 7)$ **f** $7x(1 + 2x)$
5 a $4t(t + 4s)$ **b** $4ab(3b + 7a)$ **c** $5cd(3d + 2cd + 5)$
6 $9x + 15$
7 $28 - x$
8 $5x^2 + 3x$

Are you ready? (B)

1 a $6y^2$ **b** $-12t^2$ **c** $10g^2$
2 a $2x(x + 2)$ **b** $8x(2x - 3)$ **c** $4x(2x + 5)$
3 a $-x + 7y$ **b** $17 - 3x$ **c** $5 + 10x$

Practice (B)

1 a $x^2 + 7x + 12$ **b** $x^2 + 4x + 3$
 c $x^2 - 4x - 12$ **d** $x^2 - 2x - 24$
 e $x^2 - 9x + 20$ **f** $x^2 - 2x + 1$
2 a $2x^2 - x - 36$ **b** $4x^2 + 6x - 4$
 c $15x^2 - 7x - 36$ **d** $8x^2 + 14x + 3$
 e $x^2 - 8x + 16$ **f** $x^2 + 14x + 49$
3 a $x^2 - 9$ **b** $x^2 - 36$
 c $9x^2 - 1$
4 a $x^2 + 12x + 35$ **b** $x^2 + 5x + 6$
5 a $8x^2 + 12x$
 b $8x^2 + 12x = 100 \rightarrow 8x^2 + 12x - 100 = 0 \rightarrow$
 $2x^2 + 3x - 25 = 0$
6 a $x^3 + x^2 - 22x - 40$
 b $2x^3 + 5x^2 - 4x - 3$

What do you think?

1 -23
2 Compare expansions as a class.
3 Compare expansions as a class.

Are you ready? (C)

1 a $\frac{7}{6}$ **b** $\frac{9}{35}$ **c** $\frac{20}{9}$ (or equivalent)
2 a $3x - 9$ **b** $4x + 20$ **c** $2x^2 - 10x$
3 a $5x + 5$ **b** $2x - 16$ **c** $18x - 21$

Practice (C)

1 a $\frac{23x}{56}$ **b** $\frac{4k}{5}$ **c** $\frac{9t + 10}{20}$

 d $\frac{5g + 54}{36}$ **e** $\frac{13h - 9}{42}$

2 a $\frac{7p + 4}{p(p + 1)}$ **b** $\frac{-2c - 4}{c(c - 2)}$ **c** $\frac{6v + 2}{(v - 1)(v + 1)}$

 d $\frac{11t + 16}{(t + 1)(t + 2)}$ **e** $\frac{-8e + 4}{(3e + 1)(2e + 4)}$

3 a $\frac{14x^2 + 6x}{(3x + 1)(x + 1)}$ **b** $\frac{3x^2 - 45x}{(x + 10)(x - 5)}$ **c** $\frac{20x^2 + 5x}{(x + 2)(2x - 1)}$

Consolidate

1 a $5x - 20$ **b** $7x + 35$
 c $12x + 36$ **d** $2a^2 + 10a$
 e $24b - 4b^2$ **f** $12xy + 12x^2$

2 a $14x - 9$ **b** $11x - 33$
 c $14x - 20$
3 a $2(2g + 9)$ **b** $6(h + 8)$ **c** $3t(t + 9)$
 d $5c(c + 5)$ **e** $4ab(1 + 4a - 8b)$
4 a $x^2 - 5x - 24$ **b** $x^2 - 49$ **c** $x^2 + 2x + 1$
 d $2x^2 - 7x - 15$ **e** $2x^2 + 2x - 12$
5 $x^3 + 4x^2 + x - 6$

6 a $\frac{13x}{21}$ **b** $\frac{17x}{45}$

 c $\frac{(14x + 9)}{36}$ **d** $\frac{6x}{(x + 2)(x - 1)}$

Stretch

1 Using Pythagoras: $(x - 1)^2 + (x + 3)^2 = (x + 5)^2$
 Expanding and rearranging: $x^2 - 2x + 1 + x^2 + 6x + 9 = x^2 + 10x + 25$
 Simplifying gives the required result.

2 a $\frac{2h^2 + 3h - 17}{(h + 3)(h + 4)}$ **b** $\frac{5}{(x + 2)(x - 3)}$

3 $a = 1$, $b = -12$

4 time $= \frac{\text{distance}}{\text{speed}}$

 $\frac{500}{x + 2} + \frac{600}{x + 1} = 30 \times 60$ (to convert into seconds)

 $500(x + 1) + 600(x + 2) = 1800(x + 1)(x + 2)$
 Dividing by 100 gives: $5(x + 1) + 6(x + 2) = 18(x + 1)(x + 2)$
 Expanding and simplifying gives the required result.

Chapter 6.3

Are you ready? (A)

1 a $3(k + 4)$ **b** $f(f - 5)$ **c** $6t(t - 2)$
2 a $4x - 7$ **b** $3x^2 - 2x - 5$ **c** $4x^2 - 3x + 12$
3 a 5 and 1 **b** -8 and 4 **c** -7 and -3

Practice (A)

1 a $(p + 12)(p + 2)$ **b** $(q + 16)(q + 2)$
 c $(r - 3)(r - 5)$ **d** $(s - 4)(s + 3)$
 e $(u + 3)(u - 5)$ **f** $(v - 3)(v - 3)$
 g $(w + 2)(w - 10)$ **h** $(x + 5)(x - 2)$
 i $(y - 2)(y + 7)$ **j** $(z + 6)(z + 6)$
2 a $(x - 3)(x + 3)$ **b** $(z - 4)(z + 4)$
 c $(w - 10)(w + 10)$ **d** $(2x - 5)(2x + 5)$
 e $(2u - 3)(2u + 3)$ **f** $(3t + 7)(3t - 7)$
 g $(2k + p)(2k - p)$ **h** $(3w + 2u)(3w - 2u)$
 i $(4x + 2y)(4x - 2y)$ **j** $(2x^2 + 3)(2x^2 - 3)$
3 a $(2a + 1)(a + 2)$ **b** $(3b + 2)(b - 6)$
 c $(5c + 4)(c + 3)$ **d** $(2d + 1)(2d + 7)$
 e $(3e + 8)(e - 3)$ **f** $(2f + 3)(f - 5)$
 g $(3g + 5)(g - 7)$ **h** $(2h - 1)(h + 8)$
 i $(2i - 1)(2i - 3)$ **j** $(4j - 5)(j - 1)$
4 a $(x - 3)(2x - 5)$ **b** $(x - 6)(x + 7)$
 c $(2x + 3)(2x - 1)$ **d** $(x + 13)(x - 13)$
 e $(x - 4)(x - 10)$ **f** $(2x - 9)(2x + 9)$
5 Can be factorised to $(x + y)^2$ so it is always something squared.
6 $x + 5$
7 $(9g + 8h)(9g - 8h)$
8 $5(2x + 1)(2x - 1)$
9 a Both expand to the given expression
 b $6(x + 3)(x + 1)$

What do you think? (A)

1 $(2x - 1)(2x + 1)$, letting $x = 10$, you get 19×21.
 19 is prime. 21 factorises to 3×7.
 So the prime factorisation is $3 \times 7 \times 19$
2 $(x^2 - 8)(x^2 + 4)$

Are you ready? (B)

1 a $\frac{1}{3}$ **b** $\frac{2}{7}$ **c** $\frac{7}{10}$

2 a $6(x + 4)$ **b** $(x - 3)(x + 3)$
 c $(x + 2)(x + 3)$ **d** $x(x + 1)$

Practice (B)

1 a $\frac{a}{3}$ **b** $\frac{2b}{5}$ **c** $\frac{5}{c}$

   d $\frac{y}{2}$ **e** $4y$

2 a $\frac{2(2d+1)}{3}$ **b** $\frac{5(3e-2)}{3}$ **c** $\frac{2f+1}{f-3}$

   d $\frac{g+3}{2g-1}$ **e** $\frac{2(2h-3)}{h+3}$

3 a $x-4$ **b** $x-2$

   c $\frac{x+3}{x-2}$ **d** $\frac{x+5}{x+4}$

What do you think? (B)

1 a $\frac{1}{2a}$ **b** ab **c** $\frac{b}{2c}$

2 Compare answers as a class.

Consolidate

1 a $(x+1)(x+7)$ **b** $(x-5)(x+1)$

   c $(x-5)(x+4)$ **d** $(x-5)(x-5)$

2 a $(2x+5)(x+1)$ **b** $(2x+6)(x-2)$

   c $(3x-2)(x-3)$

3 a $\frac{5a}{3}$ **b** $\frac{8b}{7}$ **c** $\frac{1}{2k}$

4 a $\frac{3g+2}{g+3}$ **b** $\frac{4-m}{1+3m}$ **c** $\frac{3(b+2)}{4(b-1)}$

5 a $\frac{c+2}{c}$ **b** $\frac{d+5}{d-2}$ **c** $\frac{g+3}{g+5}$

6 a $x+3$

   b $x+5$

Stretch

1 The given expression simplifies to $\frac{12a+24}{4a+8}$

   By factorising this simplifies further: $\frac{12(a+2)}{4(a+2)} = \frac{12}{4} = 3$

2 a $12n+12$

   b $2(6n+6)$, which is always a multiple of 2 and hence always even

3 $n(n-1)$. n and $n-1$ are consecutive integers. This means one will always be odd and the other even. Even × odd always equals an even number. So n^2-n can never be odd.

Understanding algebra: exam practice

1 61

2 a $12a^3b$ **b** $3ab^5$

3 a Formula   **b** Identity

   c Equation   **d** Expression

4 $nx\,$cm

5 $x^2+5x-14$

6 a $8(a+3b)$ **b** $3q^2(2p-5t)$

7 $(y-3)(y-4)$

8 a $(x+y)(x-y)$ **b** $(4p+5q)(4p-5q)$

9 n^2 Either

   $2n+1$ Odd

   $n+7$ Either

10 $x^3+3x^2-10x-24$

11 $\frac{7x-2}{(x+1)(x-2)}$

12 $\frac{4x}{x-2}$

Block 7 Functions and linear equations

Chapter 7.1

Are you ready? (A)

1 a $x=8$ **b** $x=7$

   c $x=30$ **d** $x=52$

2 a $x=5$ **b** $x=-6$

   c $x=-3$ **d** $x=-18$

Practice (A)

1 a $x=5$ **b** $x=21$ **c** $x=6$

   d $x=6$ **e** $x=5$ **f** $x=8$

2 a $x=-2$ **b** $x=-5$

   c $x=-7$ **d** $x=0$

3 a $a=20$ **b** $b=80$

   c $c=3$ **d** $d=-15$

4 a $x=5$ **b** $x=7$

   c $x=12$ **d** $x=3$

5 a $7x+5=68$ **b** $x=9$

6 $x=46$

7 a $x=8$ **b** $140\,$cm^2

8 78°

9 $5x+2$

10 $x=5$

11 12

Are you ready? (B)

1 a $x=3$ **b** $x=9.5$

   c $x=-4$ **d** $x=90$

2 a $-x-3$ **b** $7x-12$

   c $4x-11$ **d** $-14x+5$

Practice (B)

1 a $x=3$ **b** $x=1$ **c** $x=4$

   d $x=8$ **e** $x=2$ **f** $x=6$

   g $x=3$ **h** $x=-6$

2 a $x=4.5$ **b** $x=5$

   c $x=3$ **d** $x=4.5$

3 a $x=-2$ **b** $x=-1$

   c $x=-3$ **d** $x=1$

4 a $x=2$ **b** 36 units squared

5 44 units

6 $x=-4\frac{1}{3}$

Consolidate

1 a $x=8.5$ **b** $x=7$

   c $x=-4.5$ **d** $x=2$

2 a $x=6$ **b** $x=6.5$

   c $x=-7$ **d** $x=-4$

3 a 24   **b** 105

   c 118   **d** 0.5

4 a $x=13$ **b** $x=4$

   c $x=13$ **d** $x=-2$

5 a $x=37$ **b** $x=37.5$

6 25

Stretch

1 a $x=11$ **b** $x=-4.5$

2 a $\frac{2x+3}{4} + \frac{x-4}{5} = 300$ **b** $1285\,$m

Chapter 7.2

Are you ready? (A)

1 a 7   **b** 13   **c** 12   **d** −16   **e** 4

2 a 4.5   **b** 4   **c** 41.5   **d** −0.5   **e** 2.5

Practice (A)

1 a 18 **b** 6 **c** 10.5 **d** $x = 1$
2 a −8 **b** −5 **c** −6.75 **d** $x = 5$ or −5
3 a 3 **b** $-\frac{7}{3}$ **c** $-\frac{1}{3}$ **d** $x = 18.5$
4 $x = 3$
5 a −4 **b** 2 **c** $\frac{4}{3}$
6 $x = 12$
7 $x = 0.6$
8 a $4x + 4$ **b** $2x + 8$

Are you ready? (B)

1 a $x = \frac{y-1}{2}$ **b** $x = \frac{3-y}{4}$ **c** $x = \sqrt{(3-y)}$

d $x = \sqrt{\left(\frac{y}{5}\right)}$ **e** $x = \frac{4y}{2-y}$

2 a 3 **b** 13 **c** −25 **d** −50 **e** −26

Practice (B)

1 a $f^{-1}(x) = \frac{x-3}{4}$ **b** $f^{-1}(x) = \sqrt{(x+7)}$

c $f^{-1}(x) = 6x - 1$ **d** $f^{-1}(x) = \frac{12-x}{3}$

e $f^{-1}(x) = \frac{4-2x}{x}$ **f** $f^{-1}(x) = \frac{5-3x}{x}$

2 a $\sqrt[3]{\left(\frac{x+1}{3}\right)}$ **b** $x = 23$

3 $6 - x$ and $\frac{3}{x}$; they are their own inverses.
4 $x = -68$
5 $h^{-1}(x) = \frac{4+6x}{5x-3}$
6 a $a = 3$ **b** $f^{-1}(x) = \frac{3x}{1-x}$

Consolidate

1 a −16 **b** $k = 7$ **c** $f^{-1}(x) = \frac{x-12}{7}$

2 a 189 **b** $k = 3$ or −3 **c** $f^{-1}(x) = \sqrt{\left(\frac{x-9}{5}\right)}$

3 a 0.625 **b** $k = -\frac{1}{3}$ **c** $f^{-1}(x) = \frac{4x-1}{3}$

4 $x = \frac{10}{3}$
5 $x = \frac{4}{7}$
6 $9x^2 + 6x$
7 $f^{-1}(x) = \frac{1+x}{x-2}$

Stretch

1 $2x^2 + 11x + 12$
2 a $\frac{7+8x}{2x-3}$ **b** $\frac{dx-b}{a-cx}$
3 $p = 3$ and $q = -5$

Chapter 7.3

Are you ready?

1 a 3 **b** 7 **c** −29 **d** 1
2 a $4x + 1$ **b** $2x + 3$
   c $9x^2 - 9$ **d** $x^2 - 2x - 8$
3 a $7x - 1$ **b** $6x^2 - 6x$
   c $x^2 - 2x - 3$ **d** $x^2 - 4x + 4$

Practice

1 a 45 **b** 60 **c** 37 **d** 101
2 a 1 **b** $-\frac{2}{27}$ **c** $\frac{109}{3}$ **d** $\frac{1}{32}$
3 a $5x^2 + 13$ **b** $25x^2 - 70x + 53$
   c $25x - 42$ **d** $26 - 10x$
   e $53 - 10x$
4 a $x^2 - 15$ **b** $x^2 - 24x + 141$ **c** $x = 6.5$
5 a $f^{-1}(x) = \frac{x-4}{3}$ **b** $a = 3$

6 a $f^{-1}(x) = \frac{x-8}{a}$ **b** $a = 5$
7 a $f^{-1}(x) = \pm\sqrt{(x+1)}$ **b** $g^{-1}(x) = \frac{x-5}{4}$
   c $fg(x) = 3gf(x)$
   $(4x + 5)^2 - 1 = 3[4(x^2 - 1) + 5]$
   $16x^2 + 40x + 24 = 12x^2 + 3$
   $4x^2 + 40x + 21 = 0$ as required

What do you think?

1 $f^{-1}f(x)$ is always the same as the input x.
2 a $\frac{x-4}{2}$ and $\frac{x+8}{3}$ **b** $6x - 12$
   c $\frac{x+12}{6}$ **d** No. $\frac{x-4}{6}$
   e Yes. $\frac{x+12}{6}$

Consolidate

1 a 52 **b** 169 **c** 492 **d** 1296
2 a $12x^2 - 1$ **b** $72x^2 - 168x + 99$
   c $36x - 49$ **d** $8x^4 + 8x^2 + 3$
3 a $f^{-1}(x) = \frac{x-12}{3}$ **b** $g^{-1}(x) = \sqrt[3]{(x-5)}$
   c $3x^3 + 27$ **d** $9x + 48$
4 $x = 6.5$
5 $k = 5$ and $m = 11$
6 Both give $18x + 17$

Stretch

1 a $\frac{1+x}{x}$ **b** $\frac{2x-1}{x-1}$ **c** $\frac{x-1}{x-2}$
2 $\frac{3x+1}{8}$

Functions and linear equations: exam practice

1 $x = 10.5$
2 $y = 16$
3 $p = 8$
4 $a = 7$
5 $t = 94$
6 $m = 2.5$
7 18
8 a 1 **b** $x = -2$
9 $w = 5$
10 $4x^2 + 6x - 1$
11 a 19 **b** $\frac{x-1}{2}$
12 $(x + 3)(x + 1)$
13 $x = -2$
14 $2x^2 - 1$

Block 8 Identities, formulae and proof

Chapter 8.1

Are you ready?

1 a $5t - 7$ **b** $-8y - 8$
   c $x^2 - 4x - 21$ **d** $2p^2 - 9p - 5$
2 a $a = 9$ **b** $k = 6$ **c** $d = 3$ **d** $n = -11$
3 a $\frac{3}{4}$ **b** $\frac{4}{11}$ **c** $\frac{1}{7}$ **d** $\frac{1}{3}$
4 a $\frac{5}{8}$ **b** $\frac{27}{20}$ **c** $\frac{18}{35}$ **d** $\frac{2}{27}$

Practice

1 Example answer: both expressions are equal to $14p + 42$
2 Example answer: both expressions simplify to
   $28x + 14y + 70$
3 $(a - b)(a + b) \equiv a^2 + ab - ab - b^2 \equiv a^2 - b^2$

4 a $p = 5, q = 9$ **b** $p = -1, q = -12$
   c $p = 7, q = -28$ **d** $p = 4, q = 0$
   e $p = 4, q = -21$

5 $(a + b)^2 \equiv a^2 + 2ab + b^2, (a^2 + 2ab + b^2)(a + b)$
$\equiv a^3 + 2a^2b + ab^2 + a^2b + 2ab^2 + b^3 \equiv a^3 + 3a^2b + 3ab^2 + b^3$

6 a $\dfrac{7x}{4} \times \dfrac{3}{2x} \equiv \dfrac{21}{8}$

   b $\dfrac{4x}{9} \times \dfrac{2\overset{1}{x}}{\underset{3}{6}} \equiv \dfrac{4x^2}{27}$

   c $\dfrac{6x}{21x + 7} \times \dfrac{3x + 1}{5x} \equiv \dfrac{6}{7(3x + 1)} \times \dfrac{3x + 1}{5} = \dfrac{6}{35}$

   d $\dfrac{5y}{7x - 7} \times \dfrac{6x - 6}{7} \equiv \dfrac{5y}{7(x - 1)} \times \dfrac{6(x - 1)}{7} = \dfrac{30y}{49}$

7 a $\dfrac{7}{10x} \times \dfrac{9}{5x} \equiv \dfrac{63}{50x^2}$

   b $\dfrac{5x}{7} \times \dfrac{5}{9x} \equiv \dfrac{25}{63}$

   c $\dfrac{5x + 20}{\underset{3}{6}} \times \dfrac{4}{9x + 27} \equiv \dfrac{5(x + 4)}{3} \times \dfrac{2}{9(x + 3)} \equiv \dfrac{10(x + 4)}{27(x + 3)}$

   d $\dfrac{4x}{3x - 12} \times \dfrac{2x - 8}{7} \equiv \dfrac{4x}{3(x - 4)} \times \dfrac{2(x - 4)}{7} = \dfrac{8x}{21}$

8 a $\dfrac{(x - 3)(x + 1)}{4} \times \dfrac{3}{(x + 1)(x + 1)} \equiv \dfrac{3x - 9}{4x + 4}$

   b $\dfrac{(x + 2)(x - 4)}{2} \times \dfrac{3}{(x - 3)(x - 4)} \equiv \dfrac{3(x + 2)}{2(x - 3)}$

Consolidate

1 Example answer: both expressions are equal to $18x + 60$
2 $(x + 2y)(x + 2y) \equiv x^2 + 2xy + 2xy + 4y^2 \equiv x^2 + 4xy + 4y^2$
3 a $a = 1, b = 12$ **b** $a = -12, b = 4$
   c $a = -5, b = -33$ **d** $a = 8, b = -40$

4 a $\dfrac{\overset{3}{9}}{10x_2} \times \dfrac{5\overset{1}{x}}{\underset{3}{9}} \equiv \dfrac{3}{2}$

   b $\dfrac{7}{6x} \times \dfrac{5}{3x} \equiv \dfrac{35}{18x^2}$

   c $\dfrac{3}{5(x - 2)} \times \dfrac{3(x - 2)}{7} \equiv \dfrac{9}{35}$

   d $\dfrac{2x}{3(2x - 1)} \times \dfrac{2(2x - 1)}{7} \equiv \dfrac{4x}{21}$

5 a $\dfrac{6\overset{2}{x}}{5} \times \dfrac{4x}{\underset{1}{3x}} \equiv \dfrac{8x^2}{5}$

   b $\dfrac{\overset{3}{9}}{\underset{1}{2x}} \times \dfrac{4\overset{2}{x}}{3x} \equiv \dfrac{6}{x^2}$

   c $\dfrac{3(2x + 3)}{\underset{1}{4}} \times \dfrac{12\overset{3}{}}{(2x + 3)} \equiv 9$

   d $\dfrac{5(3x - 1)}{2\underset{}{10x}} \times \dfrac{3x}{2(3x - 1)} \equiv \dfrac{3}{4}$

Stretch

1 $a = 2, b = 2, c = -13$
2 $a = 5, b = -7, c = -10$
3 Simplifies to $\dfrac{2x + 1}{3x + 5}$, so $a = 2, b = 1, c = 3$ and $d = 5$
4 Simplifies to $4x$, so $p = 4$

Chapter 8.2

Are you ready? (A)

1 a $3v + 6$ **b** $4t^2 + 7t$ **c** $5r - 2rs$ **d** $-2e - 25$
2 a $8(3t - 2)$ **b** $4h(2 + h)$ **c** $8x(2y + x)$
   d $8ab(4a - 1)$
3 a $x = 10$ **b** $x = 20$ **c** $x = 53$ **d** $x = -4$

Practice (A)

1 a $x = \dfrac{c - b}{a}$ **b** $x = 3(d - b)$

   c $x = a(4 + b)$ **d** $x = 3y - b$

2 a $x = a(b - 2)$ **b** $x = \dfrac{2b}{a}$

   c $x = \dfrac{b^2 - a^2}{2}$ **d** $x = \dfrac{8b}{5a}$

3 a $y = \sqrt{x}$ **b** $y = \sqrt{x - 4}$ **c** $y = \sqrt{\dfrac{3x - z}{2}}$
   d $y = x^2 - a$ **e** $y = (x - 2a)^2$ **f** $y = ax^2$

4 D

5 When square rooting, he hasn't square rooted the $2b$.
Correct answer: $\sqrt{y - 2b}$

6 a $u = v - at$ **b** $u = \sqrt{v^2 - 2as}$ **c** $\dfrac{2s}{t} - v$

7 a $38.9°$ **b** $\dfrac{9C}{5} + 32 = F$

   c Compare approaches as a class.

8 a $r = \sqrt{\dfrac{3V}{\pi h}}$ **b** $h = \dfrac{3V}{\pi r^2}$ **c** $8.8\,\text{cm}$

What do you think?

1 a $u = v - at$ **b** $s = vt - \dfrac{at^2}{2}$

Are you ready? (B)

1 a $6(3t + 10)$ **b** $2k(3k + 1)$
   c $10t(r - 10t)$ **d** $5gh^2(3 - 5g^2)$
2 a $x = 23$ **b** $x = 8$ **c** $x = 17$ **d** $x = 2$

Practice (B)

1 a $x = \dfrac{4}{1 - a}$ **b** $x = \dfrac{c}{a - b}$

   c $x = \dfrac{3 + c}{b - a}$ **d** $x = \dfrac{d - b}{a - c}$

2 a $t = \dfrac{q - 2p}{2 - p}$ **b** $t = \dfrac{2q + 3p}{p - 1}$

   c $t = \dfrac{4q + 2p}{q - p}$ **d** $t = \dfrac{-q}{p + 4 - q}$

3 a $w = \dfrac{4a}{a - 2}$ **b** $w = \dfrac{-5a}{2}$

   c $w = \dfrac{-2b}{2b - 1}$ **d** $w = \dfrac{a + bc}{c - 1}$

4 a $x = \dfrac{y}{z^2 - 1}$ **b** $x = \dfrac{3z}{3y - 1}$

5 $m = \sqrt{\dfrac{3p}{q^2 - 1}}$

6 D

7 $a = \sqrt{\dfrac{c^2}{2\pi^2} - b^2}$

Consolidate

1 a false   **b** true   **c** true
   d false   **e** true

2 a $t = \dfrac{3 - w}{2}$ **b** $t = \dfrac{v}{3} + s$ **c** $t = \dfrac{-3s - 2v}{2}$
   d $t = 4(v - u)$ **e** $t = 6v - u$ **f** $t = uw + 5$

3 a $c = \sqrt{3 - d}$ **b** $c = \sqrt{\dfrac{e + 2}{3}}$ **c** $c = \sqrt{a} - e$
   d $c = d^2$ **e** $c = (2d + 6)^2$ **f** $c = de^2 - 3$

4 When multiplying by 4, Faith has forgotten to multiply the b as well.
Correct answer $a = 4c - 4b$

5 $r = \sqrt[3]{\dfrac{3V}{4\pi}}$

6 a $n = \dfrac{rq + 4p}{r - 4}$ **b** $n = \dfrac{4 - 6p}{r - 2}$ **c** $n = \dfrac{p - rq}{r - 1}$

   d $n = \dfrac{4 + pq}{q - 1}$ **e** $n = \dfrac{pq}{p - q}$ **f** $n = \dfrac{p}{q - 3}$

Stretch

1 $x = \dfrac{ay(b + c)}{b(c - y)}$

2 $u = \dfrac{fv}{v - f}$

3 $g = \dfrac{4l\pi^2}{T^2}$

Chapter 8.3

Are you ready?

1 $6n - 12$

2 a $3n^2 + 14n - 10$ **b** $n^2 - 8n + 15$ **c** $4n^2 + 38n$

3 a $2n(3n + 4)$ **b** $4n(10 - 3n)$ **c** $n(n - 1)(n + 1)$

Practice

1 a always true   **b** sometimes true
   c always true   **d** sometimes true
   e never true   **f** always true

2 Compare approaches as a class.

3 $(n + 4)^2 - (n + 2)^2 \equiv n^2 + 8n + 16 - (n^2 + 4n + 4) \equiv 4n + 12$
   $\equiv 4(n + 3)$ and so a multiple of 4

4 $(2n + 1)^2 - (2n - 1)^2 \equiv (4n^2 + 4n + 1) - (4n^2 - 4n + 1) \equiv 8n$,
   which is a multiple of 8

5 Let consecutive integers be n, $n + 1$, $n + 2$
   $n + n + 1 + n + 2 \equiv 3n + 3 \equiv 3(n + 3)$, which is a multiple
   of 3

6 Let the odd numbers be $2n + 1$ and $2m + 1$
   $2n + 1 - (2m + 1) \equiv 2n - 2m \equiv 2(n - m)$, which is always
   even.

7 Let consecutive integers be n and $n + 1$
   $(n + 1)^2 - n^2 \equiv n^2 + 2n + 1 - n^2 \equiv 2n + 1 \equiv n + (n + 1)$

8 $(2n + 1)^2 \equiv 4n^2 + 4n + 1 \equiv 4(n^2 + n) + 1$, which is 1 more
   than a multiple of 4

9 $\dfrac{8x(x^2 - 3)}{4(x^2 - 3)} \equiv 2x$, which is always even.

10 a $\dfrac{x + 1}{3x - 12}$ **b** $\dfrac{2x - 1}{x - 4}$ **c** $\dfrac{2x - 1}{x - 7}$

What do you think?

1 $n^3 - n$ can be factorised to $n(n - 1)(n + 1)$, which are
   three consecutive integers. In any set of three
   consecutive integers, one will be even and one will be
   a multiple of 3. Hence the expression is a multiple of
   both 2 and 3, and so a multiple of 6.

2 Let the even number lengths be $2n$ and $2m$.
   By Pythagoras' theorem:
   $(2n)^2 + (2m)^2 = \text{hypotenuse}^2$
   $= 4n^2 + 4m^2$
   $= 2(2n^2 + 2m^2)$
   Hence the hypotenuse^2 is even and the square root of
   an even number is also even.

Consolidate

1 a false   **b** false   **c** true   **d** true

2 $(5n + 2)^2 - (5n - 1)^2 \equiv (25n^2 + 20n + 4) - (25n^2 - 10n + 1)$
   $\equiv 30n + 3 \equiv 3(10n + 1)$, which is a multiple of 3

3 Let the even integers be $2n$, $2n + 2$ and $2n + 4$
   $2n + 2n + 2 + 2n + 4 \equiv 6n + 6 \equiv 6(n + 1)$, which is a
   multiple of 6

4 Let the integers be n, $n + 1$, $n + 2$, $n + 3$
   $n + n + 1 + n + 2 + n + 3 \equiv 4n + 6$, which cannot be
   written in the form $4(n + \text{something})$, therefore not a
   multiple of 4

5 Let the integers be n and $n + 1$
   $n(n + 1) + n + 1 \equiv n^2 + 2n + 1 \equiv (n + 1)^2$, which is always
   a square number.

6 Let the integers be $2n + 1$, $2n + 3$ and $2n + 5$
   $(2n + 1)^2 + (2n + 3)^2 + (2n + 5)^2 \equiv 12n^2 + 36n + 35$
   $\equiv 12(n^2 + 3n + 2) + 11$, which is 11 more than a
   multiple of 12

7 a $\dfrac{x + 6}{x - 4}$ **b** $\dfrac{2x + 1}{x - 3}$

Stretch

1 $k = 5n$, $a = 5n + 1$, $b = 5n - 1$
   $a^2 - b^2 \equiv (5n + 1)^2 - (5n - 1)^2 \equiv 20n$, which is a multiple
   of 20

2 a $5n - 1$
   b $(5n - 1)^2 + 4 \equiv 25n^2 - 10n + 5 \equiv 5(5n^2 - 2n + 1)$

3 $x^2 - 10x + 25$ can be written as $(x - 5)^2$ and square
   numbers cannot be negative.

4 Sometimes. For $x = 0$ to 40 the answers are prime but
   when $x = 41$, 41 itself is a factor of each term and so
   the number is composite.

Identities, formulae and proof: exam practice

1 $3(2a + 5) + 2(4a - 3) \equiv 6a + 15 + 8a - 6 \equiv 14a + 9$

2 $\dfrac{5}{6x} \div \dfrac{2}{3x^2} \equiv \dfrac{5}{\overset{2}{\cancel{6}}\cancel{x_1}} \times \dfrac{\overset{1}{\cancel{3}}x\cancel{x^x}}{2} \equiv \dfrac{5x}{4}$

3 $q = \dfrac{p + 1}{4}$

4 $t = \sqrt{k + 5}$

5 $p = 3m^2 - 2$

6 Example answer: As p is even, $p + 1$ is odd,
   so $(p + 1)^2 = \text{odd} \times \text{odd}$, which is odd.

7 $p = 3$, $q = 2$

8 $2t(6t + 1) - 3(2t + 1)(2t - 1) \equiv 12t^2 + 2t - 3(4t^2 - 1)$
   $\equiv 12t^2 + 2t - 12t^2 + 3$
   $\equiv 2t + 3$

9 $b = \dfrac{3a + 1}{2a - 1}$

10 Example answer: $2n + 2n + 2 \equiv 4n + 2 = 2(2n + 1) =$
   $2 \times \text{odd number}$

11 $(2n + 1)^2 - (2n - 1)^2 \equiv 4n^2 + 4n + 1 - (4n^2 - 4n + 1) \equiv 8n$

12 $(2n + 1)^2 \equiv 4n^2 + 4n + 1 \equiv 4n(n + 1) + 1$
   Either n or $n + 1$ must be even, so $4n(n + 1) = 4 \times \text{even}$
   $= \text{multiple of 8}$
   So $4n(n + 1) + 1$ is 1 more than a multiple of 8

Block 9 Linear graphs

Chapter 9.1

Are you ready? (A)

1 A $(3, 4)$ B $(5, -4)$ C $(0, 3)$ D $(-2, 2)$ E $(-3, -1)$

2 a Any of the form $(4, k)$ **b** Any of the form $(k, -3)$

3 a -6 **b** -13 **c** 9   **d** -7 **e** 21

Practice (A)

1

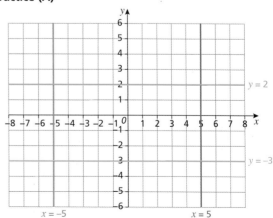

2 Amina is wrong. All the points on the x-axis have y values of 0, so the equation is $y = 0$

3 a $y = 6$ **b** $y = -4$ **c** $x = 9$ **d** $x = -7$

4 a Any of the form (p, p)

b

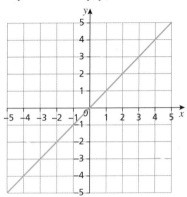

c $y = x$

d

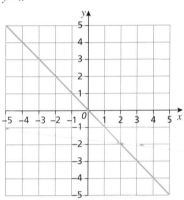

5 a

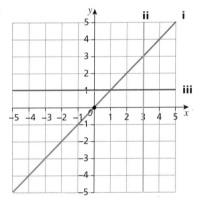

b 2 square units

6

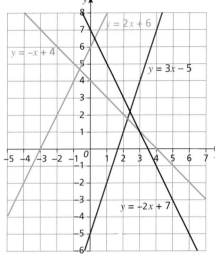

7 a Yes   **b** No   **c** Yes   **d** No

What do you think?

1 The graphs of the first three equations are straight lines. Compare methods as a class, e.g. rearranging the formulae or the 'cover up' method.

2 Ed is correct, but it is good to plot a third point as a check.

Are you ready? (B)

1 a 18   **b** 18   **c** -12
   d 43   **e** -49

2 a $y = 8 - x$ **b** $y = 10 - 2x$ **c** $y = 5x - 17$
   d $y = 7 - 4x$ **e** $y = \frac{2}{3}x + 4$ **f** $y = \frac{3}{4}x - \frac{15}{4}$

3

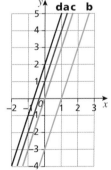

They all have the same steepness/slope and so they are parallel.

Practice (B)

1 a 2   **b** -3   **c** 4   **d** $-\frac{1}{2}$

2 a gradient: 5, y-intercept: (0, 3)
   b gradient: 1, y-intercept: (0, -3)
   c gradient: 4, y-intercept: (0, 0)
   d gradient: -4, y-intercept: (0, 3)
   e gradient: -3, y-intercept: (0, -7)
   f gradient: $\frac{1}{2}$, y-intercept: (0, 6)

3 **a** gradient: –2, *y*-intercept: (0, 7)
  b gradient: –1, *y*-intercept: (0, 8)
  c gradient: 1, *y*-intercept: (0, –10)
  d gradient: 3, *y*-intercept: (0, 0)
  e gradient: 3, *y*-intercept: (0, –2)
  f gradient: $\frac{1}{2}$, *y*-intercept: (0, $\frac{1}{4}$)

4 **a** $y = 2x + 6$ **b** $y = 3x - 2$
  c $y = -4x + 1$ **d** $y = \frac{1}{2}x + 3$

5 **a** $y = 3x + 4$ **b** $y = 6x - 3$
  c $y = 2x + 7$ **d** $y = -x + 5$

Are you ready? (C)

1 **a** 2   **b** 3   **c** 2   **d** $-\frac{1}{2}$

2 **a** 1   **b** 2   **c** –7   **d** 18

Practice (C)

1 **a** 3   **b** $-\frac{1}{4}$ **c** $\frac{1}{2}$ **d** 1

2 **a** $y = 4x - 7$ **b** $y = 3x - 16$ **c** $y = -2x + 14$
  d $y = -4x - 26$ **e** $y = \frac{1}{2}x - \frac{7}{2}$

3 **a** $y = x + 3$ **b** $y = -3x + 18$
  c $y = -2x + 5$ **d** $y = -2x + 12$

4 **a** $y = 3x + 4$ **b** $y = -5x - 6$
  c $y = 3 - 4x$ **d** $y = 4x - 2$

Consolidate

1
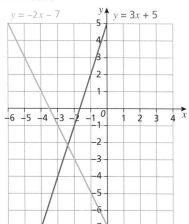

2 **a** Yes   **b** No
3 **a** gradient: 3, *y*-intercept: (0, –8)
  b gradient: –1, *y*-intercept: (0, 9)
  c gradient: –3, *y*-intercept: (0, 4.5)
  d gradient: 2, *y*-intercept: (0, 5)
4 **a** $y = x + 8$ **b** $y = 5x - 3$
  c $y = -4x + 2$ **d** $y = -\frac{1}{2}x + 7$
5 **a** $y = 6x - 34$ **b** $y = -\frac{1}{2}x + 7$
6 **a** $y = -x + 4$ **b** $y = 3x - 25$

Stretch

1 $p = 3$
2 **a** (0, 7) and (–3.5, 0)   **b** (0, –9) and (1.8, 0)
  c (0, 3) and (–9, 0)
3 **a** (0, 5) and ($\frac{15}{4}$, 0)   **b** $\frac{75}{8}$ square units

Chapter 9.2

Are you ready?

1 **a** $\frac{1}{2}$ **b** $-\frac{1}{5}$ **c** 3
  d $\frac{3}{2}$ **e** $-\frac{7}{5}$

2 **a** (4, –1)   **b** (5, 8)   **c** (1, –2)

3 **a** gradient: 3, *y*-intercept: (0, 6)
  b gradient: –2, *y*-intercept: (0, 7)
  c gradient: –5, *y*-intercept: (0, 9)
  d gradient: –6, *y*-intercept: (0, 8)
  e gradient: $\frac{4}{3}$, *y*-intercept: (0, –4)

Practice

1 **a** $y = 4x \pm k$ **b** $y = -3x \pm k$
  c $y = \frac{1}{2}x \pm k$ **d** $y = \frac{2}{5}x \pm k$

2 **a** $y = 6x + 4$ **b** $y = -1.5x - 1$
  c $y = -3x + 22$ **d** $y = \frac{1}{3}x + 5$

3 **a** $y = -\frac{1}{5}x \pm k$ **b** $y = -4x \pm k$
  c $y = \frac{1}{3}x \pm k$ **d** $y = \frac{1}{2}x \pm k$

4 **a** $y = -\frac{1}{4}x + 3$ **b** $y = 2x - 3$
  c $y = \frac{1}{3}x - 6$ **d** $y = -5x - 8$

5 They are parallel as they both have a gradient of –3
6 One line has gradient –2 and the other has gradient 4
  $-2 \times 4 \neq -1$, therefore not perpendicular.
7 **a** 6   **b** $y = 6x - 19$
  c (3, –1)   **d** $y = -\frac{1}{6}x - \frac{1}{2}$

8 l_7 has gradient 2 and l_8 has gradient $\frac{1}{2}$
  $2 \times \frac{1}{2} \neq -1$, therefore they are not perpendicular.

9 $y = -2x + 18$

Consolidate

1 **a** $y = 6x + 1$ **b** $y = \frac{1}{2}x + 1$
  c $y = -4x - 8$ **d** $y = -\frac{1}{2}x + 1$
2 **a** $y = \frac{1}{3}x + 2$ **b** $y = -5x - 1$
  c $y = 2x + 3$ **d** $y = -3x - 2$

3 One line has gradient 4 and the other has gradient $-\frac{1}{4}$
  $4 \times -\frac{1}{4} = -1$, therefore they are perpendicular.

4 **a** l_1 and l_2 **b** l_3 and l_4 **c** l_1 and l_3
5 **a** 1   **b** $y = x + 4$
  c (2, 6)   **d** $y = -x + 8$
6 l_5 has gradient –2 and l_6 has gradient $\frac{1}{2}$
  $-2 \times \frac{1}{2} = -1$, therefore they are perpendicular.

Stretch

1 $a = 10$
2 $q = \frac{2p + 13}{3}$
3 AB has gradient $\frac{1}{2}$, so BC needs to have gradient –2
  $a = 10$
4 **a** $4 \times -2 + 8 \times 3 = 16$ so lies on the line
  b $y = 2x + 7$
  c (4, 0) and (–3.5, 0)
  d $(7.5 \times 3) \div 2 = 11.25$ units squared

Chapter 9.3

Are you ready?

1 a 4   **b** 2   **c** $-\frac{1}{3}$ **d** $\frac{1}{2}$

2 a 6   **b** 3   **c** $-\frac{2}{3}$ **d** -2

3 a $v = 22$ **b** $u = 12$ **c** $a = 2.5$

Practice

1 a 30   **b** 30 litres per minute

2 a £6

   b Example answer: find the cost of 4 kg and multiply by 6

   c Cost of potatoes per kilogram

   d $C = 1.2m$

3 a £7.10

   b £15.50

   c 0.7, it costs an extra 70 pence per mile

   d 5, £5 is the fixed cost before any extra is added

   e $C = 0.7d + 5$

4 a

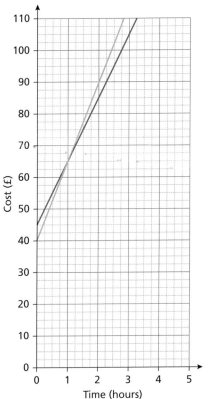

   b Pipes Are Us (£85 versus £90)

   c Cheaper to use Pipes Are Us after 1 hour

5 a i 9.6 km   **ii** 4.4 m   **b** 1.6   **c** $y = 1.6x$

6 a £12.50   **b** £13.75   **c** £1.25

Consolidate

1 a $15   **b** £16

2 a £10

   b Ali is cheaper until 10 miles. After 10 miles Flo is cheaper.

3 a

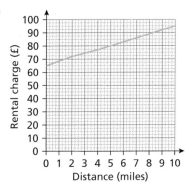

   b y-intercept is £65, the fixed fee. Gradient is 3, the extra cost per mile.

4 a Approx. £375   **b** Answers in range £6.43 to £6.81

Stretch

1 $y = 6x$, $y = 2x + 20$ and $y = 1.2x + 28$

2 a 30 000 litres

   b Approx. 9000 litres

   c -4.5. The pool is draining at a rate of 4500 litres per hour

   d Approx. 6.7 hours

Linear graphs: exam practice

1 (4, 3)

2 Yes, when $x = 4$, $y = 2 \times 4 + 1 = 9$

3 $y = 3x - 2$

4 Answer could show that both lines have gradient $\frac{1}{2}$

5 $y = 3x + 7$

6 $\frac{1}{2}$ and (0, 3)

7 $y = 2x + 4$

8 Gradient is -4, so water is leaving the tank at a rate of 4 litres per hour.

9 $y = \frac{1}{3}x - 5$

10 $y = -\frac{2}{5}x + 6$

Block 10 Non-linear functions

Chapter 10.1

Are you ready?

1 a 4   **b** 4   **c** 2   **d** -2 **e** 8

2 a -9 **b** 9   **c** 5   **d** 6   **e** 15

3 a $x = -2$, $x = -4$ **b** $x = -6$, $x = 1$ **c** $x = -3$

   d $x = 7$, $x = 9$ **e** $x = -0.5$, $x = -4$

Practice

1 a Table completed from the left: 15, 8, 3, 0, -1, 0, 3, 8, 15

   b

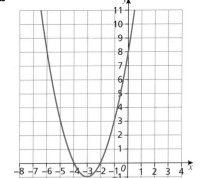

c −0.75   **d** $x = -1.25$, $x = -4.75$
e −2, −4: these are the points where the graph
      crosses the x-axis.
2 The points have been joined with straight lines
    instead of a curve. (As a result there should be a
    turning point when $x = 1.5$, not a flat section
    between $x = 1$ and $x = 2$).
3 a Table completed from the left:
      16, 8, 2, −2, −4, −4, −2, 2, 8, 16
b

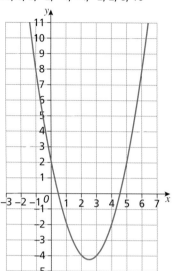

c 0.2, 4.8   **d** (2.5, −4.25)   **e** $x = 2.5$
4 a Table completed from the left: −6, 0, 4, 6, 6, 4, 0, −6
b

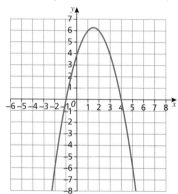

c 6.25
d (0, 4) since 4 is the constant term in the equation
e (1.5, 6.25)   **f** $x = -0.55$, $x = 3.55$
g If k is equal to 6.25, there will be one solution,
    and if k is greater than 6.25, there will be no
    solutions.
5 Graph 1 is $y = x^2 + 4x - 1$
    Graph 2 is $y = x^2 + 4$
    Graph 3 is $y = -x^2 - 3x - 5$
    Graph 4 is $y = -x^2 + 2x + 5$
6 Graphs drawn accurately and points of intersection
    as follows:
    a (−5, 0) and (2, 14)
    b (−1, −3) and (4, 12)

Consolidate

1 a Table completed from the left:
      4, −2, −6, −8, −8, −6, −2, 4

b

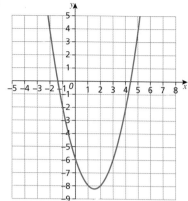

c −7.4   **d** (1.5, −8.25)   **e** $x = -1.4$, $x = 4.4$
2 a Table completed from the left: −7, −1, 3, 5, 5,
      3, −1, −7
b

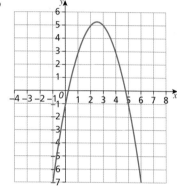

c 1.2, 3.8   **d** 5.25
3 a Table completed from the left: 9, −1, −7, −9, −7, −1, 9
b

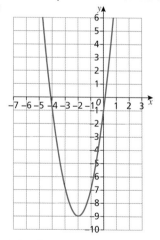

(−2, −9)
c $x = -2$ **d** $x = -4.1$, $x = 0.1$
4 (2, 7) and (−1, −2)

Stretch

1 She has drawn the graph upside down. It should have
    a maximum point rather than a minimum point.
2 $a = 2$, $b = -5$ and $c = -12$
3 Find the points of intersection with the lines:
    a $y = 1$ **b** $y = 2$
    c $y = x$ **d** $y = 3 - 2x$

Chapter 10.2

Are you ready?

1 a 4   **b** 10   **c** 9   **d** 23

2 a 18   **b** −31   **c** −40   **d** −40

3 a $\frac{1}{3}$ **b** $\frac{1}{2}$ **c** $-\frac{1}{7}$

   d 2   **e** 4

Practice

1 a Table completed from left: −8, −4, 4, −2, 8

   b

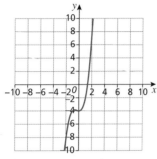

   c 0.75   **d** $x = 1.3$

2 a Table completed from the left: −4, 2, 0, −4, −4, 6

   b

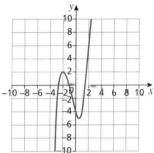

   c −2, −1.7, 1.7

3 a Table completed from the left: −1, −2, −4, 4, 2, 1

   b

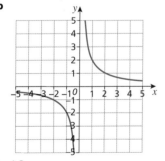

   c 1.3

   d

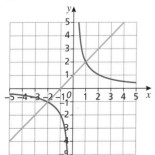

   e (1, 2) and (−2, −1)

4 a Table completed from the left: 1.5, 3, 6, −6, −3, −1.5

   b

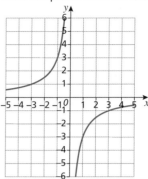

   c The two branches are in different quadrants.

5 $y = -\frac{1}{x}$ is graph 3

    $y = x^3$ is graph 4

    $y = 3x − 5$ is graph 1

    $y = 5 − 3x$ is graph 5

    $y = x^3 − 9x^2 + 23x − 15$ is graph 6

What do you think?

1 Graphs of the form $y = \frac{1}{x + a}$ are horizontal

   translations of $y = \frac{1}{x}$, whereas graphs of the form

   $y = \frac{1}{x} + a$ are vertical translations of $y = \frac{1}{x}$

Consolidate

1 a Table completed from the left: −2, 2.125, 3, 2.875, 4

   b

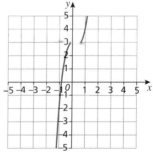

   c −0.8   **d** −0.4

2 a Table completed from the left: −4, 12, 14, 8, 0, −4, 2

   b

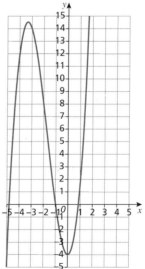

   c 1.2, −1.6, −4.6

   d No. For example, when y is greater than 15, there is only one x value for each y value.

3 a Table completed from the left: −1.5, −3, −6, −12, 12, 6, 3, 1.5

b
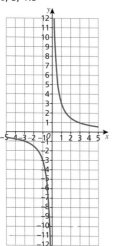

c 1.5

d
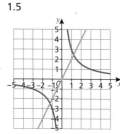

e (1.2, 2.5) and (−1.2, −2.5)

4 Graph 1 is $y = 2x + 4$ Graph 2 is $y = -x^2 + 6$
Graph 3 is $y = -x^3$ Graph 4 is $y = \frac{1}{x}$

Stretch

1 a
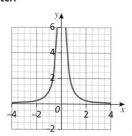

b There are no negative y values, so both branches are above the x-axis.

2 a
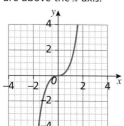

b $f^{-1}(x) = \sqrt[3]{(x)}$

c
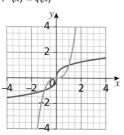

d The graphs are mirror images of each other, reflected in the line $y = x$

3 a
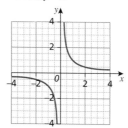

b $f^{-1}(x) = \frac{1}{x}$

c It would be the same graph; the function $\frac{1}{x}$ is self-inverse.

4 $y = x$ and $y = -x$

Chapter 10.3

Are you ready?

1 a 64   **b** 16   **c** 81   **d** $\frac{1}{16}$

2 a $\frac{1}{4}$ **b** $\frac{1}{9}$ **c** 4   **d** 25

3 a 1   **b** 1   **c** $\frac{1}{9}$
d $\frac{1}{4}$ **e** $\frac{1}{125}$

Practice

1 a Table completed from the left: 0.25, 0.5, 1, 2, 4, 8, 16, 32

b
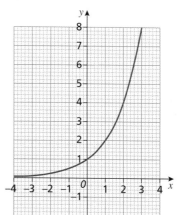

c 2.3   **d** −0.7

2 a Table completed from the left: 16, 8, 4, 2, 1, 0.5, 0.25, 0.125

b
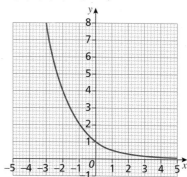

c −2.8   **d** 1.3

3 a 20   **b** Approx. 55

c The model suggests the number of rabbits will grow forever.

4 a 200 mg   **b** Approx. 33 mg   **c** 2 hours
   d Because the graph never reaches the x-axis, this model suggests that the medicine will never fully leave the body.
5 $a = 4$ and $b = 3$
6 The graph should go through (0, 1) and not (0, 0).

Consolidate

1 a Table completed from the left: $\frac{1}{16}, \frac{1}{4}$, 1, 4, 16, 64
   b

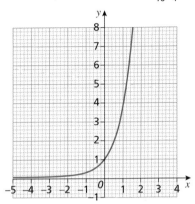

   c 1.4   **d** −0.08

2 a Table completed from the left: 27, 9, 3, 1, $\frac{1}{3}, \frac{1}{9}, \frac{1}{27}$
   b

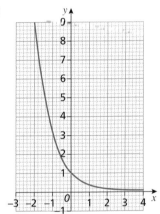

   c −1.8   **d** 0.8

3 a 50
   b Approx. 104
   c Approx. 3 m 45 s
   d This suggests the bacteria population will grow forever, which is not realistic as it will eventually run out of space/food.

4 a £20 000   **b** £8000
   c A little over 6 years

Stretch

1 a 9 years   **b** 150%
   c 25%   **d** $y = 1000 \times 1.08^x$
   e a represents the initial investment
     $b = 1.08$ shows that the investment is growing by 8% per year

2 Graph 1 is $y = 2^x - 1$
   Graph 2 is $y = 4 \times 2^x$
   Graph 3 is $y = 2^x$
   Graph 4 is $y = -2^x$
   Graph 5 is $y = \left(\frac{1}{2}\right)^x$

Chapter 10.4

Are you ready?

1 a −2   **b** −4   **c** −2
2 a $-\frac{1}{3}$ **b** $\frac{1}{2}$ **c** −3   **d** 5
3 a $y = 4x - 19$ **b** $y = -3x + 11$

Practice

1 a $x^2 + y^2 = 1$ **b** $x^2 + y^2 = 49$ **c** $x^2 + y^2 = 100$
2 a

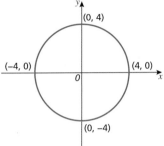

   b

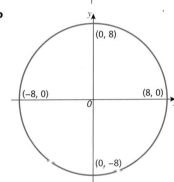

3 a $x^2 + y^2 = 100$
   b i Inside   **ii** On   **iii** Inside
     iv Inside   **v** Outside
4 a (0, 0)   **b** 4
   c No, because $4^2 + 2^2$ is not equal to 16.
5 a $2\sqrt{10}$ **b** $k = \pm 6$
6 $x^2 + y^2 = 81$
7 $x^2 + y^2 = 34$
8 a $4^2 + (-2)^2 = 16 + 4 = 20$ **b** $-\frac{1}{2}$
   c 2   **d** $y = 2x - 10$
9 a $y = \frac{4}{3}x - \frac{25}{3}$ **b** $y = -3x + 20$ **c** $y = \frac{1}{4}x + \frac{17}{2}$

What do you think?

1 $(x - h)^2 + (y - k)^2 = r^2$

Consolidate

1 a $x^2 + y^2 = 25$ **b** $x^2 + y^2 = 16$
2

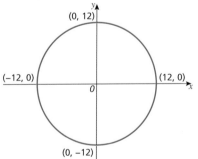

3 a (0, 0)   **b** $2\sqrt{5}$ **c** Yes
4 a 10   **b** $m = \pm 8$

5 $y = -x + 12$

6 $y = \frac{1}{3}x - \frac{10}{3}$

Stretch

1 **a** $c = 26$ **b** $x^2 + y^2 = 208$

2 $x^2 + y^2 = 36$

3 $(17, 0)$

4 **a** $26\frac{1}{24}$ square units   **b** 10.4 units (1 d.p.)

Non-linear functions: exam practice

1 **a** Table completed from the left: 5; −3; −1; 5

b

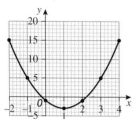

2 **a**

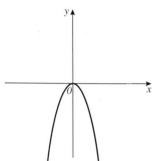

b

3 **a**

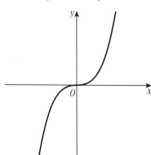

b

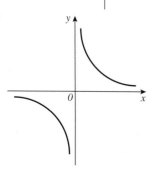

4 $x = -1.7$, $x = 4.2$

5 4

6 $y = -3x + 10$

Block 11 More quadratics

Chapter 11.1

Are you ready? (A)

1 **a** $x = -3$ **b** $x = 5$ **c** $x = 2$

    d $x = -\frac{1}{2}$ **e** $x = \frac{2}{3}$

2 **a** $(x + 3)(x + 5)$ **b** $(x + 4)(x - 9)$

    c $(x + 7)(x - 2)$ **d** $(x - 3)(x - 7)$

3 **a** $(x + 4)(x - 4)$ **b** $(x + 5)(x - 5)$

    c $(2x + 1)(2x - 1)$ **d** $(2x + 3)(2x - 3)$

4 **a** $(2x + 1)(x - 4)$ **b** $(2x - 1)(x + 3)$

    c $(3x - 2)(x + 7)$ **d** $(2x - 5)(x + 2)$

Practice (A)

1 **a** $x = -4$, $x = -5$ **b** $x = -9$, $x = -2$

    c $x = 7$, $x = -3$ **d** $x = 11$, $x = -6$

    e $x = 2$, $x = 3$ **f** $x = 8$, $x = 4$

    g $x = -1$, $x = 7$ **h** $x = -10$, $x = 3$

2 **a** $x = 7$, $x = -7$ **b** $x = 9$, $x = -9$

    c $x = 13$, $x = -13$ **d** $x = \frac{1}{2}$, $x = -\frac{1}{2}$

    e $x = \frac{4}{3}$, $x = -\frac{4}{3}$ **f** $x = \frac{5}{4}$, $x = -\frac{5}{4}$

3 **a** $x = -\frac{3}{2}$, $x = -1$ **b** $x = -\frac{1}{2}$, $x = -4$

    c $x = -\frac{7}{2}$, $x = 6$ **d** $x = \frac{9}{2}$, $x = 2$

    e $x = -\frac{1}{3}$, $x = 1$ **f** $x = -\frac{7}{3}$, $x = -2$

    g $x = -\frac{5}{3}$, $x = 8$ **h** $x = -\frac{3}{2}$, $x = -\frac{5}{2}$

4 **a** $x = -1$, $x = -7$ **b** $x = -3$, $x = -9$

    c $x = 10$, $x = -10$ **d** $x = \frac{1}{2}$, $x = -5$

    e $x = 8$, $x = 2$ **f** $x = 6$, $x = -\frac{3}{2}$

5 **a** Factorises to $(x - 7)(x - 7)$ so $x = 7$

    b Factorises to $(x + 3)(x + 3)$ so $x = -3$

    c Factorises to $(2x - 5)(2x - 5)$ so $x = \frac{5}{2}$

Are you ready? (B)

1 **a** $(x + 1)(x + 5)$ **b** $(x + 2)(x - 2)$

    c $(x - 3)(x + 4)$ **d** $(x - 6)(x - 7)$

2 **a** $(2x + 1)(x + 3)$ **b** $(x - 2)(2x - 5)$

    c $(2x + 7)(x - 7)$ **d** $(3x + 7)(x - 8)$

3 **a** $\dfrac{3x + 2}{(x + 3)(x - 4)}$ **b** $\dfrac{7x - 13}{(x + 2)(x - 7)}$

    c $\dfrac{-x + 4}{(x + 1)(x + 2)}$ **d** $\dfrac{4x - 1}{(x - 9)(x - 4)}$

Practice (B)

1 The rectangle measures 7 cm by 3 cm

2 **a** $\frac{1}{2}(a + b) \times h = 36$

$$\frac{1}{2}(x + 2 + x + 6)(x - 5) = 36$$

$$\frac{1}{2}(2x + 8)(x - 5) = 36$$

$$(x + 4)(x - 5) = 36$$

$$x^2 - x - 20 = 36$$

$$x^2 - x - 56 = 0$$

    b $x = 8$

3 $n = 4$

4 **a** $x^2 + 8x = 84$

    b Faith is 14 (and Seb is 6)

5 $x = 2$

6 a $x = \frac{11}{2}, x = -1$ **b** $x = 0, x = 1$

   c $x = \frac{4}{3}, x = 4$ **d** $x = 7, x = -7$

   e $x = 2, x = -\frac{10}{3}$ **f** $x = 5, x = -\frac{1}{2}$

Consolidate

1 a $x = 1, x = 7$ **b** $x = 11, x = -2$
   c $x = 3, x = -4$ **d** $x = 5, x = -5$
   e $x = -8, x = 9$ **f** $x = -10, x = -6$

2 a $x = \frac{1}{2}, x = 10$ **b** $x = -\frac{9}{2}, x = -2$

   c $x = \frac{4}{3}, x = 5$ **d** $x = \frac{5}{2}, x = -\frac{5}{2}$

   e $x = -\frac{7}{2}, x = 7$ **f** $x = -\frac{10}{3}, x = \frac{10}{3}$

3 a $\frac{1}{2}x(x + 4) = 30$, leading to $x^2 + 4x - 60 = 0$

   b $x = 6$

4 24 cm

5 a $n^2 + (n + 1)^2 = 145$

   b 8 and 9

6 a $x = 3, x = -2$ **b** $x = 1, x = -\frac{1}{8}$

   c $x = 0, x = 4$ **d** $x = \frac{1}{2}, x = 4$

   e $x = -4, x = 5$

Stretch

1 $x = 4, x = -1$

2 a $x = 1, x = 9$
   h i $y = \pm1, y = \pm3$ **ii** $z = 1, z = 81$ **iii** $w = -1, w = 7$

3 a Compare approaches as a class.
   b 60 mph and 80 mph

Chapter 11.2

Are you ready?

1 a $x^2 + 8x + 16$ **b** $x^2 - 4x + 4$
   c $x^2 + 24x + 144$ **d** $x^2 - 16x + 64$

2 a $x = 4, x = -8$ **b** $x = -3, x = 9$
   c $x = -1, x = -7$ **d** $x = 10, x = 4$

3 a $2\sqrt{5}$ **b** $10\sqrt{2}$
   c $3\sqrt{3}$ **d** $4\sqrt{5}$

Practice

1 A and C

2 a $(x + 4)^2 - 24$ **b** $(x + 5)^2 - 28$ **c** $(x - 2)^2 + 7$

   d $(x - 1)^2 + 2$ **e** $\left(x + \frac{5}{2}\right)^2 - \frac{29}{4}$ **f** $\left(x - \frac{7}{2}\right)^2 - \frac{41}{4}$

3 a $-2 \pm \sqrt{3}$ **b** $-1 \pm 2\sqrt{2}$ **c** $2 \pm 2\sqrt{3}$

   d $-6 \pm \sqrt{11}$ **e** $1 \pm \sqrt{6}$ **f** $\frac{1}{2} \pm \frac{\sqrt{17}}{2}$

4 $2(x + 2)^2 + 2$

5 a $2(x - 5)^2 - 57$ **b** $3(x + 1)^2 - 8$ **c** $5(x - 3)^2 - 34$

6 $-10 + 10\sqrt{11}$ and $10 + 10\sqrt{11}$

What do you think?

1 Rewriting as $(x + 6)^2 - 11$, the minimum value is -11, which occurs at $x = -6$

2 The expression can be written as $(x + 1)^2 + 4$
   $(x + 1)^2 \geqslant 0$ and adding 4 means that the expression will always be positive.

Consolidate

1 a $(x + 3)^2 - 10$ **b** $(x + 2)^2 + 3$ **c** $(x - 4)^2 - 34$

   d $(x - 6)^2 - 32$ **e** $\left(x + \frac{1}{2}\right)^2 - \frac{13}{4}$

2 a $1 \pm \sqrt{2}$ **b** $-2 \pm \sqrt{2}$ **c** $-5 \pm \sqrt{23}$ **d** $-6 \pm \sqrt{41}$

Stretch

1 a $-(x - 3)^2 + 14$ **b** $-2(x + 3)^2 + 25$ **c** $(x + 3\sqrt{2})^2 - 19$

2 $p = 8$ and $q = -5$

Chapter 11.3

Are you ready?

1 a $x = 6, x = -2$ **b** $x = -1, x = -4$
   c $x = 2, x = 1$ **d** $x = 9, x = -11$

2 a $(x + 2)^2 - 1$ **b** $(x + 4)^2 - 18$
   c $(x - 3)^2 + 4$ **d** $(x - 1)^2 + 8$

Practice

1 a (4, 0) (7, 0) (0, 28)   **b** (−8, 0) (−2, 0) (0, 16)
   c (6, 0) (−9, 0) (0, −54)   **d** (−1.5, 0) (1, 0) (0, −3)

2 a (−3, −20)   **b** (2, −3)
   c (−1, −8)   **d** (−1.5, −7.25)

3 a (0, −20)   **b** (−10, 0) (2, 0)   **c** (−4, −36)
   d

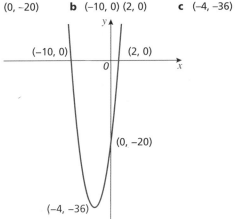

4 a $y = x^2 - 10x + 24$

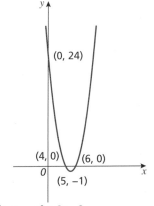

   b $y = -x^2 + 6x - 8$

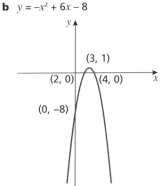

5 a (0, 9)   **b** $x = 3$
c Only one solution; graph touches the x-axis at $x = 3$
d

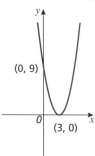

6 a Can't be solved as you end up needing to square root a negative number.
b Doesn't have any real roots so doesn't cross the x-axis.
c (−2.5, 3.75)
d

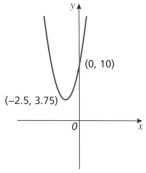

7 $b = -4$, $c = 1$

Consolidate

1 a (3, 0) (−3, 0) (0, −9)   **b** (8, 0) (5, 0) (0, 40)
c (5, 0) (−2, 0) (0, −10)   **d** (−2.5, 0) (−1, 0) (0, 5)
2 a (−4, −29)   **b** (1, 2)
c (−2, −13)   **d** (−2.5, −12.25)
3 a

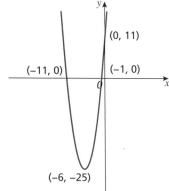

b

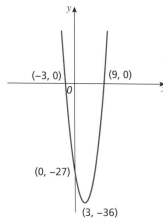

4

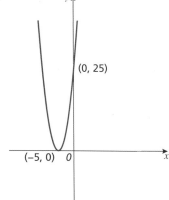

Stretch

1 a (−3, −22)   **b i** (−3, −23) **ii** (−3, −12)   **iii** (−3, −8)
2 a (−b, −b^2 + c)
b i (−b, −b^2 + 3c) **ii** (−2b, −4b^2 + c) **iii** (−2b, −4b^2 + 2c)
3 $2x^2 + 7x - 15$
4 a 1.2 m   **b** 2.45 m
c

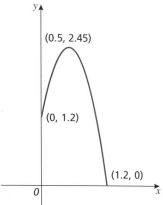

Chapter 11.4

Are you ready?

1 a $x = 3$, $x = -7$ **b** $x = -2$, $x = 9$
c $x = -\frac{3}{2}$, $x = -6$ **d** $x = 4$, $x = \frac{1}{3}$
2 a 40   **b** 19   **c** 4.69   **d** 9.5

Answers

Practice

1 a $x = 1.77, x = -2.27$ **b** $x = -0.81, x = -6.19$
   c $x = -1.39, x = -2.27$ **d** $x = 1, x = -0.6$
   e $x = 0.36, x = -0.79$ **f** $x = -0.23, x = -1.43$

2 a $-3 \pm 2\sqrt{5}$ **b** $-1 \pm 2\sqrt{2}$ **c** $\dfrac{-5 \pm \sqrt{21}}{2}$

   d $\dfrac{3 \pm \sqrt{229}}{10}$ **e** $\dfrac{-1 \pm \sqrt{13}}{6}$ **f** $\dfrac{3 \pm 2\sqrt{6}}{5}$

3 a $x = 1.89, x = 0.106$ **b** $x = 1.37, x = -4.37$
   c $x = 1.32, x = -0.687$ **d** $x = 0.777, x = -0.920$

4 a $x \times (x + 2) = 21$
   $x^2 + 2x = 21$, so $x^2 + 2x - 21 = 0$
   b 3.7

5 a $\frac{1}{2}(x + 3 + 3x - 1) \times (x + 2) = 64$

   $\frac{1}{2}(4x + 2)(x + 2) = 64$

   $(2x + 1)(x + 2) = 64$
   $2x^2 + 5x + 2 = 64$
   $2x^2 + 5x - 62 = 0$
   b 4.46

6 a $x^2 + 4^2 = (2x + 3)^2$
   $x^2 + 16 = 4x^2 + 12x + 9$
   $3x^2 + 12x - 7 = 0$
   b $\dfrac{-6 + \sqrt{57}}{3}$

7 4.24

Consolidate

1 a $x = 1.24, x = -3.24$ **b** $x = 8.39, x = -2.39$
   c $x = 2.87, x = 0.46$ **d** $x = 1.52, x = -2.19$

2 a $\dfrac{-7 \pm \sqrt{33}}{4}$ **b** $\dfrac{-5 \pm \sqrt{105}}{4}$

   c $\dfrac{3 \pm \sqrt{2}}{7}$ **d** $\dfrac{9 \pm \sqrt{401}}{4}$

3 a $x \times (x + 5) = 400$
   $x^2 + 5x = 400$
   $x^2 + 5x - 400 = 0$
   b $x = 17.6556...$ Perimeter = 80.6 m

4 30.0

Stretch

1 $4x^2 + 24x + 33$
2 6.5 cm by 10.5 cm

3 $x^2 + \dfrac{bx}{a} + \dfrac{c}{a} = 0$

  $x^2 + \dfrac{bx}{a} = -\dfrac{c}{a}$

  $\left(x + \dfrac{b}{2a}\right)^2 - \left(\dfrac{b}{2a}\right)^2 = -\dfrac{c}{a}$ (completing the square)

  $\left(x + \dfrac{b}{2a}\right)^2 = \left(\dfrac{b}{2a}\right)^2 - \dfrac{c}{a}$

  $\left(x + \dfrac{b}{2a}\right)^2 = \dfrac{b^2}{4a^2} - \dfrac{c}{a}$

  $\left(x + \dfrac{b}{2a}\right)^2 = \dfrac{b^2}{4a^2} - \dfrac{4ac}{4a^2}$

  $x + \dfrac{b}{2a} = \pm\sqrt{\dfrac{b^2 - 4ac}{4a^2}}$

  $x + \dfrac{b}{2a} = \dfrac{\pm\sqrt{b^2 - 4ac}}{2a}$

  $x = \dfrac{-b}{2a} \pm \dfrac{\sqrt{b^2 - 4ac}}{2a} = \dfrac{-b \pm \sqrt{b^2 - 4ac}}{2a}$

More quadratics: exam practice

1 $(x + 3)(x + 2)$
2 $x = -3, x = 6$
3 $x = 2, x = -12$
4 $(0, -20), (-4, 0)$ and $(5, 0)$
5 a $x = -1, x = 7$
   b $(3, -16)$
6 $x = -4.28, x = 1.28$
7 $(x - 2)^2 + 8$
8 $3 \pm \sqrt{19}$

Block 12 Inequalities

Chapter 12.1

Are you ready?

1 a $x = 6$ **b** $x = -2$ **c** $x = 4$ **d** $x = 22$
2 a Any numbers less than 4
   b Any numbers greater than or equal to 7
   c Any numbers between 4 and 10 (not including 4 and 10)
   d Any numbers less than or equal to –3
3 a >   **b** >   **c** <   **d** <

Practice

1 a

   b

   c

   d

2 a $x \geqslant -3$ **b** $x < 4$
   c $-3 < x < 2$ **d** $-4 \leqslant x < -1$
3 a 5, 6, 7   **b** –1, 0, 1, 2
   c –3, –2, –1, 0, 1   **d** 6, 7, 8
4 a $x \geqslant 1$ **b** $x \leqslant -2$
   c $x \geqslant 2.5$ **d** $x < 2$
   e $x > 10$ **f** $x < 3$
   g $x \geqslant -2$
5 a $x \geqslant 1$ **b** $x \leqslant 1.5$ **c** $x < -3$
   d $x \leqslant 2$ **e** $x \leqslant -3$ **f** $x \geqslant 2$
   g $x \leqslant -4$
6 a $4 < x < 7$ **b** $2 < x < 4$
   c $-1 \leqslant x \leqslant 5$ **d** $-6 < x \leqslant -2$
7 a $x \leqslant 6.5$

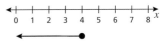

   b $x \leqslant -5$

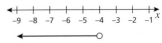

   c $x < 4$

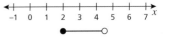

   d $-2 \leqslant x < 0.5$

8 a $4x + 12 < 40$ **b** $0 < x < 7$
9 $15 < \pi r^2 < 25$, so $2.19 < r < 2.82$

Consolidate

1 a
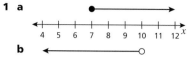

b
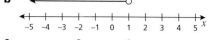

c
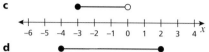

d

2 a $x < 5$ **b** $x \geq 2$
c $-5 < x < -2$ **d** $3 \leq x \leq 7$
3 a $x \leq 6$ **b** $x > 4$
c $x < 6$ **d** $x \leq 2$
4 a $x > 8$ **b** $x < 3$
c $-17 < x < -7$ **d** $2.5 \leq x \leq 4.5$
5 $3 \leq x < 6$
6 $-1, 0, 1, 2$

Stretch

1 a $-0.5 < x < 4$ **b** $3 < x \leq 4$
2 $0, 1, 2, 3$
3 a 16 **b** 16
c -5 **d** -8

Chapter 12.2

Are you ready?

1
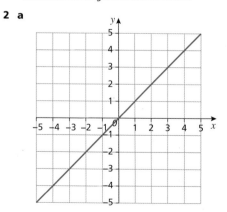

2 a

b
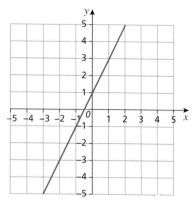

c
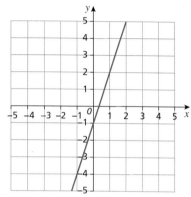

d
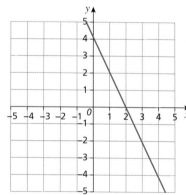

Practice

1 a
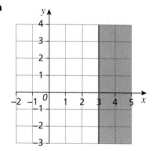

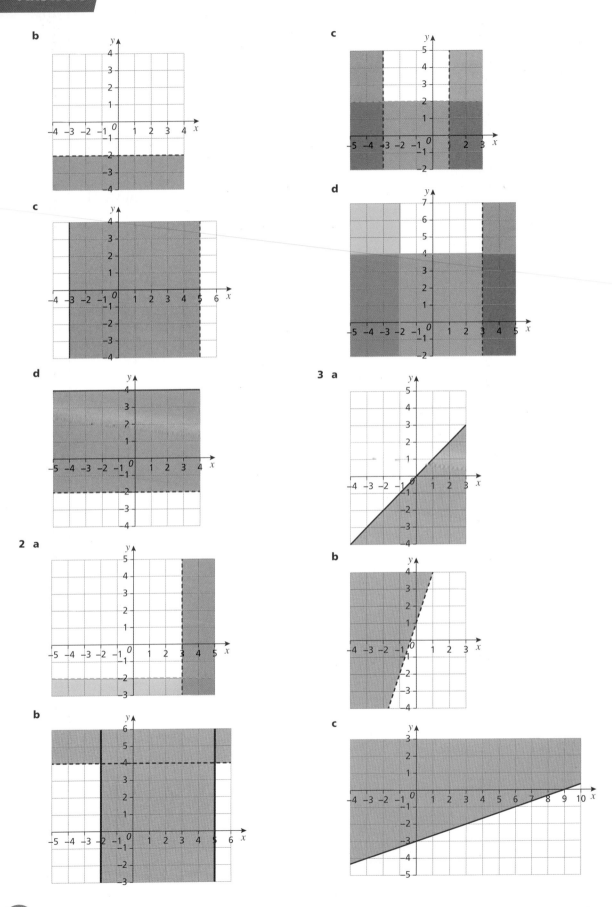

b

c

c

d

d

3 a

2 a

b

b

c

d

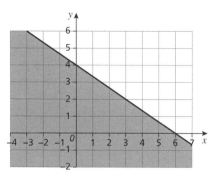

4 a

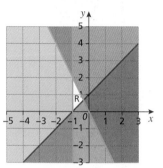

b

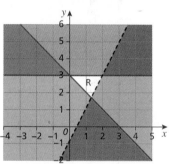

c

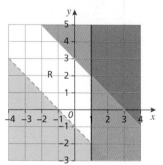

d

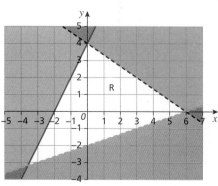

5 a $x > 3$ **b** $y \leqslant 4$
  c $y < 2x + 1$ **d** $y \geqslant 3x - 2$

6 a $y < 2$, $x \geqslant -2$ and $y > 2x$
  b $x \geqslant 3$, $y > -2$ and $y \leqslant 4 - x$

7 a $y > x - 2$, $x \geqslant 1$ and $y \leqslant 3 - x$

  b $y < 2x + 1$, $y \leqslant -\frac{1}{2}x + 3$ and $y \geqslant \frac{2}{3}x - 4$

8 a

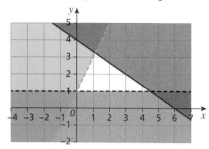

b (1, 2), (3, 2) and (2, 2)

Consolidate

1 a

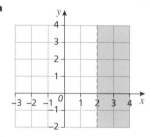

b

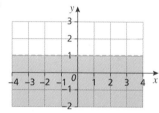

c

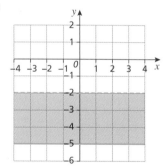

d

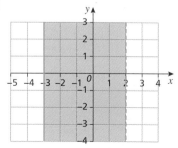

2 a

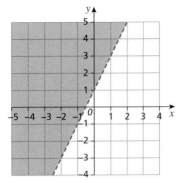

b

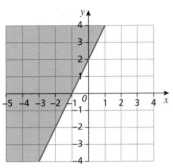

c

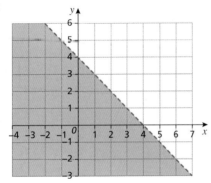

d

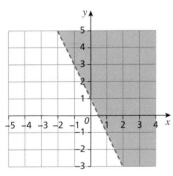

3 a

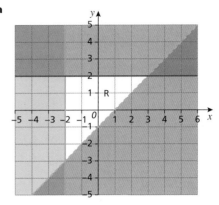

b

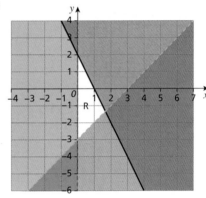

c

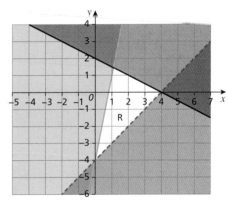

4 a $x \geqslant 3$ **b** $-1 \leqslant y < 4$
  c $y < 3x - 1$ **d** $y \geqslant 2x + 4$

5 a $y > x - 2$, $x + y \leqslant 10$ and $x > 3$
  b $y < 2x + 1$, $y > 1$ and $7x + 4y \geqslant 28$

Stretch

1

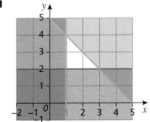

Benji can have 2 choco-wafer, 2 cocoa-caramel;
1 choco-wafer, 3 cocoa-caramel; 1 choco-wafer,
2 cocoa-caramel; 1 choco-wafer, 4 cocoa-caramel;
2 choco-wafer, 3 cocoa-caramel; 3 choco-wafer,
2 cocoa-caramel

2 a i The number of rides on the water slide has to be positive and cannot be more than 5, as 5 rides would cost £20.

ii The number of rides on the Ferris wheel would have to be positive and cannot be more than 6, as 7 rides would cost £21.

iii The total cost of water slide rides would be $4 \times W$; the total cost of Ferris wheel rides would be $3 \times F$. The sum of these must be less than or equal to the £20 that Flo has.

b

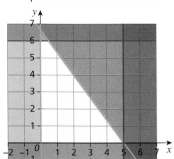

c

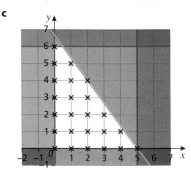

d 2 rides on the water slide and 4 rides on the Ferris wheel

Chapter 12.3

Are you ready?

1 a $x = -5, x = -6$ **b** $x = 2, x = -3$ **c** $x = 7, x = -4$

2 a $-2 \pm \sqrt{3}$ **b** $3 \pm 2\sqrt{3}$ **c** $-1 \pm 2\sqrt{3}$

3 a

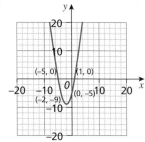

b

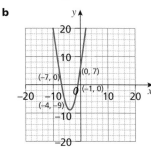

c

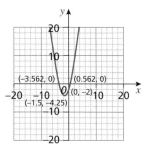

Practice

1 a $x < -7, x > 5$ **b** $x \leqslant -5, x \geqslant 6$

c $-8 < x < 6$ **d** $x < -2, x > 8$

e $x < -8, x > 4$

2 a $\{x: 2 \leqslant x \leqslant 7\}$ **b** $\{x: x \leqslant -4\} \cup \{x: x \geqslant -3\}$

c $\{x: -3 < x < 8\}$ **d** $\{x: x < 4\} \cup \{x: x > 5\}$

e $\{x: 3 < x < 4\}$

3 a $1 - \sqrt{2} < x < 1 + \sqrt{2}$

b $x \leqslant -2 - \sqrt{6}, x \geqslant -2 + \sqrt{6}$

c $x < -6 - \sqrt{41}, x > -6 + \sqrt{41}$

d $-5 - \sqrt{23} \leqslant x \leqslant -5 + \sqrt{23}$

e $3 - 2\sqrt{3} < x < 3 + 2\sqrt{3}$

4 a $\{x: 1.5 < x \leqslant 10\}$ **b** $\{x: x < -3\} \cup \{x: 8 < x < 12\}$

c $\{x: 3 < x \leqslant 6.5\}$

5 a $\{x: < -2.5\} \cup \{x: x > -1\}$

b $\{x: -3 \leqslant x \leqslant \frac{1}{3}\}$

c $\{x: \frac{1}{2} < x < 4\}$

6 a -0.35 and -5.65 **b** $-5, -4, -3, -2, -1$

Consolidate

1 a $\{x: x > -3\} \cup \{x: x < -4\}$

b $\{x: 4 \leqslant x \leqslant 6\}$

c $\{x: x \geqslant 5\} \cup \{x: x \leqslant -4\}$

d $\{x: 1 < x < 2\}$

e $\{x: -2 < x < 6\}$

2 a $-4 - \sqrt{5} \leqslant x \leqslant -4 + \sqrt{5}$

b $x < \frac{-11 - 3\sqrt{13}}{2}, x > \frac{-11 + 3\sqrt{13}}{2}$

c $\frac{-7 - \sqrt{61}}{2} < x < \frac{-7 + \sqrt{61}}{2}$

d $x \leqslant \frac{-1 - \sqrt{13}}{2}, x \geqslant \frac{-1 + \sqrt{13}}{2}$

e $-1 - 2\sqrt{2} < x < -1 + 2\sqrt{2}$

3 a $-4 < x < -\frac{1}{3}$ **b** $-\frac{1}{2} \leqslant x \leqslant 3$ **c** $x \leqslant -6, x \geqslant 6$

4 a $-1, -2, -3, -4, -5, -6$

b $-2, -1, 0$

c $-1, 0, 1$

5 $\{x: -12 \leqslant x < -10\}$

6 a $x(x - 3) > 10, x^2 - 3x > 10, x^2 - 3x - 10 > 0$

b $x > 5$

Stretch

1 a $2x^2 - x - 1 < 90$ and $6x > 20$

b $\frac{10}{3} < x < 7$

2 Example answer: $x^2 + 3x - 4 < 0$

3 Example answer: $x^2 - 4x - 32 \geqslant 0$

Inequalities: exam practice

1

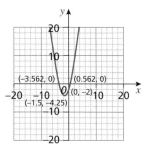

2 $-4, -3, -2, -1, 0, 1$

3 $x \leqslant 7.5$

4 $x < 5$

5

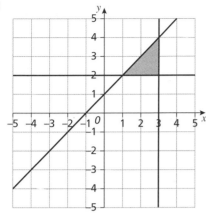

6 $x \le 3$, $y \ge 0$, $y \le \frac{1}{2}x + 2$

7 a

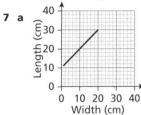

b No, you cannot have a length or width of 0 cm.

8 $-2 \le x \le 6$

9 $2, -1, 0, 1$

10 $x \le -4$ or $x \ge -\frac{3}{2}$

Block 13 Simultaneous equations

Chapter 13.1

Are you ready? (A)

1 a 3   **b** 5   **c** 1.5   **d** 4.5

2 a 11   **b** 0   **c** 8.5   **d** $\frac{4}{3}$

3 a Gradient 6; y-intercept 3
   b Gradient -3; y-intercept 4
   c Gradient 2; y-intercept -1.5
   d Gradient -1; y-intercept 17

Practice (A)

1 a $x = -2$, $y = -2$ **b** $x = 4$, $y = 4$ **c** $x = 1.5$, $y = 5.5$
2 a $x = 5$, $y = 2$ **b** $x = 1$, $y = 4$ **c** $x = 3$, $y = -2$
   d $x = 4$, $y = 3$ **e** $x = 4$, $y = 2$
3 a $x = 4$, $y = 1$ **b** $x = -1$, $y = 4$ **c** $x = 1.5$, $y = -2$
   d $x = 3$, $y = 1$ **e** $x = -2$, $y = 4$
4 a $x = 4$, $y = 2$ **b** $x = 2$, $y = -1$ **c** $x = 2$, $y = -1$
   d $x = -1$, $y = 5$ **e** $x = 6$, $y = 2$
5 Tea £2, coffee £1.50
6 Child £6, adult £10
7 Seb is 14, Rob is 19
8 $x = 6$, $y = 14$
   Perimeter is 66

Are you ready? (B)

1 a -9 **b** -15 **c** 11
2 a $5x + 8$ **b** $3x - 11$ **c** $5x + 18$

Practice (B)

1 a $x = 1$, $y = 5$ **b** $x = 8$, $y = 1$ **c** $x = -2$, $y = -4$
   d $x = 1$, $y = 3$ **e** $x = 6$, $y = 5$
2 Ali has £9.75, Benji has £5.25
3 9 and 18

Consolidate

1 a $x = 3.2$, $y = 2.2$ **b** $x = 3$, $y = 5$
2 a $x = 2$, $y = 1$ **b** $x = 2.5$, $y = -2$
   c $x = 7$, $y = -2$ **d** $x = 8$, $y = 9$
3 a $x = -3$, $y = 3$ **b** $x = 1.5$, $y = -3$
4 a $x = 3$, $y = 1$ **b** $x = 4$, $y = 1$
5 Pastry £2.50, cake £3
6 $x = 12$ and $y = 15$
   Perimeter is 304 units

Stretch

1 a To get 13, you do $4 \times a$ and add b.
   b $13a + b = 31$
   c $a = 2$, $b = 5$
   d 67
2 a One solution
   b (Infinitely) many solutions (graphically it is the same line)
   c No solutions; the lines are parallel and never meet
3 The intersections are $(-1, 6)$, $(2.6, -1.2)$ and $(5, 0)$.
Two of the lines are perpendicular so it is a right-angled triangle.
Area = 10.8 (1 d.p.)

Chapter 13.2

Are you ready?

1 a $x = 3$, $y = 8$ **b** $x = 7$, $y = 2$ **c** $x = 10$, $y = 3$
2 a $x = -4$, $x = 6$ **b** $x = 5$, $x = 8$ **c** $x = -1.5$, $x = -5$

Practice

1 a $x = 4$, $y = 12$ and $x = -1$, $y = -3$
   b $x = 2$, $y = 7$ and $x = -1$, $y = -2$
   c $x = -1$, $y = 3$ and $x = -2$, $y = 6$
   d $x = 4$, $y = 2$ and $x = -4$, $y = -2$
2 a $x = -5$, $y = 0$ and $x = 2$, $y = 14$
   b $x = -\frac{1}{3}$, $y = -\frac{11}{3}$ and $x = 1$, $y = -1$
   c $x = -2$, $y = -3$ and $x = 2$, $y = 5$
3 a $x = 5$, $y = -1.5$ and $x = -1$, $y = 1.5$
   b $x = 2$, $y = 4$ and $x = 4$, $y = 2$
   c $x = 4$, $y = 3$ and $x = 3$, $y = 4$
4 a $x = 4$, $y = 6$ and $x = -6$, $y = -4$
   b $x = \frac{10}{3}$, $y = -\frac{1}{6}$
   c $x = 5$, $y = 3$ and $x = -1.5$, $y = -10$
5 a $x = 0.44$, $y = -0.12$ and $x = 4.56$, $y = 8.12$
   b $x = -1.56$, $y = -2.56$ and $x = 2.56$, $y = 1.56$
6 a $x = 1$, $y = 0$
   b There is only one solution, so the graphs don't intersect, but instead touch at one point.
   c

$y = x^2 + 3x - 4$

$(-4, 0)$ $(1, 0)$

$y + 5 = 5x$

Consolidate

1 a $x = -1$, $y = 0$ and $x = 1$, $y = 2$
   b $x = 0.5$, $y = 1$ and $x = 5$, $y = 19$
   c $x = -3$, $y = 4$ and $x = -4.8$, $y = -1.4$
2 a $x = -1$, $y = -2$ and $x = 5$, $y = 4$
   b $x = 3$, $y = 3$ and $x = 1$, $y = -1$
   c $x = 2$, $y = 12$ and $x = -1$, $y = 3$
3 a $x = -1$, $y = 2$ and $x = -2$, $y = -1$
   b $x = 2$, $y = 3$ and $x = 3$, $y = -2$
   c $x = 0$, $y = -5$ and $x = 4$, $y = 3$
4 a $(-1, 4)$ and $(2, 7)$ **b** $(2, 0)$ and $(4, -3)$

Stretch

1 a (−1, 6) and (3, 2)   **b** $4\sqrt{2}$ units
2 a (2, 1) and (5, 7)   **b** (3.5, 4)
3 a You cannot solve the quadratic.
  b The graphs don't meet.
  c

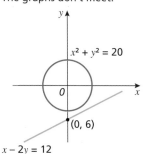

Simultaneous equations: exam practice

1 $x = 14.5$, $y = 3.5$
2 $x = 3$, $y = −4$
3 Coffee = £2.45, tea = £1.90
4 $x = 1$, $y = 4$
5 78 cm
6 $x = 5$, $y = 4$ and $x = −4$, $y = −5$
7 $x = 2$, $y = −3$ and $x = 3$, $y = −2$

Block 14 Sequences

Chapter 14.1

Are you ready?

1 a 14, 17   **b** −8, −13   **c** 19, 25   **d** −0.4, −1
2 a 15   **b** 6   **c** −5   **d** 5.5
3 a $n > 2$ **b** $n < 5$ **c** $n > 5$ **d** $n > 13.25$

Practice

1 a 5, 9, 13   **b** 2, 5, 8   **c** −1, −6, −11   **d** 7.5, 8, 8.5
2 a Yes   **b** No   **c** No   **d** Yes
3 a $3n + 6$ **b** $9n + 2$ **c** $10n − 4$ **d** $7n − 17$
4 a $8 − n$ **b** $12 − 0.5n$ **c** $4 − 6n$ **d** $2 − 3n$
5 a 207   **b** 411   **c** −294   **d** 802
6 a i $6n − 7$ **ii** $4n + 1$ **iii** $13n + 5$ **iv** $8n − 3$
  b i 503   **ii** 501   **iii** 512   **iv** 501
7 a

  b 17   **c** Pattern 25   **d** No   **e** $2n + 1$
8 a

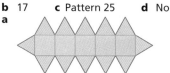

  b 8   **c** 22   **d** $n + 1$ **e** $2n + 4$

What do you think?

1 a $\dfrac{2n + 1}{n + 2}$ **b** $\dfrac{3n − 1}{4n + 3}$
2 a 4   **b** 4

Consolidate

1 a $3n + 7$ **b** $6n − 4$ **c** $8n − 1$ **d** $5n$
  e $4 − 5n$ **f** $3 + 0.5n$ **g** $2 − 0.5n$
2 a Yes   **b** No   **c** Yes   **d** Yes
3 a 907   **b** 440   **c** −754   **d** −440
4 a i $8n + 9$ **ii** $10n − 12$ **iii** $11n + 6$ **iv** $3n + 15$
  b i 1001   **ii** 1008   **iii** 1007   **iv** 1002

5 a

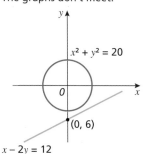

b 17   **c** 12
  d 20   **e** $n + 2$ **f** $2n + 1$

Stretch

1 $11n − 9$
2 a $2.5n + 14.5$ **b** Yes
3 a $m = 6$ **b** 1

Chapter 14.2

Are you ready?

1 a $3n + 1$ **b** $4n − 6$ **c** $5 − 2n$ **d** $8 − n$
2 a 11, 17, 23   **b** −2, −7, −12   **c** 5, 8, 13   **d** 4, 14, 30

Practice

1 a 35, 48   **b** 39, 52   **c** 50, 65
2 a 5, 8, 13, 20   **b** 3, 12, 27, 48
  c 0, 2, 6, 12   **d** 2, 7, 14, 23
3 a $n^2 + 1$ **b** $2n^2$ **c** $n^2 − 3$ **d** $(n + 1)^2$
4 a $n^2 + 7$ **b** $n^2 + 3n + 1$ **c** $n^2 + 4n − 3$
  d $n^2 + 2n + 17$ **e** $n^2 − 3n − 8$
5 a $2n^2 + 6$ **b** $2n^2 + n − 3$ **c** $3n^2 + 4n$
  d $3n^2 − 3n + 1$ **e** $−n^2 + 2n + 4$
6 a 3rd term   **b** 10th term
  c Neither solution of $n^2 + 2n − 30 = 0$ is a positive integer
7 a 25   **b** $2n^2 − 2n + 1$ **c** Yes (7th pattern)
8 a 10, 15   **b** $\frac{1}{2}n^2 + \frac{1}{2}n$
9 Lots of possibilities, such as $b = 1$ and $c = −14$ or $b = 2$ and $c = −18$

What do you think?

1 Cubic sequences have rules of the form $an^3 + bn^2 + cn + d$, and they have constant third differences.

Consolidate

1 a −2, 1, 6, 13   **b** 2, 6, 12, 20   **c** 5, 11, 21, 35
  d −1, 5, 13, 23   **e** 4, 9, 16, 25
2 57
3 a $n^2 + 2n + 4$ **b** $n^2 − 3n − 1$
  c $n^2 + n + 7$ **d** $n^2 − 2n + 3$
4 a $2n^2$ **b** $2n^2 − 2n + 5$
  c $3n^2 + n − 1$ **d** $4n^2 + n − 2$
5 a 8   **b** 42   **c** $n + 1$ **d** $n^2 + 3n + 2$ **e** 90
6 5th term

Stretch

1 $b = 3$ and $c = −2$
2 8th term
3 234 and 266, the 13th and 14th terms

Chapter 14.3

Are you ready?

1 a 16, 19   **b** 80, 160   **c** 25, 36   **d** −10, −14
2 a $4a + 2b$ **b** $6a + 2b$ **c** $6a + 3b$ **d** $8a + 11b$

Practice

1 a Geometric   **b** Quadratic
  c Linear   **d** Fibonacci
2 a 81, 243   **b** 40, 80
  c 64, 128   **d** 1.25, 0.625
3 a 3, 9, 27   **b** 2, 4, 8   **c** 10, 100, 1000
  d 2, 8, 32   **e** 3, 15, 75

4 a 5^n **b** 6^n **c** 3^n **d** $\left(\frac{1}{2}\right)^n$
e $2 \times 10^{n-1}$ **f** $3 \times 4^{n-1}$ **g** $5 \times 2^{n-1}$

5 a 2.4m **b** 1.23m
c 3×0.8^n **d** 0.000 0428m
e Not realistic as it suggests the ball will never stop bouncing as the sequence will never include 0

6 a 11, 18 **b** 19, 31 **c** 42, 68
d $8a$, $13a$ **e** $a + 4b$, $2a + 6b$

7 a $a + 2b$ **b** $3a + 10b$

8 The numbers in the sequence are being halved each time, so they will get smaller and smaller but they will never be negative.

9 a 8, $8\sqrt{2}$ **b** $144\sqrt{3}$, 864

10 a 130
b Because each time you are multiplying by 1.3 as the common ratio
c 6.27

Consolidate

1 a 4, 16, 64 **b** 2, 12, 72
c $\frac{1}{3}, \frac{1}{9}, \frac{1}{27}$ **d** 3, 12, 48

2 a 4^n **b** $\left(\frac{1}{5}\right)^n$ **c** 7^n
d $3 \times 6^{n-1}$ **e** $5 \times 2^{n-1}$

3 a 33, 53 **b** 0, 1
c $3a + 4b$, $6a + 6b$ **d** $4a - 4b$, $8a - 6b$

4 a 9, 12 **b** 12, 24 **c** 9, 15

5 a £525
b £638.14
c Because each time you are multiplying by 1.05 which is the common ratio
d 15 years

Stretch

1 a 12 **b** Any two numbers with a sum of 12
2 $a = 4$
3 $9 + 6\sqrt{3}$, $18 + 9\sqrt{3}$
4 Third term: $a + b$
First term: $-2a + 2b$

Chapter 14.4

Are you ready? (A)

1 a 15, 19 **b** 48, 96 **c** 47, 95 **d** −38, −86
2 a 10.9 **b** 5.4 **c** 2.37 **d** 4.29

Practice (A)

1 a 10, 13, 16, 19 **b** 8, 20, 56, 164
c 3, 8, 63, 3968
2 $u_{n+1} = u_n + 9$
3 $t_{n+1} = 2t_n$
4 No, because you start with 6, which isn't a multiple of 10, and then add 10, so it will still not be a multiple. Each time you are adding 10 to a number that isn't a multiple of 10.
5 a 39 **b** 110
c The number of rabbits keeps increasing forever, so it is not a suitable model.

What do you think?

1 $u_{n+1} = 2u_n + 3$

Are you ready? (B)

1 a 1.71 **b** −1.91
c 2.18 **d** −2.02
2 a $x = \frac{2y - 4}{3}$ **b** $x = \frac{10 + y}{4}$
c $x = \sqrt{b - a}$ **d** $x = \sqrt[3]{\frac{3d + c}{2}}$

Practice (B)

1 $x = 0$ gives 3 and $x = 1$ gives −2; there is a sign change, hence the root lies between 0 and 1

2 $x = 2$ gives −14 and $x = 3$ gives 63; there is a sign change, hence the root lies between 2 and 3
3 2.49, 2.44, 2.44
4 1.6303, 1.6370, 1.6373
5 a Compare approaches as a class. **b** 2.05
6 a Compare approaches as a class. **b** 0.45
7 a $x = 3$ gives −5 and $x = 4$ gives 25; there is a sign change, hence the root lies between 3 and 4
b Compare approaches as a class.
c 3.23071, 3.22565, 3.22647

Consolidate

1 a 8, 44, 260, 1556
b −3, 15, 447, 399 615
2 a 10% **b** 72.9g **c** 34.9g
d The model suggests that the amount of substance will never disappear as it cannot reach zero.
3 $x = 1.35$ gives −0.42925 and $x = 1.40$ gives 0.088; there is a sign change, hence the root lies between 1.35 and 1.40
4 a Compare approaches as a class.
b 2.43
5 a Compare approaches as a class.
b 2.7778, 2.9072, 2.8282

Stretch

1 a Compare approaches as a class.
b Compare approaches as a class.
c Compare approaches as a class.
d Compare approaches as a class.
e Equation **a** diverges and gets bigger, equation **b** gives one solution (0.219) and equations **c** and **d** both give another solution (2.28)
2 a $a = 3$, $b = 5.5$ **b** 2.32

Sequences: exam practice

1 Example answer: $3n + 2 = 1000$ does not have an integer solution
2 50
3 81
4 130
5 The 7th term
6 25, 12
7 $9\sqrt{3}$
8 $\frac{3}{(n + 1)(n + 2)}$
9 $2n^2 + 3n + 5$
10 775.06

Block 15 More graphs

Chapter 15.1

Are you ready?

1 a 0.77 **b** 0.57 **c** −0.75 **d** 0.39
2 a 66.4° **b** 50.2° **c** 53.1°

Practice

1 a

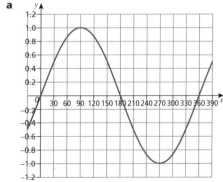

b

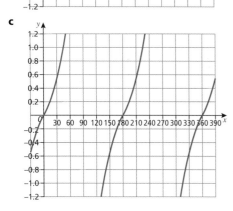

c

2 a 325° **b** 113° **c** 258°
3 a 17.5°, 162.5° **b** 45.6°, 314.4° **c** 50.2°, 230.2°
  d 233.1°, 306.9° **e** 107.5°, 252.5° **f** 115.5°, 295.5°
4 Possible answers: 63.3°, 296.7°, 423.3°, 656.7°
5 Possible answers: 139.6°, 319.6°, 499.6°, 679.6°
6 a True **b** False **c** False **d** True
  e False **f** True **g** True

What do you think?

1 The graphs are the same shape in a different place on the horizontal axis.
2 a sin x = sin(360° − x) is false; the others are all true.
  b Compare answers as a class.

Consolidate

1 A (90, 1) B (180, 0)
2 A (90, 0) B (180, −1)
3 a 44.4°, 135.6° **b** 78.5°, 281.5° **c** 63.4°, 243.4°
  d 244.2°, 295.8° **e** 95.7°, 264.3° **f** 149.0°, 329.0°
4 a k = 0 **b** k = −1
  c She is correct as all the angles between 0° and 90° have a cosine value between 0 and 1
  d No, he is not correct as angles between 270° and 360° also have a cosine value between 0 and 1

Stretch

1 a cos 25° **b** tan 35°
2 a 10.5 m **b** 2:23 am
3 a 60°, 300° **b** 66.8°, 246.8°
  c 30°, 150°, 210° and 330°
4 30°, 150°, 210° and 330°

Chapter 15.2

Are you ready?

1 a

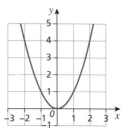

b

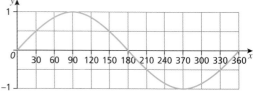

c

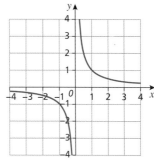

d

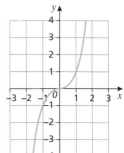

e

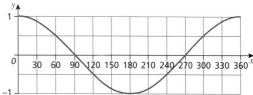

Practice

1 a Translation 1 unit to the left
  b Translation 1 unit up
  c Translation 1 unit to the right
  d Translation 1 unit down

2 a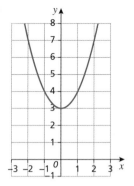

b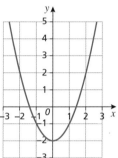

c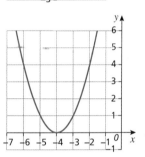

d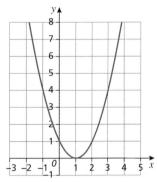

3 a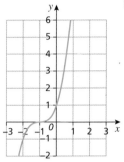

b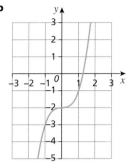

c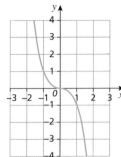

d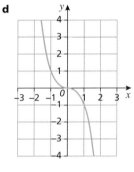

4 a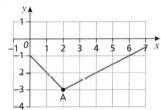

b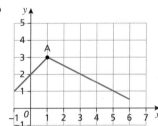

c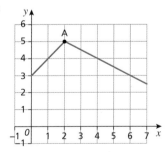

d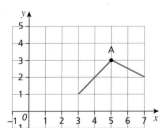

5 a (3, –4)   **b** (–1, –2)   **c** (1, –4)   **d** (–1, 4)
6 a $y = f(x - 2)$ **b** $y = f(x) - 4$ **c** $y = f(-x)$
7 a

b

8 a

b

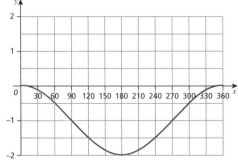

What do you think?
1 (7, 7)
2 $y = x^2 + 12x + 25$

Consolidate
1

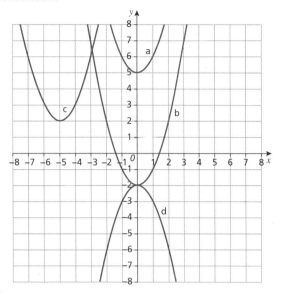

2

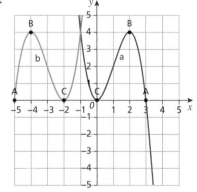

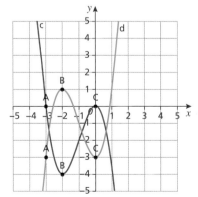

3 a (6, –1)   **b** (4, 2)   **c** (–4, –1)   **d** (4, 1)
4 $y = f(x - 3)$
5 $y = g(-x)$

Stretch
1 D
2 a Translation 2 units to the left and 7 units down
(–2, –7)
  b Translation 3 units to the right and 7 units down
(3, –7)
  c Translation 1 unit to the left and 3 units down
(–1, –3)

3 $y = -x^2 - 5x - 3$

4 a

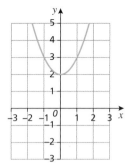

b

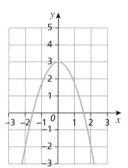

c

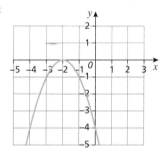

d

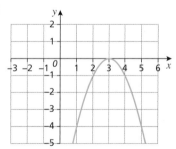

Chapter 15.3

Are you ready?

1 a 2   **b** −3   **c** 4   **d** $\frac{1}{2}$

2 a −3   **b** 2   **c** −6   **d** 1

Practice

1 a 1 m/s   **b** −1.75 m/s

  c The rocket is instantaneously at rest since speed equals 0

2 a 9 m/s   **b** 7.8 m/s

3 a 1.4 m/s²

  b −0.5 m/s²; the answer is negative as the car is slowing down.

  c Speed is instantaneously constant as the car goes from accelerating to decelerating.

4 a 85°C   **b** −4.5°C per minute

  c −8; rate of temperature loss per minute is 8°C.

  d No, because eventually it would reach room temperature and stop decreasing

5 a −3   **b** 1

  c It is an estimate because the gradient of the curve is being approximated by the gradient of a straight line tangent to the curve.

Consolidate

1 a 35 m/s   **b** 5.6 m/s

2 a 4.9; the rate of change of the depth of water at that moment is 4.9 cm/s

  b 3 cm/s

3 a 2.3; acceleration at 6 s is 2.3 m/s²

  b $a = 0$ so the cyclist is travelling at constant speed at that moment, before decelerating.

  c −0.9 m/s²

4 a 3.75   **b** −2.25

Stretch

1 a The gradient is always 4

  b The gradient is $4 + k$

Chapter 15.4

Are you ready?

1 a 12 cm²   **b** 48 cm²   **c** 12 cm²   **d** 75 cm²

Practice

1 1660 m

2 a 34 m   **b** Use more strips.

3 a 29.85 m

  b Underestimate as all the trapezia/triangles are below the curve so the real area is greater.

4 a 21 m

  b Overestimate as all the trapezia are above the curve, so the real area is less.

  c Use more strips.

5 17.5 square units

6 18.5 square units

Consolidate

1 a 16.25 m   **b** Use more strips.

2 The first 4 seconds (6.5 m versus 6.125 m)

3 a 11.25 m   **b** 11.925 m

  c Part **b**, because more strips were used.

4 a 4   **b** 11.1%

Stretch

1 a 150 m   **b** 14.5 s

2 a i 675 m   **ii** 600 m

  b Both vehicles will have covered the same distance.

  c 75 s, having covered 900 m

More graphs: exam practice

1 197° and 343°

2 a (1, 0) and (7, 0)   **b** (−2, 7)

3 a 4   **b** 96 square units

Algebra: exam practice

1 $x = 3, y = -1$

2 $6n - 4$

3 $x = 2, x = 4$

4 $y = -3x + 10$

5 $r = \frac{C}{2\pi}$

6 a 343   **b** $2x - 10$

7 $x = 1, x = -\frac{3}{2}$

8 $\frac{x - 4}{x - 3}$

9 $x^2 + y^2 = 12$

10

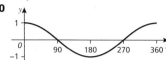

11 a (−6, 5) **b** (−3, 8)

12 a $(x − 2)^2 + 5$

b Example answer: $(x − 2)^2$ is at least zero, so $(x − 2)^2 + 5$ is at least 5, so always positive.

Block 16 Ratio

Chapter 16.1

Are you ready? (A)

1 a 4 : 3 **b** 7 : 8 **c** 8 : 7 **d** 7 : 5 : 3

2 10 benches

3 3 : 8

4 a 14 **b** 10.5 **c** 31.5 **d** 63 **e** 54

Practice (A)

1 a £150 : £210 **b** 20 bananas : 15 bananas **c** 3 cm : 18 cm : 21 cm **d** 280 ml : 420 ml

2 a Ali gets £14, Flo gets £49 **b** £18 **c** £283.50 **d** £113.40

3 60 g of flour

4 12

5 1250 kg of timber, 625 kg of sand and 3125 kg of steel

6 600 g : 650 g

7 £51 562.50

What do you think? (A)

1 90 tulips

2 a 1, 2, 3, 7, 9, 12, 17, 27 or 57

b Because 3 + x needs to sum to a factor of 60; for example, you can't share 60 sweets in the ratio 3 : 4 as 7 is not a factor of 60

3 £64

4 180° ÷ (1 + 5 + 6) = 15°

1 × 15° : 5 × 15° : 6 × 15° = 15° : 75° : 90°

Are you ready? (B)

1 a 6 **b** 8

2 a $\frac{1}{3}$ **b** $\frac{3}{4}$ **c** $\frac{4}{7}$ **d** $\frac{3}{10}$

3 a 5 : 1 **b** 5 : 6 **c** 3 : 5 **d** 7 : 3

4 a 3500 g **b** 142 cm **c** 6070 ml

Practice (B)

1 a 4 : 1 **b** 3 : 8 **c** 1 : 15 **d** 1 : 3 **e** 4 : 3

2 a 5 : 1 **b** 5 : 2 **c** 2 : 3 : 4 **d** 1 : 10

3 a 3 : 10 **b** 2 : 1 **c** 10 : 1 **d** 5 : 1

4 a 7 : 10 **b** 7 : 2 **c** 100 : 1 **d** 2 : 1

What do you think? (B)

1 3 : 2

2 2 : 1 : 7

3 200 : 4 : 1

4 2 : 3 : 1

5 80 counters

Consolidate

1 a £120 : £280

b 40 strawberries : 24 strawberries

c 14 mm : 28 mm : 84 mm

d 200 g : 1000 g

2 a Sven gets £90, Huda gets £150

b £400 **c** £384 **d** £960

3 a 6 : 1 **b** 5 : 12 **c** 4 : 15 **d** 1 : 4 **e** 5 : 3

4 a 5 : 2 **b** 25 : 2 **c** 2 : 1 : 12 **d** 1 : 3

5 a 7 : 10 **b** 5 : 3 **c** 14 : 1 **d** 7 : 3

6 a 5 : 1 **b** 12 : 7 **c** 1 : 5 **d** 1 : 1

Stretch

1 a 60 **b** 50 square milk chocolates

2 30 seconds

3 9 : 4

Chapter 16.2

Are you ready?

1 a 5 : 1 **b** 5 : 6 **c** 3 : 5

2 a 1 : 2.5 **b** 0.4 : 1

3 a 2 : 5 **b** $\frac{2}{7}$ **c** $\frac{5}{7}$

Practice

1 $\frac{1}{5}$

2 $\frac{2}{5}$

3 a $\frac{3}{10}$ **b** $\frac{4}{5}$

4 a Example answer: Because 5 : 2 boys to girls would mean $\frac{2}{7}$ of his class are girls.

b 3 : 2

5 a 3 : 1 **b** 3 : 4

6 $\frac{6}{25}$

What do you think?

1 $\frac{x}{x + y}$

2 $\frac{10}{27}$

Consolidate

1 $\frac{3}{8}$

2 $\frac{2}{7}$

3 a $\frac{3}{14}$ **b** $\frac{5}{14}$

4 11 : 9

5 a 4 : 1 **b** 1 : 5

6 $\frac{11}{25}$

Stretch

1 $b = a\left(1 − \frac{2}{k}\right)$

2 $m = \frac{3}{8}$

3 6 : 5

4 1110 litres

Chapter 16.3

Are you ready? (A)

1 48

2 $a = 25$

3 1 : 2

Practice (A)

1 4 : 10 : 3

2 15 : 5b : 3g

3 $m = 4$

4 $y = 10$

5 21

6 $\frac{32}{105}$

What do you think? (A)

1 $5 : 6 : 7 : 8$
2 $k = 61$

Are you ready? (B)

1 $\frac{2a}{5}$

2 $2x - x^2$

3 $k = \frac{3}{2x}$

4 $\frac{r}{y + r}$

Practice (B)

1 20 years old
2 44 lemons and 36 limes
3 36 raspberries
4 Class A: 27 chairs and Class B: 36 chairs
5 $y = 8$
6 $w = 5$ or $w = 2$
7 $x = 8$
8 $19 : 15$

What do you think? (B)

1 $1 : 2$
2 $2 : 1$
3 $p = 6$ and $q = 4$
4 $k = 7$

Consolidate

1 $24 : 20 : 15$
2 66
3 16 years old
4 72
5 $10 : 7$

Stretch

1 $3 : 4 : 9 : 6$
2 76
3 $66 : 65$
4 $a = 2$ and $b = 6$, $a = 3$ and $b = 3$, $a = 6$ and $b = 2$

Ratio: exam practice

1 $1 : 2.625$

2 $\frac{2}{5}$

3 £1680
4 6.15 km
5 $6 : 15 : 35$

6 $y = \frac{3}{2}x$

7 £1050

Block 17 Proportion

Chapter 17.1

Are you ready? (A)

1 a £14.58   **b** £21.87   **c** £10.94
2 a 1125 ml of yoghurt and 312.5 g of flour
 b 28 flatbreads
3 a 522 kr   **b** £38.31

Practice (A)

1 a Table completed from the left: 12.5, 25, 62.5
 b 48 000 bees **c** 1.25 ml

2 a Table completed from the left: 4, 8, 12, 20, 24
 b

 c i £10   **ii** 125 litres
3 Table 1
4 A
5 a ₺450   **b** €2   **c** ₺1500   **d** $1 = ₺27
6 a 2950 m²   **b** 2 minutes
7 40 miles
8 Graph 1 is not a straight line.
 Graph 2 is a straight line but does not go through the origin.

What do you think? (A)

1 a $16 : 1$ **b** $P = 64$

2 $\frac{5}{2}m$

3 $125a + 350b$

Are you ready? (B)

1 $j = 15$
2 $(2, 10)$ $(0.7, 3.5)$ $(c, 5c)$

Practice (B)

1 a $a = 40.5$ **b** $b = 50$
2 B and F
3 a $w = 6z$ **b** $w = 306$ **c** $z = 37$
4 a 15.75 N   **b** 2.67 m/s²
5 a $d = 2.2$ **b** $f = 759$
6 a $j = 36$ **b** $l = 552$
7 a $y = 105$
 b Example answer: By substitution in the equation
  $y = 3x$

8 a $p = \frac{q}{8}$ **b** $p = 1\frac{3}{4}$ **c** $q = 976$

What do you think? (B)

1 a $t = \frac{2}{3}u$ **b** $t = \frac{1}{2}$ **c** $u = 1\frac{4}{9}$

2 $x = 15z$

3 a $r = 50$ **b** $y = \frac{3}{8}x$ **c** $x = \frac{8y}{3}$

Consolidate

1 a Table completed from the left: 2200, 4400, 11 000
 b 22 kg   **c** 228 trees
2 Table completed from the left: 9, 18, 27, 45, 54
3 a $e = 29.4$ **b** $f = 60$
4 a $g = 18$ **b** $h = 54$
5 a $w = \frac{3}{16}x$ **b** $w = 30$ **c** $x = 656$

Stretch

1 $q = 12$
2 Assuming fuel consumption is constant throughout
 the journey, $n = 50$ and $n = 350$

Chapter 17.2

Are you ready? (A)

1 a 3 cm   **b** 1.5 cm   **c** 15 cm
    d 4.8 cm   **e** 120 cm
2 a 32 mph   **b** 18.75 mph

Practice (A)

1 a Table completed from the left: 6, 3, 1.5, 1, $\frac{2}{3}$ or $0.\dot{6}$, 0.5
    b The time is halved.
    c

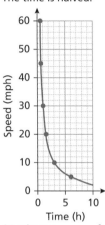

Time (h)

2 a 30 minutes   **b** 3 hours
3 a B The number of kitchen workers is inversely
    proportional to the time taken.
    b 8 minutes
    c 360 workers
4 a 3.2 minutes   **b** 128 mph
5 a 37.5 minutes
    b The number of people and the time taken are
    inversely proportional. One person would take five
    times as long as 5 people.
6 a 120 builders   **b** 18 months
    c Example answers: weather; work rate of builders
7 a Table completed from the left: 16, 30, $\frac{1}{15}$ or $0.0\dot{6}$
    b $xy = 24$
8 a i 1.25 hours or 1 hour and 15 minutes
    ii 25 minutes
    b 10 cleaners
    c Example answers: same cleaning equipment; offices
    all identically sized; work rate of each cleaner
    being the same

What do you think? (A)

1 a 16.5 days
    b Example answer: the work rate of all workers
    was equal
2 a Inversely   **b** Directly   **c** Neither   **d** Directly

Are you ready? (B)

1 $k = 833.49$
2 $x = \frac{7}{3}$

Practice (B)

1 a Example answer: If w is halved then v must be
    doubled or $64 \times 3.75 = 32 \times 7.5 = 240$
    b $v = 0.6$ **c** $w = 300$
2 a $e = 33.6$ **b** $d = 0.7$
3 a $r = 2.4$ **b** $t = 0.24$
4 a $j = 60$ **b** $j = 9.6$
5 a 135°C   **b** 20 minutes and 15 seconds
    c If the oven is not hot enough, the potato will not
    cook. If the oven is too hot, the outside of the
    potato may be burnt and the inside overcooked.

6 16 hours
7 $c = 3703$

What do you think? (B)

1 $x = 88$
2 a $n = 0.175$ and $p = 0.6$
    b n is directly proportional to p.
3 Example answer, using $\theta = 30°$, opposite side length
    10 cm and hypotenuse 20 cm. Triangle not possible if
    opposite side doubles and hypotenuse halves.

Consolidate

1 a 40 workers   **b** 16 days
2 a 2.5 minutes   **b** 5 people
3 a $q = \frac{2}{3}$ **b** $p = 250$
4 a $d = \frac{2}{15}$ **b** $b = 7$

Stretch

1 a $x = 2880$ and $y = \frac{1}{64}$
    b y is inversely proportional to z.
    c $y = \frac{5}{32z}$
2 Discuss answers as a class. Chloe is correct based on
    $\cos \theta = \frac{\text{adj}}{\text{hyp}}$ where the adjacent is the constant of
    proportionality.
3 a $a = \frac{3}{2b}$
    b $a = 3$, $b = 0.5$ and $a = 0.5$, $b = 3$

Chapter 17.3

Are you ready?

1 48
2 $b = 4.8$
3 $v = 196$

Practice

1 a $z = \frac{4\sqrt{y}}{5}$ or $z = 0.8\sqrt{y}$ **b** $z = 6.4$
2 a He has not squared t.
    b $t = 3$ **c** $r = \frac{2}{3}$
3 A
4 a $g = 512$ **b** $h = 320$
5 $q = 600$
6 Final column completed from top to bottom: B, D, A, C

What do you think?

1 a True   **b** False   **c** True   **d** True
2 a

x	p	pq	p^3	$\frac{p}{q^2}$
y	q^2	q^3	p^2q^2	1

    b

x	p	pq	$\frac{1}{p}$	pq^2
y	q^2	q	p^2q^2	1

Consolidate

1 a $l = 10$ **b** $m = 512$
2 a $d = 2.5$ **b** $f = \frac{1}{8}$
3 A: y is inversely proportional to x^2
    B: y is directly proportional to x^2
    C: y is directly proportional to $\sqrt{x}$
    D: y is directly proportional to x

Stretch

1 $\left(\frac{9a}{16}\right)^{\frac{1}{3}}$
2 $x(x - y)$

3 a

p	$x - 1$	$x^2 - 1$	$2x^2 - 2$	x
q	$x + 1$	1	$\frac{1}{2}$	$x - \frac{1}{x}$

b

p	$x - 1$	$\frac{x-1}{x+1}$	$2x^2 - 2$	$\frac{(x-1)^2}{x}$
q	$x + 1$	1	$2(x + 1)^2$	$x - \frac{1}{x}$

Proportion: exam practice

1 £6
2 £1.20
3 $n = 3$
4 No.
Example justification: When $a = 45$, $\frac{a}{b}$ is not the same as for the other values of a.
5 $p = \frac{216}{q}$
6 16 minutes
7 12, assuming all workers work at the same rate.
8 0.96

Block 18 Rates

Chapter 18.1

Are you ready?

1 2 hours and 20 minutes
2 $k = 25$
3 a $r = df$ **b** $r = \frac{f}{d}$ **c** $r = \frac{f}{d}$
4 a 14.1   **b** 10 100   **c** 0.0294

Practice

1 a 87.5 mph   **b** 116 km/h   **c** 258 mph   **d** 6.4 m/s
2 a 117 miles   **b** 177.1 km
   c 630 miles   **d** 82 800 m or 82.8 km
3 a 2.5 hours   **b** 5 hours
   c 25 seconds   **d** 3 minutes or equivalent
4 a

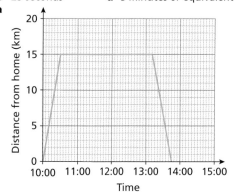

b 25.7 km/h
5 a

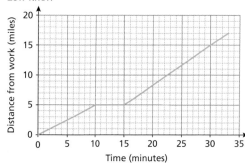

b 30.9 mph

6 AB
7 Example answer: Monday and Tuesday's journeys both cover the same total distance and time.
8 8.70 m/s
9 a 86.4 mph   **b** 45 mph   **c** 30 mph   **d** 330 mph
10 3.16 km/h
11 Example answer: 33.15 m/s = 1989 m/min = 119 340 m/h
    = 119.34 km/h
    119.34 km/h > 115 km/h

What do you think?

1 a 10 minutes   **b** 8.3%

Consolidate

1 a 75 mph   **b** 100 km/h   **c** 400 mph   **d** 8 m/s
2 a 8.82 miles   **b** 875 km
   c 76.2 miles (3 s.f.)   **d** 278 m (3 s.f.)
3 a 1.5 hours   **b** 3 hours
   c 40 seconds   **d** 15 minutes
4 a

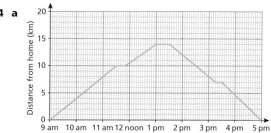

b 3.43 km/h

Stretch

1 Let the train journey be 1 mile long.
This means the total time taken is $\frac{0.5}{60} + \frac{0.5}{30} = \frac{1.5}{60}$
So the average speed is $1 \div \frac{1.5}{60} = 40$ mph
2 a 100 m
   b The distance travelled by the car
   c 215 m
3 Let the van journey be 1 mile long.
Total time taken = total distance ÷ average speed
$$= \left(\frac{1}{3} \div a\right) + \left(\frac{1}{3} \div 2a\right) + \left(\frac{1}{3} \div 4a\right) = \frac{7}{12a} \text{ hours}$$
Average speed = total distance ÷ total time taken
$$= 1 \div \frac{7}{12a} = \frac{12a}{7}$$

Chapter 18.2

Are you ready?

1 a 1.75 kg   **b** 17 500 g
2 a 216 cm³   **b** 0.125 m³
3 a $a = 16$ **b** $b = 360$ **c** $c = 6.85$

Practice

1 8.52 g/cm³
2 a 0.625 g/cm³   **b** 2.25 g
3 Table completed from top as follows: 7.75, 1.96, 2.72, 5.5
4 The mass and the volume of the paper are halved so the density stays the same.
5 a 2.18 litres   **b** 0.545 litres   **c** 1.36 litres
6 a 10.53 g   **b** 16.848 g   **c** 3 510 000 g
7 a 0.538 kg/cm³   **b** 0.008 kg/cm³
   c 0.061 kg/cm³   **d** 18.189 kg/cm³
8 22 200 kg/m³
9 A and C
10 538 kg/m³

Consolidate

1 $4.5\,\text{g/cm}^3$
2 Table completed from top as follows: 244, 328, 152
3 $1.76\,\text{cm}^3$
4 $2\,\text{g/cm}^3$
5 $18.39\,\text{g/l}$

Stretch

1 $163.4\,\text{kg/m}^3$
2 $8.94\,\text{g/cm}^3$
3 $0.968\,\text{g/cm}^3$
4 75%
5 Volume $= \dfrac{x(x+1)}{2} \times 2x = x^2(x+1)$

Density $= \dfrac{\text{Mass}}{\text{Volume}} = \dfrac{3x^2}{x^2(x+1)} = \dfrac{3}{x+1}$

6 Tin (B), $410\,543\,\text{kg} > 189\,333\,\text{kg}$

Chapter 18.3

Are you ready?

1 **a** €92   **b** £500
2 5 hours
3 **a** 2700 potatoes   **b** 10 hours

Practice

1 **a** 75 shirts   **b** 6 minutes and 40 seconds
2 Rob
3 X: 5 litres per minute
  Y: 3.75 litres per minute
4 **a** 190   **b** 3 830 400
  c 87 hours and 43 minutes
5 **a**

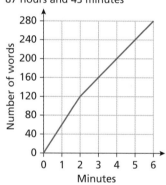

  b Correct average is $280 \div 6 = 46.\dot{6}$ words per minute.
6 **a** 0.8 passports per minute
  b 36 passengers
  c 125 minutes
  d Example answers: The security officer works at the same constant rate / doesn't tire / doesn't take a break.
7 **a** 147 rupees   **b** $0.34y$
8 288 minutes

Consolidate

1 **a** 16   **b** 22.5 minutes
  c Huda keeps chopping at a constant rate, however many onions she chops.
2 **a** 800 carrots per minute   **b** 3 456 000 carrots
  c 7.5 seconds
3 **a** Between 3 and 5 minutes
  b i 20 litres per minute   **ii** 10 litres per minute
    iii 35 litres per minute
  c 24 litres per minute
4 79 minutes

Stretch

1 4 hours
2 10 hours and 28 minutes
3 Assuming they continue to work at exactly the same rate when working together as they did working individually, $\dfrac{cd}{c+d}$

Rates: exam practice

1 25 mph
2 350 g
3 2.4 hours or 2 hours 24 minutes
4 10 mph
5 149 yen
6 67.5 mph
7 75 g
8 0.96 g/ml

Ratio, proportion and rates of change: exam practice

1 The pressure decreases.
2 £98
3 $2.5\,\text{g/cm}^3$
4 $7:5$
5 49 cm
6 $1:2.5$
7 **a** 1 hour 21 minutes
  b 6 cm
8 £108
9 15 km/h
10 41
11 $3:4:6$
12 112
13 $20:12:6:21$

This page has deliberately been left blank

This page has deliberately been left blank

This page has deliberately been left blank